Undergraduate Topics in Computer Science

Series Editor

Ian Mackie, University of Sussex, Brighton, France

'Undergraduate Topics in Computer Science' (UTiCS) delivers high-quality instructional content for undergraduates studying in all areas of computing and information science. From core foundational and theoretical material to final-year topics and applications, UTiCS books take a fresh, concise, and modern approach and are ideal for self-study or for a one- or two-semester course. The texts are authored by established experts in their fields, reviewed by an international advisory board, and contain numerous examples and problems, many of which include fully worked solutions.

The UTiCS concept centers on high-quality, ideally and generally quite concise books in softback format. For advanced undergraduate textbooks that are likely to be longer and more expository, Springer continues to offer the highly regarded *Texts in Computer Science* series, to which we refer potential authors.

Robertas Damaševičius

Human-Centred Scientific Data Visualisation

Making Complex Data Accessible for Everyone

 Springer

Robertas Damaševičius
Department of Software Engineering
Kaunas University of Technology
Kaunas, Lithuania

ISSN 1863-7310 ISSN 2197-1781 (electronic)
Undergraduate Topics in Computer Science
ISBN 978-3-032-01605-8 ISBN 978-3-032-01606-5 (eBook)
https://doi.org/10.1007/978-3-032-01606-5

This Springer imprint is published by the registered company Springer Nature Switzerland AG
The registered company address is: Gewerbestrasse 11, 6330 Cham, Switzerland

If disposing of this product, please recycle the paper.

Preface

When I first embarked on my journey as a scientist, I was struck by the gulf that often separates complex data from clear insight. Time and again, I witnessed researchers drowning in tables and formulas, yearning for a way to see their discoveries take shape before their eyes. It was this experience—this frustration and this spark of possibility—that inspired me to write *Human-Centred Scientific Data Visualisation: Making Complex Data Accessible for Everyone.*

This book originates from more than a decade of working at the intersection of user experience design and scientific research. Across disciplines, one truth emerged: powerful visualization techniques must serve the human mind, guiding our eyes and our intuition toward understanding. Throughout these pages, I share not only theory but hands-on techniques and real-world examples, distilled from my years of practice and experimentation.

At its heart, this work is driven by a singular purpose: to break down barriers between data and discovery. Whether you are a graduate student struggling to present your thesis results, a seasoned researcher seeking new ways to explore large datasets, or an educator aiming to make complex concepts intuitive for your students, you will find guidance here. The book's plan is straightforward: each chapter introduces a technique, explains its rationale and best practices, illustrates its implementation with clear Python or R examples, and demonstrates its impact through compelling case studies.

I have designed this book for a broad audience: scientists, engineers, data analysts, UX and visualization practitioners, and anyone curious about turning numbers into narratives. No prior experience in design is required—only a willingness to see data through the lens of human perception and interaction. As you read, you will be encouraged to experiment, to challenge assumptions, and to craft visualizations that speak directly to the people who rely on your insights.

It is my hope that by the end of this book, you will not only understand these techniques but also feel empowered to invent your own.

Kaunas, Lithuania
Robertas Damaševičius

Competing Interests The author has no competing interests to declare that are relevant to the content of this manuscript.

Contents

1 Importance of Accessibility and Inclusivity in Visualizing Scientific Data 1
   1.1   Introduction to Human-Centered Visualization 2
   1.2   Understanding the Needs of Diverse Audiences 3
   1.3   Accessibility in Scientific Visualization 6
   1.4   Inclusivity in Data Representation 9
   1.5   Cognitive Load and User-Centered Design 11
   1.6   Guiding UX Design Principles 14
   1.7   Emerging Trends and Future Directions 15
   1.8   Summary: Creating Inclusive and Accessible Scientific Visualizations 17
   Suggested Readings 19

2 Designing Marvelous Line Plots 21
   2.1   Introduction to Line Plots 22
   2.2   Essential Components of a Line Plot 24
   2.3   Choosing the Right Scale and Range 28
   2.4   Aesthetic Considerations in Line Plots 31
   2.5   Enhancing Readability with Smart Labels 35
   2.6   Plotting Multiple Series and Overlays 38
   2.7   Advanced Techniques for Line Plot Design 41
   2.8   Avoiding Common Pitfalls in Line Plot Design 43
   2.9   Case Studies and Real-World Examples 47
   2.10   Design Tips and Tools for Creating Line Plots 53
   2.11   Conclusion and Future Trends 58
   Suggested Readings 61

3 Designing Stunning Bar Charts 62
   3.1   Introduction to Bar Charts and Box Plots 63
   3.2   Design Principles for Bar Charts 66
   3.3   Creating Stunning Box Plots 69
   3.4   Data Storytelling with Bar Charts and Box Plots 73
   3.5   Going Beyond Bar Charts and Box Plots 75

3.6   Tools and Best Practices for Bar Chart Design 79
3.7   Case Studies and Examples 83
3.8   Conclusion 85
Suggested Readings 87

4   Designing Amazing Scatterplots 88
4.1   Introduction to Scatterplots 89
4.2   Choosing the Right Data for Scatterplots 91
4.3   Design Aesthetics of Scatterplots 94
4.4   Advanced Scatterplot Techniques 101
4.5   Storytelling Through Scatterplots 105
4.6   Practical Examples and Case Studies 109
4.7   Tools for Creating Scatterplots 112
4.8   Evaluating the Effectiveness of Your Scatterplot 117
4.9   Concluding Remarks 121
Suggested Readings 124

5   Designing Effective Histograms 126
5.1   Introduction to Histograms 127
5.2   Histogram Design Fundamentals 130
5.3   Enhancing Histogram Readability 134
5.4   Advanced Techniques in Histogram Design 139
5.5   Avoiding Common Pitfalls 145
5.6   Accessibility in Histogram Design 148
5.7   Best UX Practices in Histogram Design 152
5.8   Real-World Examples of Effective Histograms 157
5.9   Tools for Creating Visually Appealing Histograms 160
5.10   Histogram Visualization Tips and Tricks 166
Suggested Readings 169

6   Designing Astonishing Pie Charts 170
6.1   Introduction to Pie and Donut Charts 171
6.2   Principles of Effective Pie and Donut Design 174
6.3   Enhance User Engagement with Color and Contrast 177
6.4   Labeling and Annotations for Maximum Clarity 180
6.5   Donut Charts for Modern Data Storytelling 183
6.6   Animation and Interaction for Dynamic Charts 186
6.7   Best Practices 189
6.8   Tools and Libraries for Pie and Donut Charts 191
6.9   Summary and Key Takeaways 194
Suggested Readings 196

7   Creating Enchanting Spider Plots 198
   7.1   Introduction to Spider Plots 199
   7.2   Design Principles for Effective Spider Plots 202
   7.3   Choosing the Right Tools for Spider Plots 206
   7.4   Advanced Design Techniques for Spider Plots 216
   7.5   Accessibility Considerations 219
   7.6   Troubleshooting Spider Plot Design 222
   7.7   Common Mistakes in Spider Plot Design 226
   7.8   Summary and Best Practices 229
   Suggested Readings 231

8   Creating Vibrant Ridgeline Plots 232
   8.1   Introduction to Ridgeline Plots 233
   8.2   Design Principles for Effective Ridgeline Plots 237
   8.3   Creating Ridgeline Plots in Python with `seaborn` and `plotly` 240
   8.4   Customizing Ridgeline Plots in R with `ggplot2` and `ggridges` ... 245
   8.5   Enhancing Ridgeline Plots with Aesthetic Design Techniques 252
   8.6   Accessibility Considerations 255
   8.7   Case Studies and Practical Applications 257
   8.8   Troubleshooting Common Design Issues 262
   8.9   Summary and Best Practices 266
   Suggested Readings 269

9   Creating Breathtaking Density Plots 270
   9.1   Introduction to 2D Density Plots 271
   9.2   Design Principles for Effective 2D Density Plots 274
   9.3   Creating 2D Density Plots in Python with `Matplotlib` and `Seaborn` 278
   9.4   Enhancing 2D Density Plots with Interactive Elements Using `Plotly` 280
   9.5   Creating and Customizing Contour Plots 283
   9.6   Heatmaps for Density Visualization 285
   9.7   Advanced Techniques for 2D Density Plots 287
   9.8   Accessibility Considerations for 2D Density Plots 291
   9.9   Common Pitfalls in 2D Density Plot Design and How to Avoid Them 293
   9.10   Accessibility Considerations for 2D Density Plots 295
   9.11   Summary and Best Practices 297
   Suggested Readings 299

10   Designing Other Beautiful Charts 301
   10.1   Outline of Advanced Charts 302
   10.2   Treemaps 304
   10.3   Network Graphs 311
   10.4   Sankey Diagrams 316
   10.5   Chord Diagrams 322

10.6   Streamgraphs 326
10.7   Hexbin Plots. 332
10.8   Radial Bar Charts 337
10.9   Insights and Takeaways. 343
Suggested Readings 344

11   Advanced Visualization Techniques 346
11.1   Introduction 347
11.2   Zoom-Out. 350
11.3   Panning. 353
11.4   Brushing. 357
11.5   Drill-Down. 363
11.6   Filtering 368
11.7   Linking. 374
11.8   Annotation 378
11.9   Animation 384
11.10   Small Multiples 391
11.11   Focus + Context (Fisheye Views) 396
11.12   Summary and Concluding Thoughts. 400
Suggested Readings 402

Appendix A: Visualization Tools and Libraries 403
A.1   Python Libraries 403
A.2   JavaScript Frameworks. 405
A.3   Business Intelligence Platforms 405
A.4   Emerging and Specialized Tools 406

Appendix B: Accessibility and Inclusive Design 407
B.1   Principles of Inclusive Visualization 407
B.2   Color and Contrast 407
B.3   Keyboard Navigation and Focus Management 408
B.4   Screen Reader Support 409
B.5   Accessible Charts and Graphs 409
B.6   Testing and Validation. 410

Glossary 411

Index 419

Acronyms

AI	Artificial Intelligence
API	Application Programming Interface
AR	Augmented Reality
CPU	Central Processing Unit
CSS	Cascading Style Sheets
CSV	Comma-Separated Values
EDA	Exploratory Data Analysis
GIS	Geographic Information System
GPU	Graphics Processing Unit
GUI	Graphical User Interface
HCD	Human-Centered Design
HTML	HyperText Markup Language
JPG	Joint Photographic Experts Group
JSON	JavaScript Object Notation
KDE	Kernel Density Estimation
LOD	Level of Detail
PDF	Probability Density Function
PNG	Portable Network Graphics
PPI	Pixels Per Inch
RGB	Red Green Blue
SVG	Scalable Vector Graphics
UI	User Interface
UX	User Experience
VR	Virtual Reality
XAI	Explainable Artificial Intelligence
XML	EXtensible Markup Language

1 Importance of Accessibility and Inclusivity in Visualizing Scientific Data

Abstract

This chapter introduces the principles of human-centered scientific visualization, emphasizing the importance of accessibility and inclusivity in representing complex data. It explores how visualizations can be designed to accommodate diverse audiences, from experts to non-experts, by reducing cognitive load and aligning with users' mental models. Strategies for designing accessible visualizations are discussed, including considerations for users with visual impairments, such as colorblindness and low vision, and the use of inclusive color schemes and assistive technologies. The chapter also addresses ethical issues in data representation, such as reducing bias, ensuring cultural sensitivity, and avoiding stereotypes, to foster fairness and clarity. Emerging trends, including augmented and virtual reality, personalized visualization, and AI-driven adaptive designs, are highlighted as transformative opportunities for making data representation more interactive and inclusive. By focusing on user-centered design, the chapter underscores the role of scientific visualization in promoting data accessibility and scientific literacy, paving the way for a more inclusive, informed, and engaged society.

After reading this chapter you should be able to:

➤ Understand the fundamental principles of human-centered visualization and its importance in scientific communication.

➤ Identify the challenges and strategies for making scientific data accessible to diverse audiences, including non-experts.

➤ Apply best practices for designing accessible visualizations, with considerations for colorblindness, low vision, and other disabilities.

➤ Analyze the impact of cognitive load on user understanding and design visualizations that minimize complexity while maximizing clarity.

➤ Evaluate real-world case studies demonstrating the principles of accessibility and inclusivity in scientific visualization.

➤ Apply a human-centered approach to visualizations, ensuring they are both inclusive and accessible for diverse user groups.

© The Author(s), under exclusive license to Springer Nature Switzerland AG 2026
R. Damaševičius, *Human-Centred Scientific Data Visualisation*,
Undergraduate Topics in Computer Science,
https://doi.org/10.1007/978-3-032-01606-5_1

1.1 Introduction to Human-Centered Visualization

1.1.1 Definition and Concept of Human-Centered Visualization

Human-centered visualization is an approach to data representation that places the user at the core of the design process. Rather than focusing purely on the technical accuracy or aesthetic aspects of a visual, human-centered visualization seeks to ensure that data is accessible, understandable, and meaningful to diverse audiences. It emphasizes the need to consider user experience, cognitive abilities, and context in which the data is being consumed. The goal is to create visualizations that communicate complex scientific data in a way that is not only informative but also engaging and inclusive, catering to users with varying levels of expertise, including non-experts.

1.1.2 History and Evolution of Visualization Practices

The practice of visualizing data has evolved significantly over time. Early visualization techniques, such as charts, maps, and diagrams, can be traced back centuries, used primarily to convey simple patterns or relationships in information. Over time, as datasets grew larger and more complex, so did the need for sophisticated visual representations. The advent of computers in the 20th century marked a transformative period in the field, allowing for dynamic, interactive, and highly detailed visualizations. Today, with the rise of big data and advances in machine learning and AI, the field continues to evolve, pushing the boundaries of how we visualize and interact with data. Despite these advancements, the core principles of effective visualization–clarity, accuracy, and the ability to convey a message–remain the same. However, the focus has shifted more towards making these visualizations accessible and meaningful to a broader audience, which is central to the concept of human-centered design.

1.1.3 Visualization in Modern Scientific Communication

In the context of modern scientific communication, visualization plays an indispensable role in transforming complex datasets into comprehensible formats. As scientific research continues to produce vast amounts of data, researchers rely on visualization not only to make sense of their findings but also to effectively communicate their results to the broader community. A well-designed visualization can bridge the gap between complex scientific information and diverse audiences, making it easier for non-experts to grasp essential insights without needing in-depth technical knowledge. Visualizations can highlight patterns, trends, and relationships that might not be immediately apparent through raw data analysis. In today's interdisciplinary and global research environment, visualization serves as a universal language that transcends barriers of discipline, culture, and expertise, facilitating collaboration and innovation. The human-centered approach ensures that these visualizations are not just scientifically accurate but also intuitive, accessible, and inclusive, helping to democratize data and scientific knowledge.

1.2 Understanding the Needs of Diverse Audiences

1.2.1 Scientific Data for Non-Experts: Challenges and Opportunities

One of the key challenges in scientific visualization is presenting complex data to non-experts in a way that is both comprehensible and engaging. Non-experts, such as policymakers, educators, and the general public, often lack the technical background to interpret raw scientific data or highly specialized visualizations. This creates a significant communication gap, as critical findings may remain inaccessible or misunderstood by a broad audience. To address this challenge, visualizations must be designed with simplicity, clarity, and context in mind, avoiding overwhelming the viewer with unnecessary details. However, this presents an opportunity for creative approaches in design–using storytelling, interactive elements, and user-friendly interfaces to make scientific data more approachable and informative for non-expert audiences. By adopting such strategies, we not only enhance public understanding of science but also foster greater engagement and informed decision-making.

1.2.2 Visualizing for Experts vs. Non-Experts: Bridging the Gap

The needs of expert and non-expert audiences in scientific visualization (Table 1.1) differ significantly, yet visualizations must often cater to both. Experts typically seek precise, detailed representations of data that allow for deep analysis and insights. They require the ability to drill down into complex datasets, identifying specific trends and anomalies that inform their research or decision-making processes. Non-experts, on the other hand, may need broader overviews that prioritize high-level summaries, key findings, and simplified visuals. Bridging the gap between these audiences requires a flexible, adaptive approach to visualization design. One solution is to create layered visualizations (Figure 1.1), where the initial view presents a simplified, high-level overview, while additional layers offer more detailed information that experts can explore as needed. By integrating both broad overviews and detailed data insights, designers can create visualizations that cater to a wider range of audiences, ensuring that both experts and non-experts can derive value from the same visualization without being overwhelmed or underserved.

Table 1.1: Comparison of Visualization Needs: Experts vs. Non-Experts

Visualization Characteristic	Experts	Non-Experts
Level of Detail	High granularity	Summarized overview
Data Interaction	Drill-down exploration	Simplified interaction
Terminology	Technical jargon accepted	Layman-friendly language
Purpose	In-depth analysis	Quick understanding
Design Priority	Precision	Clarity

Fig. 1.1: A layered visualization approach showing a simplified overview for non-experts (left) and a detailed exploration layer for experts (right).

1.2.3   Importance of User Personas in Scientific Visualization

User personas play a critical role in the design of human-centered scientific visualizations. These personas represent archetypal users who embody the goals, needs, and challenges of real-life users interacting with visualizations. By defining user personas–such as scientists, policymakers, educators, or the general public–designers can create visualizations that are tailored to the specific needs of different audience segments (Figure 1.2). Personas help in making informed decisions about the complexity of the visualization, the level of detail required, the type of interaction, and the context in which the data will be consumed. For instance, a visualization aimed at a policy-making persona might focus on actionable insights, such as trends that can influence policy decisions, whereas a visualization aimed at a researcher persona may prioritize detailed data exploration tools. Incorporating user personas ensures that visualizations are not one-size-fits-all but are instead designed with a clear understanding of the users' varying expertise levels, goals, and expectations.

Fig. 1.2: Persona-based design: aligning visualization features with user needs (e.g., scientists prefer detailed data; policymakers prefer high-level insights).

1.2.4   Common Misconceptions in Scientific Visualization

Several common misconceptions can arise when visualizing scientific data, particularly in the context of human-centered design. One major misconception is that more data or complexity in a visualization leads to better understanding. In reality, overwhelming users with too much information often hinders comprehension and reduces the effectiveness of the visualization. Another misconception is that all users have the same needs or abilities, which overlooks the importance of tailoring visualizations to different audience segments, as discussed earlier with user personas. Designers may assume that aesthetic appeal automatically translates to effective communication. While visual appeal is important, it should not come at the expense of clarity and accessibility. It's also a common error to assume that visualizations will be universally understood without providing necessary context or explanations. Ensuring that scientific visualizations are accompanied by clear legends, annotations, or guides can greatly enhance their effectiveness. Addressing these misconceptions is crucial for creating visualizations that truly meet the diverse needs of the audience.

1.3   Accessibility in Scientific Visualization

1.3.1   Defining Accessibility in Data Visualization

Accessibility in data visualization refers to the practice of designing visual representations of data in ways that allow people of all abilities, including those with disabilities, to engage with, understand, and act upon the information presented. This is a crucial aspect of human-centered visualization because scientific data often needs to be communicated to diverse audiences with varying cognitive and physical abilities. Ensuring accessibility means considering factors such as the visual clarity of charts and graphs, the readability of text, and the ease with which users can interact with the visualization. It also involves anticipating barriers that certain users might face, such as visual impairments, and proactively designing solutions that allow all individuals to access the information equally. By making visualizations accessible, we not only adhere to inclusive design principles but also expand the reach and impact of scientific communication.

1.3.2   Addressing Visual Impairments: Designing for Color-blindness, Low Vision, and Other Disabilities

When designing accessible visualizations, it's essential to account for users with visual impairments, including colorblindness, low vision, and other disabilities (Figure 1.3). Colorblindness is one of the most common visual impairments, affecting around 8% of men and 0.5% of women globally. To accommodate colorblind users, designers should avoid relying solely on color to convey information and instead use patterns, textures, or symbols to differentiate data points.

Tools like colorblind-friendly palettes (such as ColorBrewer or CVD palettes) can ensure that colors used in visualizations are distinguishable by individuals with various types of color vision deficiency. For users with low vision, designers should prioritize high contrast between text and background, as well as large, readable fonts (Figure 1.4). It's also important to allow users to magnify or adjust visualizations to suit their needs. For users with severe vision loss, audio descriptions or haptic feedback may be useful. By considering the needs of users with visual impairments, we can make scientific visualizations more inclusive and usable for all.

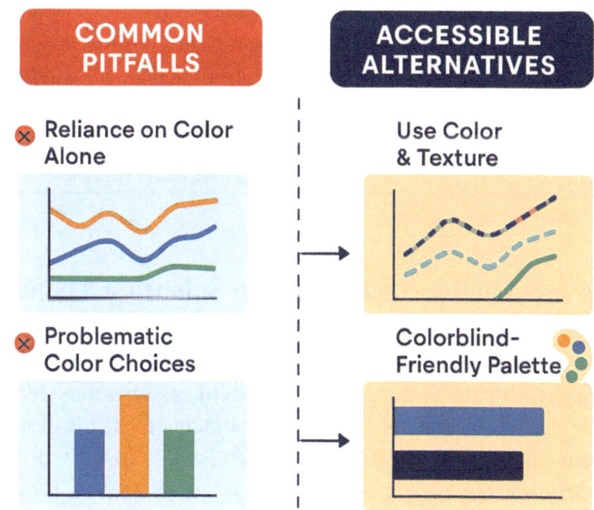

Fig. 1.3: Common visual impairments in data visualization design include color-blindness (difficulty distinguishing colors), low vision (reduced acuity or field), and other disabilities that limit visual perception. Designing with these in mind supports broader accessibility.

Fig. 1.4: Common pitfalls in using color for visualization and accessible alternatives.

1.3.3 Best Practices for Accessible Visualization

Creating accessible visualizations involves adhering to several best practices that ensure data can be consumed and understood by a wide range of users (Table 1.2). First, simplicity is key: designers should avoid clutter and unnecessary complexity in visuals, as these can be overwhelming and hinder comprehension. Clear labeling, concise legends, and informative titles help provide necessary context. Another important practice is to provide alternative text (alt text) for any images or visualizations, allowing screen readers to describe the content for users with visual impairments. Interactive visualizations should be keyboard-navigable to ensure users who cannot use a mouse can still engage with the content. Offering multiple modes of interaction–such as touch, voice, or keyboard controls–can help accommodate users with different abilities. Testing visualizations with assistive technologies, such as screen readers or voice-controlled interfaces, is also critical in identifying and addressing accessibility issues. By following these best practices, designers can create visualizations that are not only scientifically accurate but also universally accessible.

Table 1.2: Best Practices for Accessible Scientific Visualizations

Practice	Explanation
Use high contrast	Ensures readability for users with low vision
Avoid color-only encoding	Use patterns, shapes, or labels in addition to color
Add alternative text	Screen readers can convey meaning to visually impaired users
Provide keyboard navigation	Ensures accessibility for motor-impaired users
Enable zoom and scaling	Supports users needing larger text or symbols

1.3.4 Use of Assistive Technology in Scientific Visualization

Assistive technologies (Table 1.5) play an important role in making scientific visualizations accessible to individuals with disabilities. Screen readers, for instance, allow users with visual impairments to access the content of visualizations by providing audio descriptions of graphs, charts, and other visuals. To fully use the potential of screen readers, designers must include detailed alt text or long descriptions that explain the key insights of the visualization. Haptic feedback can enhance interactive visualizations by providing sensory cues to users with limited vision. Another emerging technology is eye-tracking software, which enables individuals with motor impairments to navigate and interact with visualizations using only their eye movements. These technologies greatly enhance the accessibility of scientific data and ensure that all users, regardless of ability, can engage meaningfully with visual content. Incorporating assistive technologies (Figure 1.5) into the design process is an essential step towards inclusive, human-centered visualization.

Table 1.3: Overview of Assistive Technologies for Scientific Visualization

Technology	Functionality	Target Users
Screen readers	Audio-based description of visuals	Visually impaired
Haptic feedback tools	Tactile representation of data	Blind or low vision
Eye-tracking systems	Control interfaces via eye movement	Motor-impaired users
Keyboard navigation systems	Full control via keyboard	Users unable to use mouse
Alt text/image description tools	Describes visuals in text form	Screen reader users

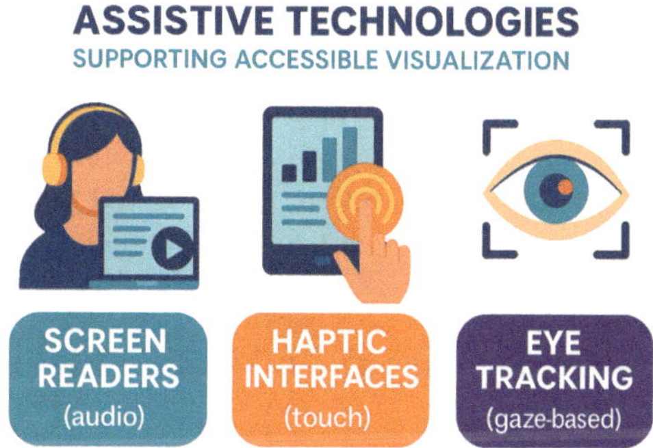

Fig. 1.5: Assistive technologies supporting accessible visualization: screen readers (audio), haptic interfaces (touch), and eye-tracking (gaze-based control).

1.4　Inclusivity in Data Representation

1.4.1　Ensuring Cultural Sensitivity in Data Depiction

Cultural sensitivity in data visualization is crucial for ensuring that visual representations of scientific data are respectful and relevant to diverse audiences. In a globalized world, scientific data is consumed by people from various cultural backgrounds, each with its own norms, values, and interpretations of visual symbols. A culturally sensitive visualization avoids using symbols, colors, or metaphors that may carry unintended or negative connotations in different cultural contexts. For instance, certain colors may signify positive emotions in one culture but be associated with mourning or danger in another. Similarly, icons or imagery that seem neutral in one region might have cultural or political implications elsewhere. To ensure cultural sensitivity, designers must consider their audience's cultural context during the design process, seeking input from local or global stakeholders to avoid misunderstandings. By doing so, data visualizations can foster greater understanding and engagement across cultures, making scientific data more accessible and inclusive.

1.4.2 Reducing Bias in Visualization: Ethical Considerations

Reducing bias in data visualization is a critical ethical responsibility, as biased visuals can mislead or misrepresent the underlying data, influencing viewers' perceptions and decisions. Bias in visualization can occur in various forms, such as through selective data representation, manipulative scaling, or misleading visual metaphors. For instance, omitting key data points or cherry-picking data to highlight a particular trend can skew the viewer's understanding of the actual situation. Ethical data visualization practices demand transparency in how data is selected, processed, and displayed. This includes clearly indicating the source of the data, avoiding distortions through inappropriate scaling, and presenting the full context of the data, including uncertainties and limitations. Ethical considerations also extend to the use of algorithms that may reflect existing biases in the data, leading to visual outputs that reinforce stereotypes or discrimination. It is essential to actively counter such biases through careful design choices, ensuring that visualizations promote fairness and accuracy, rather than perpetuate inequality or misinformation.

1.4.3 Inclusive Color Schemes and Design Elements

Designing inclusive visualizations involves using color schemes and design elements that are accessible to all users, including those with color vision deficiencies (Table 7.6). An inclusive color scheme avoids heavy reliance on color as the sole means of conveying information, instead using combinations of color, texture, patterns, or shapes to differentiate data points. For example, pairing colors with distinct shapes or textures can help colorblind users distinguish between elements in a visualization. Inclusive color schemes often employ colors that are distinguishable across various forms of color blindness, such as using blues and oranges rather than reds and greens, which are difficult for many colorblind individuals to differentiate. Tools such as color contrast checkers and colorblind simulators can be used to test the accessibility of a color scheme before finalizing a design. Inclusive design elements also should consider cultural associations with certain colors, ensuring that the chosen colors align with the context of the visualization without causing confusion or offense.

1.4.4 Avoiding Stereotypes in Data Representation

One of the key principles of inclusivity in data visualization is avoiding the use of stereotypes or generalizations in how data is represented. Visuals can sometimes unintentionally reinforce harmful stereotypes by the way they categorize or display

Table 1.4: Color Palette Accessibility for Color Vision Deficiencies

Palette Name	Color Combinations	Colorblind Safe?
ColorBrewer Diverging	Blue–Orange–Brown	Yes
Traditional Red–Green	Red–Green	No
CVD-safe Sequential	Blue–Purple–Yellow	Yes
Monochromatic Gray	Light–Dark gray	Yes
Rainbow Spectrum	Red–Violet (full)	No

data, particularly when representing demographic groups based on gender, ethnicity, socioeconomic status, or geographic location. For example, overgeneralizing or using reductive categories when visualizing data related to race or gender can perpetuate outdated or offensive stereotypes. To avoid this, designers should ensure that their visualizations respect the diversity and complexity of the subjects being represented. This can be achieved by consulting experts or stakeholders from the relevant communities, carefully choosing the language used in labels and legends, and avoiding visual cues that might reinforce biases. Data should be contextualized appropriately, and its presentation should encourage a nuanced understanding rather than oversimplified or stereotypical interpretations. By being mindful of these factors, visualizations can better reflect the diversity of the real world and contribute to more equitable, respectful data representation.

1.5 Cognitive Load and User-Centered Design

1.5.1 Reducing Cognitive Load through Simplification

Cognitive load refers to the mental effort required to process and understand information, and it plays a crucial role in how users interact with scientific visualizations. When visualizations are overly complex, they can overwhelm the viewer, making it difficult to extract meaning or retain information (Table 1.5). Reducing cognitive load involves simplifying the visualization without sacrificing the integrity of the data. This can be achieved through several strategies (Figure 1.7), such as minimizing unnecessary details, avoiding excessive use of colors or fonts, and ensuring that the most important data points are clearly highlighted. Visual hierarchy is also important, guiding the viewer's eye toward the key elements first and allowing them to absorb additional details gradually. Grouping related data points, providing clear labels and legends, and using intuitive layouts all help to lower the cognitive demands placed on the viewer. Simplified visualization does not mean oversimplification; instead, it seeks to balance clarity with the complexity of the data (Figure 1.7), enabling users to focus on the insights that matter most.

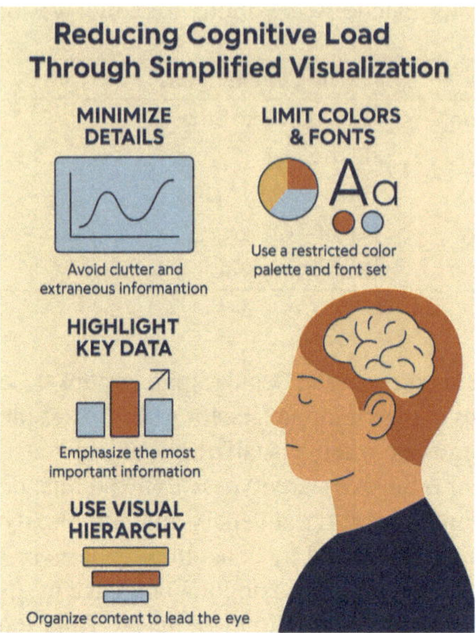

Fig. 1.6: Strategies for reducing cognitive load in scientific visualizations include minimizing extraneous details, limiting color and font usage, highlighting key data, and applying visual hierarchy. These practices help users focus on essential insights without being overwhelmed.

Table 1.5: Techniques for Reducing Cognitive Load in Visualization

Technique	Description
Minimize visual clutter	Avoid unnecessary decorative elements
Highlight key data points	Use contrast or annotations to guide attention
Apply consistent layout	Reduces need to relearn interface
Group related elements	Helps user chunk information
Use familiar metaphors	Aligns with users' mental models

1.5.2 Mental Models and the Role of User-Centered Design

Mental models refer to the internal representations that people form based on their prior knowledge and experience, which they use to interpret and make sense of new information. In the context of data visualization, understanding the mental models of users is key to effective design. A user-centered design approach (Figure 1.8) takes into account the different mental models that users may have when interacting with a visualization, ensuring that the design aligns with their expectations and cognitive

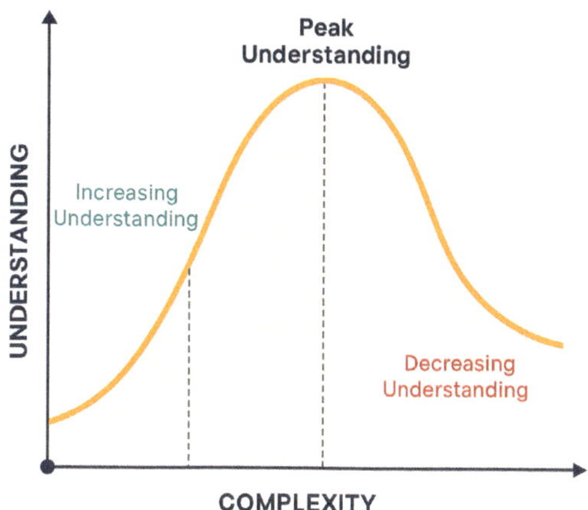

Fig. 1.7: Cognitive load curve: as complexity increases, understanding peaks and then declines, suggesting a need for balance in visualization design.

processes. For example, a user with a background in statistics might expect a more technical and detailed representation of data, while a layperson might need a simpler, more visual approach. By designing visualizations that align with users' mental models, we make the data more intuitive and easier to understand. User-centered design also emphasizes the importance of user feedback throughout the design process, allowing designers to iteratively refine their visualizations based on how real users interpret and engage with the data.

1.5.3 Engaging Visuals for Better User Retention

Engaging visuals are crucial for improving user retention and understanding of scientific data. Well-designed visualizations not only convey information clearly but also capture the viewer's attention, making it more likely that they will remember and act on the information presented. Engaging visuals often use storytelling techniques, where data is presented in a way that highlights a narrative or progression, helping users make connections and grasp the significance of the data. Interactive elements, such as clickable data points or adjustable views, allow users to explore the data more deeply, increasing their engagement and understanding. Another way to enhance engagement is through the use of familiar visual metaphors or relatable examples that resonate with the user's prior experiences. By creating an emotional or cognitive connection to the content, engaging visuals can help users retain information more

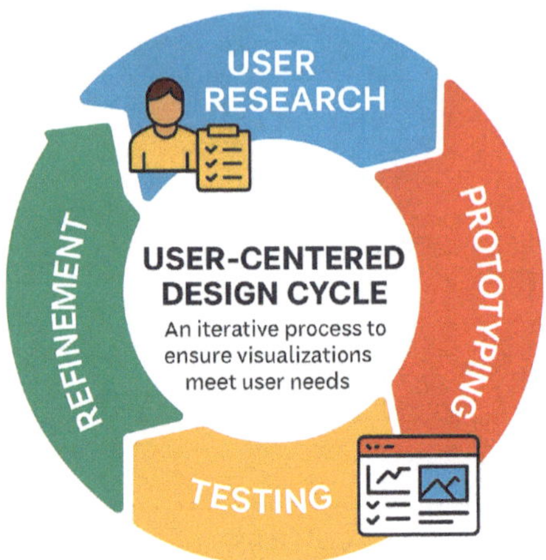

Fig. 1.8: User-Centered Design Cycle: an iterative process involving user research, prototyping, testing, and refinement to ensure visualizations meet user needs.

effectively. Ultimately, the goal is to design visualizations that not only communicate data but also foster curiosity and a deeper comprehension of the subject matter.

1.6   Guiding UX Design Principles

Data visualizations are not merely about displaying information; they are about communicating insights effectively, clearly, and accessibly. The foundation of every successful chart or diagram lies in its user-centered design. In this chapter, all visualizations have been crafted in accordance with core User Experience (UX) design principles to ensure high usability, interpretability, and aesthetic appeal.

Clarity is a fundamental principle, where the removal of visual clutter and the emphasis on essential elements enables users to focus on meaningful data patterns.

Consistency in color schemes, labeling, and layout promotes user familiarity and reduces cognitive load.

Accessibility is ensured through the use of high-contrast palettes, colorblind-friendly hues, and font choices that support legibility for diverse audiences.

Feedback and affordance are subtly embedded through interactive elements (e.g., tooltips, hover effects) in interactive diagrams, enabling users to explore without confusion.

Hierarchical organization and **progressive disclosure** guide users from high-level overviews to more detailed views without overwhelming them.

The visual design decisions made throughout this chapter align with a set of interdisciplinary best practices grounded in cognitive science, perceptual psychology, and usability engineering.

These principles provide the foundation for creating effective, accessible, and engaging data visualizations:

- **Cleveland & McGill's graphical perception framework** [4] and **Shneiderman's Information-Seeking Mantra** [19] emphasize that visual encodings should align with human perceptual strengths: "overview first, zoom and filter, then details on demand." This principle ensures efficient user navigation through visual complexity.

- **Web Content Accessibility Guidelines (WCAG) 2.1** [23] and the **Color Universal Design (CUD)** principles [18] mandate high-contrast, colorblind-safe palettes, and textual alternatives for color cues–supporting inclusive access for diverse user groups.

- **Nielsen's usability heuristic of system status visibility** [16] encourages responsive, real-time feedback through interactive and animated visualizations, reducing user uncertainty and enhancing confidence.

- **Tufte's Principle of Data-Ink Maximization** [22] guides the removal of redundant graphics, emphasizing clarity and integrity of data representation.

- The **Gestalt Principle of Proximity** and **Labeling Guidelines** ensure that related elements are perceived as groups, and that labels are integrated directly with visuals, minimizing the need for excessive eye movement.

- **Cognitive Load Theory** [20] advocates for concise labeling and progressive disclosure, minimizing mental effort while supporting interpretability.

- The **Aesthetic-Usability Effect** [16] and **Whitespace Usage Guidelines** promote visual elegance and breathing room in layout, enhancing user satisfaction and readability.

- **Fitts's Law** [7] informs spatial positioning and element sizing to improve target accessibility , aiding attention guidance in interactive interfaces.

- Finally, **Responsive Design Guidelines** [13] ensure that visualizations adapt seamlessly across screen sizes and devices, maintaining usability and visual hierarchy regardless of context.

1.7 Emerging Trends and Future Directions

1.7.1 Personalized Data Visualization: The Future of Inclusivity

Personalized data visualization is an emerging trend that focuses on tailoring visual content to meet the individual needs, preferences, and cognitive abilities of users. As scientific data becomes increasingly available to diverse audiences, one-size-fits-all visualizations are no longer sufficient to ensure inclusivity. Personalized visualizations can adapt in real-time based on user behavior, preferences, or accessibility needs. For example, data visualizations can adjust complexity based on the user's level of expertise, offering a more detailed view for professionals and a simplified version for laypersons. AI-driven systems can also personalize visualizations based on a user's past interactions, learning their preferred formats, color schemes, or interaction styles. Personalization can enhance accessibility by automatically adjusting visualizations for users with disabilities, such as colorblind-friendly palettes or simplified layouts for users with cognitive challenges. The future of personalized data visualization holds immense potential for making scientific data more inclusive, engaging, and relevant to every user.

1.7.2 Artificial Intelligence in Human-Centered Visualization

Artificial intelligence (AI) and machine learning (ML) are increasingly influencing how scientific data is visualized, with implications for human-centered design. AI-powered algorithms can automate the creation of visualizations, identifying the most relevant patterns in the data and presenting them in a format optimized for user comprehension. For instance, ML techniques can analyze user interaction with a visualization and adapt it dynamically to better suit their needs, ensuring that the most important insights are highlighted. AI can also assist in handling large datasets, by pre-processing and simplifying complex data into digestible visual forms. ML algorithms can personalize visualizations based on past user behavior, providing a more customized experience. As AI continues to evolve, its integration into visualization design will become more sophisticated, allowing for deeper insights and more adaptive, intelligent visual experiences that are aligned with user preferences and needs.

1.7.3 Collaboration and Participation in Visualization

Collaborative and participatory approaches are gaining traction in the field of scientific visualization, emphasizing the importance of co-creation and user involvement in the design process. In collaborative visualization (Figure 1.9), multiple users or stakeholders work together to explore and interpret data, often through interactive platforms that allow real-time contributions and insights. This approach is particularly valuable in interdisciplinary research, where experts from various fields contribute their unique perspectives to the data.

Participatory visualization, on the other hand, involves users in the design process itself, ensuring that the resulting visualizations meet the needs of the intended audience (Figure 1.10). This can be especially beneficial in creating inclusive visualizations for underrepresented or marginalized communities, allowing their voices to shape how data is presented and interpreted. These trends highlight a shift towards more democratic and user-centered approaches to scientific visualization, where the focus is not only on the data but also on the collective and diverse interpretations of it.

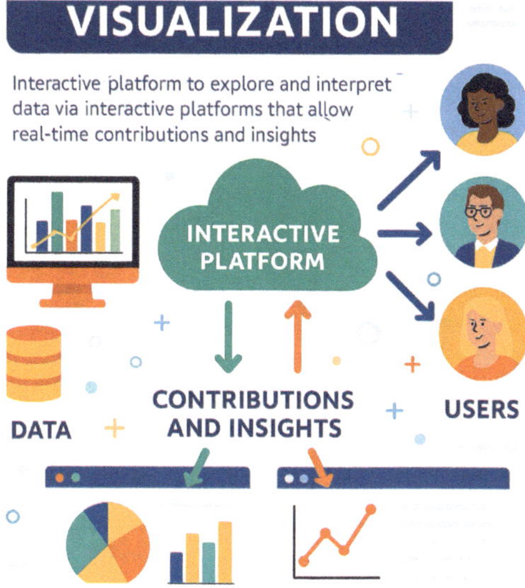

Fig. 1.9: Collaborative visualization enables multiple users or stakeholders to work together to explore and interpret data through interactive platforms that support real-time contributions and shared insights.

1.8 Summary: Creating Inclusive and Accessible Scientific Visualizations

In this chapter, we explored the essential principles of human-centered visualization, focusing on accessibility and inclusivity in the representation of scientific data (Figure 1.11). We discussed the need to address the diverse abilities and knowledge levels of users, highlighting the importance of simplifying visualizations to reduce cognitive load and align with users' mental models. We also examined strategies for designing accessible visualizations, particularly for users with visual impairments, and emphasized the ethical responsibilities involved in reducing bias and ensuring cultural sensitivity in data depiction. Key best practices, such as using inclusive color schemes and avoiding stereotypes, were also covered to promote fairness and clarity in data representation. By incorporating assistive technologies and personalized design elements, scientific visualizations can become more inclusive, allowing a broader audience to engage with complex data. Ultimately, this chapter underscored the importance of user-centered design in making scientific visualization more accessible and impactful.

Fig. 1.10: Participatory visualization involves users in the design process to ensure that visualizations are inclusive, reflect diverse perspectives, and address the needs of underrepresented communities.

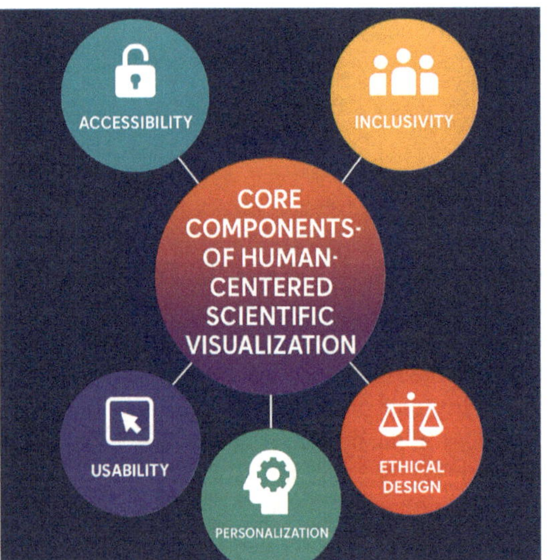

Fig. 1.11: Core components of human-centered data visualization: accessibility , inclusivity, usability, ethical design, and personalization.

As we look toward the future, the role of human-centered visualization in promoting scientific literacy will become even more vital. In an era of rapid technological advancement and increasing data complexity, visualizations must continue to evolve in ways that foster understanding and engagement across diverse audiences. Human-

Centered Design principles ensure that visualizations not only communicate data effectively but also democratize access to scientific knowledge, making it more approachable for everyone, regardless of their expertise or abilities. Emerging trends, such as augmented reality, personalized data visualization, and AI-driven adaptive interfaces, offer exciting possibilities for making data even more accessible and inclusive. Participatory and collaborative approaches in visualization design will empower users from all backgrounds to contribute to and shape how data is represented. By prioritizing inclusivity and accessibility , human-centered visualization will play a critical role in enhancing scientific literacy and fostering a more informed, data-literate society.

Suggested Readings

1. Borgo, R., Kehrer, J., Chung, D.H.S., Maguire, E., Laramee, R.S., Hauser, H., Ward, M., Chen, M.: State of the art report on visualization for climate science. Computer Graphics Forum **32**(8), 201–221 (2013). DOI 10.1111/cgf.12003
2. Brewer, C.A.: Evaluation of color sequences for data classed maps. Cartographic Journal **34**(2), 66–75 (1997). DOI 10.1179/caj.1997.34.2.66
3. Cleveland, W.S.: The Elements of Graphing Data. Wadsworth (1985)
4. Cleveland, W.S., McGill, R.: Graphical perception: Theory, experimentation, and application to the development of graphical methods. Journal of the American Statistical Association **79**(387), 531–554 (1984)
5. Few, S.: Now You See It: Simple Visualization Techniques for Quantitative Analysis. Analytics Press (2009)
6. Few, S.: Show Me the Numbers: Designing Tables and Graphs to Enlighten, 2nd edn. Analytics Press (2012)
7. Fitts, P.M.: The information capacity of the human motor system in controlling the amplitude of movement. Journal of Experimental Psychology **47**(6), 381 (1954)
8. Healey, C.G., Enns, J.T.: Visualizing data with motion. IEEE Computer Graphics and Applications **16**(4), 20–30 (1996)
9. Heer, J., Boyd, D.: A streamgraph: a form of stacked area graph for visualization of thematic changes over time. In: Proc. of InfoVis (2010)
10. International Organization for Standardization: Ergonomics of human-system interaction — part 210: Human-centred design for interactive systems (2010). ISO 9241-210
11. Itoh, M., Okabe, M.: Color universal design (cud) guidelines. Online Resource (2006). https://jfly.uni-koeln.de/color/
12. Mackinlay, J.D.: Automating the design of graphical presentations of relational information. In: ACM Transactions on Graphics (TOG), vol. 5, pp. 110–141 (1986)
13. Marcotte, E.: Responsive web design. A Book Apart (2011)
14. Moreland, K.: Diverging color maps for scientific visualization. Proceedings of the 5th International Symposium on Visual Computing (2016)
15. Munzner, T.: Visualization Analysis and Design. CRC Press (2014)
16. Nielsen, J.: Usability Engineering. Morgan Kaufmann (1994)
17. Norman, D.A.: The Design of Everyday Things. Basic Books (2013)
18. Okabe, M., Ito, K.: Color universal design (cud): How to make figures and presentations that are friendly to colorblind people. J. Sci. Commun **3**(2) (2008)
19. Shneiderman, B.: The eyes have it: A task by data type taxonomy for information visualizations. In: Proceedings 1996 IEEE Symposium on Visual Languages, pp. 336–343. IEEE (1996)
20. Sweller, J.: Cognitive load during problem solving: Effects on learning. Cognitive science **12**(2), 257–285 (1988)

21. Tufte, E.R.: Envisioning Information. Graphics Press (1990)
22. Tufte, E.R.: The visual display of quantitative information. Graphics Press (2001)
23. W3C: Web content accessibility guidelines (wcag) 2.1. World Wide Web Consortium (W3C) (2018). https://www.w3.org/TR/WCAG21/
24. Ware, C.: Information Visualization: Perception for Design, 3rd edn. Morgan Kaufmann (2012)
25. Wilke, C.O.: Fundamentals of Data Visualization. O'Reilly Media, Inc. (2019)
26. Wilkinson, L.: The Grammar of Graphics. Springer (2005)
27. Williams, R.: The Non-Designer's Design Book. Peachpit Press (2014)
28. Wong, D.M.: The Wall Street Journal Guide to Information Graphics. W. W. Norton & Company (2011)

2 Designing Marvelous Line Plots

Abstract

Line plots are one of the most versatile and widely used tools in data visualization, offering an intuitive way to represent trends, patterns, and relationships over time or across continuous variables. This chapter provides a comprehensive guide to designing effective, visually appealing, and human-centered line plots. Beginning with an introduction to their fundamental components, such as axes, gridlines, and labels, the chapter explores how thoughtful design choices enhance clarity and interpretability. Advanced techniques, including the use of shading to represent uncertainty, combining line plots with other chart types, and creating interactive plots for dashboards, are discussed in detail. The chapter emphasizes the importance of avoiding common pitfalls, such as misleading axis scaling, overcrowding plots with too many lines, and mishandling data gaps. Practical tips for using popular tools like Matplotlib, Plotly, and Tableau are provided, along with guidance on maintaining consistency through templates and themes.

Aims

After reading this chapter, you should be able to:

➤ Understand the core components and structure of line plots.

➤ Identify appropriate use cases for line plots in various contexts.

➤ Choose suitable scales and ranges for plotting complex data.

➤ Apply aesthetic considerations such as color, line style, and thickness to improve readability and impact.

➤ Label and annotate line plots effectively to highlight key information.

➤ Implement advanced techniques such as uncertainty shading and combining line plots with other plot types.

➤ Avoid common pitfalls, such as misleading scales and visual clutter in complex plots.

© The Author(s), under exclusive license to Springer Nature Switzerland AG 2026
R. Damaševičius, *Human-Centred Scientific Data Visualisation*,
Undergraduate Topics in Computer Science,
https://doi.org/10.1007/978-3-032-01606-5_2

2.1 Introduction to Line Plots

Line plots are one of the most fundamental and versatile tools in data visualization. They are widely used across various disciplines to represent data that changes over time or some continuous variable. Unlike other plot types that may focus on categorical comparisons or distributions, line plots excel at illustrating trends, patterns, and relationships in a clear and intuitive manner. They allow the viewer to quickly identify increases, decreases, plateaus, and other dynamics within the data. Through careful design choices and enhancements, line plots can be adapted to present complex data in a way that remains accessible and insightful to a broad audience.

2.1.1 What is a Line Plot?

A line plot is a type of chart that connects individual data points using straight lines. Each point represents a value at a specific location along the x-axis (often corresponding to time or a sequential measurement) and a y-axis value that reflects the magnitude or amount of that variable at the given point. Line plots are particularly useful for depicting continuous data, making them a popular choice for time-series data, trend analysis, and scientific measurements.

> **Design Tip: When to Use Line Plots**
>
> Use line plots when your data represents continuous trends or measurements over time. Avoid using them for categorical comparisons – bar or dot plots are more appropriate in those cases.

In a typical line plot (Figure 2.1), the horizontal axis (x-axis) represents an independent variable, such as time, and the vertical axis (y-axis) represents a dependent variable. The data points are connected with lines to show the trend of how the dependent variable changes as the independent variable progresses. By visualizing this relationship, line plots can reveal overall trends, fluctuations, and anomalies in data that might not be immediately apparent from raw numbers.

```python
import matplotlib.pyplot as plt

x = [1, 2, 3, 4, 5]
y = [2, 4, 3, 5, 7]

plt.plot(x, y, marker='o', linestyle='-', color='teal',
    label='Data Series')
plt.xlabel('X-Axis Label')
plt.ylabel('Y-Axis Label')
plt.title('Basic Line Plot')
plt.grid(True)
```

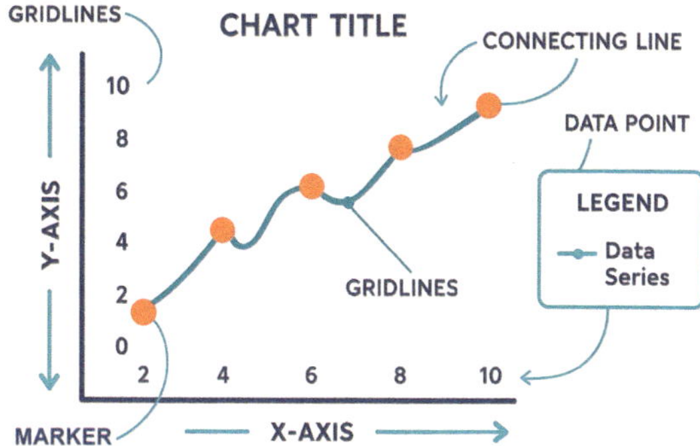

Fig. 2.1: Anatomy of a basic line plot showing labeled axes, gridlines, data points, connecting lines, and a legend.

```
11  plt.legend()
12  plt.show()
```

2.1.2 Importance of Line Plots in Data Visualization

The importance of line plots lies in their ability to convey complex data trends in an easily interpretable format. Line plots distill large datasets into a visual summary that highlights key patterns such as growth, decline, seasonality, and cyclical behavior. Because they make use of continuous lines, line plots are ideal for showing changes in variables over time or space, helping viewers grasp the pace, direction, and magnitude of changes.

From a communication standpoint, line plots are powerful because they simplify the process of identifying data trends for audiences who may not be data experts. The human eye is naturally attuned to detecting slopes and curves, making line plots an intuitive way to present data in a narrative form. Line plots facilitate comparison between multiple data sets. By overlaying several line plots on the same graph, one can easily compare trends, identify divergences or correlations, and draw insights from those relationships.

2.1.3 When to Use Line Plots

While line plots are versatile, it's crucial to understand when they are most appropriate. Line plots are best suited for data that is continuous rather than categorical. They are ideal when the primary goal is to show how a variable changes over time, or when you wish to emphasize trends, rather than individual data points. Common scenarios include visualizing stock prices, temperature fluctuations, sales data, and scientific measurements where time or another continuous variable is a critical factor.

However, line plots may not be appropriate for categorical data or when the goal is to compare discrete groups. In such cases, bar charts or pie charts might be more suitable. For datasets with large amounts of noise or many overlapping data points, line plots might create visual clutter, making it harder to extract meaningful insights. For this reason, it is important to consider the data type and the story you want to tell before choosing a line plot as your visualization tool.

2.1.4 Real-World Applications of Line Plots

Line plots find a wide range of applications across various domains, from finance and economics to scientific research and business analytics. In finance, for example, line plots are commonly used to track stock market performance over time, showing how share prices fluctuate and enabling investors to detect trends or forecast future movements. Economists use line plots to track indicators such as inflation rates, GDP growth, and unemployment levels, providing a visual summary of economic trends.

In the scientific domain, line plots are widely used to visualize experimental data. They can show the progress of a chemical reaction, track the growth of a population over time, or illustrate the output of a sensor as environmental conditions change. In business, line plots are a common way to visualize key performance indicators (KPIs) such as sales figures, web traffic, and customer acquisition rates. By visualizing these metrics over time, companies can identify patterns, assess the impact of strategic decisions, and forecast future performance.

In addition, line plots are commonly used in education, where they help students and researchers visualize data from experiments or simulations. The simplicity and clarity of line plots make them an excellent educational tool for teaching concepts related to data analysis, mathematical modeling, and trend analysis. Whether in a scientific research paper or a corporate report, line plots help to communicate data in a way that is both accessible and impactful.

In the following sections, we will focus on the specific design choices that can make a line plot not only functional but also aesthetically pleasing. From color selection to axis scaling and the use of annotations, we will explore how thoughtful design can elevate a line plot from a basic chart to a powerful storytelling tool.

2.2 Essential Components of a Line Plot

A well-designed line plot is much more than just connecting data points with lines. It involves several key components that work together to create a clear, informative, and visually appealing visualization. In this section, we will discuss the essential elements of a line plot, including the role of axes and gridlines, the importance of data points and continuous lines, how to use labels and annotations effectively, and the best practices for handling legends and multiple data series.

2.2.1 Axes and Gridlines

The axes in a line plot serve as the fundamental framework upon which the data is displayed. Typically, the x-axis represents the independent variable, while the y-axis represents the dependent variable. The clarity and precision of these axes are crucial to ensuring the accuracy and interpretability of the plot. Proper labeling of both axes is essential, and the scale of the axes should be chosen carefully to reflect the nature of the data. For instance, logarithmic scales may be appropriate for data with exponential growth, while linear scales are suitable for data with uniform changes.

Gridlines play a supportive but equally important role in line plots. They act as visual guides, helping the viewer to trace data points across the plot and compare values easily. While gridlines are not always necessary, they can improve readability, especially in cases where precise values need to be compared. However, too many gridlines can clutter the plot, making it difficult to focus on the actual data. Therefore, gridlines should be used sparingly and styled in a way that they do not dominate the visual space – often using lighter colors or dashed lines to keep them unobtrusive.

Another consideration is the range of the axes. Setting appropriate minimum and maximum values for the axes ensures that the data is neither too compressed nor too stretched. In some cases, including a break in the axis can help manage data with large gaps or extreme outliers .

> **Design Tip: Simplify the Grid**
>
> Use light gray or dashed gridlines to guide the eye without overwhelming the data. Avoid thick or dark gridlines, which can distract from the main message.

2.2.2 Data Points and Continuous Lines

The heart of any line plot lies in the data points and the continuous lines that connect them. Data points represent the individual observations or measurements at specific values of the independent variable (x-axis), while the connecting lines show the relationship or trend between these points.

The choice of whether to display the data points themselves (with markers like dots or circles) or only the connecting lines depends on the type of data and the message you want to convey. For small datasets or datasets where each point has significant meaning (e.g., an experimental result), displaying data points is essential. In contrast, for larger datasets where the overall trend is more important than individual points, the line alone may suffice. In some cases, a combination of both can be used to highlight certain critical data points along the continuous trend.

The line itself can take various forms depending on the data and design intent. Solid lines are the most common, but dashed or dotted lines can be used to represent different data series or to indicate uncertainty or prediction intervals. Line thickness is another design element that can help differentiate between different trends or highlight certain aspects of the data. Thicker lines often draw attention to key trends, while thinner lines are used for secondary or background data.

A common challenge when dealing with line plots is managing data gaps. Missing data points or measurements should be handled carefully, either by leaving gaps in the line or using interpolation techniques to connect the available data points without misleading the viewer.

2.2.3 Labels and Annotations

Labels and annotations are critical for providing context to a line plot. Without proper labeling, the viewer is left guessing what the axes, data points, or trends represent. At a minimum, both the x-axis and y-axis should be labeled clearly, and the units of measurement (if applicable) should be included. Axis labels should be concise yet descriptive enough to allow the reader to understand the variables being compared without referring back to the dataset (Table 2.1).

Annotations offer another layer of clarity and insight to a line plot. These are typically used to highlight important data points, trends, or events that might otherwise go unnoticed. For example, annotations can mark peaks, troughs, outliers , or critical turning points in the data. They can also provide explanatory notes or commentary, guiding the viewer's interpretation of the plot.

Positioning annotations is a key consideration. They should be placed close to the data they reference but in such a way that they do not overlap or obscure the lines and data points. Annotations can be simple, such as a label next to a peak, or more complex, incorporating callouts, shapes, or arrows to draw attention to specific aspects of the data.

2.2.4 Legends and Multiple Data Series

In cases where a line plot contains more than one data series, a legend becomes indispensable. The legend helps the viewer distinguish between different lines and understand what each line represents. In a simple plot with just one line, a legend may not be necessary, but in a plot with multiple lines – each representing a different category, group, or condition – the legend provides essential context (Figure 2.2).

> **Design Tip: Use Direct Labels**
>
> Whenever possible, label lines directly near their endpoints instead of using a separate legend. This improves usability, especially in static or print visualizations.

The design of the legend should be consistent with the visual style of the plot. For example, if different line colors or styles (solid, dashed, etc.) are used to differentiate between data series, the legend should reflect these distinctions clearly. It's important to position the legend in a way that it does not obstruct the plot itself, often placing it outside the plot area or in a corner where it won't interfere with the data.

Table 2.1: Best Practice Guidance for Using Labels and Annotations in Line Plots

Aspect	Best Practice	Rationale / Expert Tip
Axis Labeling	Always label both x-axis and y-axis, including units of measurement where applicable.	Provides immediate clarity about the variables being analyzed and eliminates ambiguity for the reader.
Label Clarity	Use short, descriptive labels that explain what each axis represents (e.g., "Temperature (°C)", "Year").	Ensures viewers understand the data without needing to refer back to source documentation.
Consistent Labeling Style	Follow a consistent typography, font size, and case style across all plots.	Maintains visual consistency and improves readability across multi-plot figures.
Annotation Purpose	Use annotations to highlight key data points, such as peaks, troughs, outliers , or major changes.	Draws attention to significant features and supports narrative storytelling within the data.
Annotation Placement	Position annotations near the point of interest, avoiding overlap with data lines or other labels.	Reduces clutter and maintains legibility while clearly linking annotation to data.
Annotation Type	Use callouts, arrows, or shapes for important notes; inline text for simple labeling.	Callouts provide a visual cue for importance, while inline text can be used unobtrusively for routine points.
Dynamic/ Interactive Annotations	Use hover-to-reveal tooltips or interactive callouts in web-based plots.	Improves usability in digital contexts by reducing clutter while preserving access to detail.
Avoid Over-Annotation	Limit the number of annotations to the most essential points.	Prevents visual overload and helps maintain focus on key aspects of the data.

When dealing with multiple data series (Table 2.2), one of the main challenges is avoiding visual clutter. Too many overlapping lines can make the plot difficult to interpret. In these cases, using distinct colors, line styles, or even transparency (for semi-opaque lines) can help distinguish between the different series without overwhelming the viewer. In some situations, it might also be useful to plot the data series on different axes (dual-axis plots) to ensure that each data series is represented accurately and is easy to compare.

Legends should be used in conjunction with well-chosen color schemes and line styles to ensure that the plot remains clear and that different data series can be easily identified and compared. Avoid using colors that are too similar, especially for viewers with color vision deficiencies. Tools such as color-blind friendly palettes can be helpful in ensuring that your visualizations are accessible to a broader audience.

The essential components discussed in this section – axes, gridlines, data points, continuous lines, labels, annotations, and legends – form the backbone of any well-designed line plot. Together, these elements provide structure, clarity, and meaning, turning raw data into a visual story that can be easily understood by the viewer. In the following sections, we will explore how to further enhance line plots through

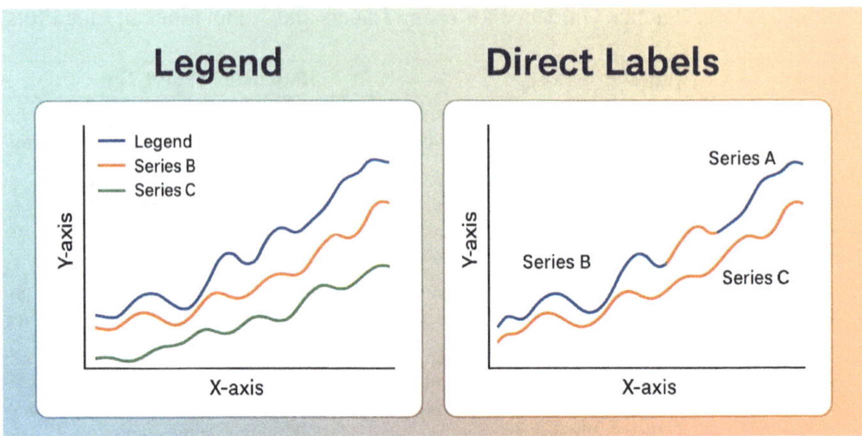

Fig. 2.2: Left: Traditional legend referencing; Right: Direct labels placed near line endpoints. Direct labeling improves usability, especially in print or static media.

Table 2.2: Best Practices for Plotting Multiple Series

Technique	Description
Color	Use clearly distinguishable, color-blind friendly palettes
Line Style	Combine styles (solid, dashed, dotted) with colors
transparency	Apply Alpha Blending to reduce visual overlap
Direct Labels	Place labels at line ends to avoid legend referencing
Interactive Legends	Enable toggling of series in web-based plots
Small Multiples	Use subplots to reduce clutter when comparing many series

aesthetic design, including color choices, line styles, and techniques for dealing with large datasets and overlapping information.

2.3   Choosing the Right Scale and Range

Choosing the correct scale and range is essential for creating accurate and meaningful line plots. The scale of the axes affects how the data is perceived, and an inappropriate choice can distort the data's true nature, leading to misinterpretation. In this section, we explore the considerations that come into play when selecting between linear and logarithmic scales, handling large ranges of data, plotting time series data, and dealing with missing data.

> **Design Tip: Choose Your Scale Wisely**
>
> Select a logarithmic scale when visualizing data that spans several orders of magnitude or grows exponentially. Always label the scale clearly and explain it in the figure caption if needed.

2.3.1   Linear vs. Logarithmic Scales

One of the most critical decisions in plotting data is whether to use a linear or logarithmic scale. The choice between these two scales depends largely on the nature of the data being visualized and the story you want the data to tell.

A linear scale is the most commonly used axis type, where equal intervals on the axis represent equal increments in value. For instance, a distance of 10 units on a linear scale represents the same change in magnitude, regardless of whether the points lie between 1 and 11 or 100 and 110. Linear scales are intuitive and appropriate for data that does not exhibit exponential growth or large disparities in values.

In contrast, a logarithmic scale represents data in multiplicative increments. This means that each equal interval on a logarithmic scale corresponds to a constant factor rather than a constant difference. Logarithmic scales are particularly useful when dealing with data that spans several orders of magnitude, or for data that grows exponentially, such as population growth, financial investments, or certain types of scientific measurements. By compressing the larger values, logarithmic scales make it easier to compare growth rates or detect patterns in data that would otherwise be lost in a linear scale.

For example, a dataset showing exponential growth will appear as a steep upward curve on a linear scale, but it may show a more manageable slope on a logarithmic scale, making it easier to compare data points. However, logarithmic scales can be less intuitive for some audiences, particularly if they are unfamiliar with this type of representation. Therefore, when using a logarithmic scale, it's important to provide clear labeling and annotations that help the viewer understand the scale and its implications for the data.

2.3.2   Handling Large Ranges of Data

When plotting data that spans a large range, the choice of scale and axis range becomes even more critical. A line plot that compresses a vast data range into a small space may obscure important details or create visual artifacts. Conversely, expanding the range too much can exaggerate minor fluctuations and mislead viewers about the data's overall trends.

One technique for managing large ranges of data is to use axis breaks. An axis break introduces a discontinuity in the axis, allowing you to focus on the most relevant part of the data while omitting outlier values or extreme ranges that are not necessary for the plot's primary message. For example, if your data contains a few extreme outliers , an axis break can help you visualize the majority of the data in more detail without letting the outliers distort the overall view.

In some cases, dual-axis plots can also be helpful when comparing data with vastly different ranges. By assigning one data series to the primary y-axis and another to a secondary y-axis on the opposite side of the plot, you can visualize both datasets without overwhelming the plot or compressing the data.

However, these techniques should be used with caution. Axis breaks or dual-axis plots can sometimes confuse viewers or lead to misinterpretation. Always ensure that the choice of axis range reflects the true nature of the data and clearly communicate any adjustments made to the scale or range.

2.3.3 Plotting Time Series Data

Time series data, which tracks changes in a variable over time, is one of the most common applications of line plots. The x-axis in such cases typically represents time, while the y-axis represents the variable of interest. When plotting time series data, it is essential to choose an appropriate time scale and ensure that the intervals on the x-axis correspond to meaningful units.

Time scales can vary widely, depending on the data and its context. For example, financial data might be plotted using daily or monthly intervals, while scientific measurements might use intervals of seconds, milliseconds, or even years. Choosing the right time interval is critical for accurately reflecting the data's trends. Too large an interval may obscure short-term fluctuations, while too small an interval can clutter the plot with excessive detail.

Another consideration when plotting time series data is the handling of irregular intervals. In many real-world datasets, measurements may not be taken at consistent time points. For instance, financial data might have missing data on weekends or holidays, and sensor data might have gaps due to outages. In such cases, it is important to maintain the integrity of the time axis. Gaps in data should not be filled in with false points, but instead, left as blank spaces or marked with annotations, ensuring the viewer understands the data's true nature.

In addition, time series data often benefits from smooth lines or interpolation techniques that help reveal overall trends. However, these should be used carefully to avoid distorting the data. Smoothing techniques like moving averages or spline interpolation can be helpful, but they should be clearly labeled, and raw data points should also be shown where appropriate.

2.3.4 Dealing with Missing Data in Line Plots

Missing data is a common challenge in real-world datasets, and handling it effectively is crucial for creating accurate and reliable visualizations. In a line plot, missing data points can disrupt the continuity of the line, leaving gaps that might confuse or mislead viewers if not handled properly.

One approach to dealing with missing data is to simply leave gaps in the line. This technique maintains the integrity of the dataset and avoids introducing false information. Gaps in the line indicate to the viewer that data is missing for those points, and they can interpret the trend based on the available data. This method is particularly useful when the missing data is sporadic or random.

In some cases, however, it may be useful to connect the points before and after the missing data, using interpolation techniques to estimate the missing values. Linear interpolation is one of the simplest methods, connecting two points with a straight line and assuming that the trend continues smoothly through the missing data. However, this technique can introduce bias if the missing data represents a significant change in the underlying trend, and it should be used with caution.

Another option is to use markers or annotations to explicitly indicate missing data points. For example, you might place a symbol (such as an asterisk or a different colored marker) at the location of missing data, alerting the viewer to the gap without distorting the overall trend. In scientific and financial plots, it is common to annotate missing data with notes explaining the cause of the gaps, such as "sensor failure" or "data unavailable."

Dealing with missing data requires careful consideration of the context and the message you want to convey. Leaving gaps, using interpolation, or marking missing data explicitly can all be valid approaches, depending on the nature of the data and the plot's purpose. Regardless of the method chosen, transparency is key – the viewer should always be made aware of missing data and how it was handled in the visualization.

Choosing the right scale and range for a line plot is a fundamental part of ensuring the accuracy, readability, and effectiveness of the visualization. By selecting appropriate axis scales, handling large ranges, carefully plotting time series data, and dealing transparently with missing data, you can create line plots that not only convey trends and patterns clearly but also maintain the integrity of the data. In the next section, we will explore how aesthetic considerations, such as color and line styles, can further enhance the design and impact of line plots.

2.4 Aesthetic Considerations in Line Plots

Aesthetics play a significant role in the design of line plots, influencing not only the visual appeal but also the clarity and effectiveness of the plot. Thoughtful use of color, line styles, thickness, and backgrounds can transform a basic plot into a powerful tool for data communication. Aesthetic choices are especially important when creating plots for presentations or reports where the audience may be less familiar with the underlying data. In this section, we explore key aesthetic considerations, such as selecting appropriate colors for lines, choosing the right line styles, adjusting line thickness for emphasis, and deciding between white and colored backgrounds.

2.4.1 Color Selection for Lines

Color is one of the most impactful design elements in a line plot. When used effectively, it can help differentiate between multiple data series, highlight important trends, and improve the plot's overall readability. However, poor color choices can lead to confusion, misinterpretation, or even make the plot inaccessible to color-blind viewers (Table 2.3).

Table 2.3: Recommended Color Schemes for Color-Blind Accessibility

Palette Name	Characteristics
ColorBrewer Set2	Soft, distinct colors, color-blind friendly
Viridis (Matplotlib)	Perceptually uniform, works in grayscale
Tableau 10	Business-friendly and color-blind safe
High Contrast (Plotly)	Clear differences in hue and brightness

When plotting multiple data series on the same line plot, choosing distinct and contrasting colors for each series is essential. Colors should be chosen in such a way that they are easy to distinguish, even in small areas or where the lines overlap. It's important to avoid using colors that are too similar in hue or saturation, as these can blur the distinction between data series. When selecting colors, consider accessibility: a significant portion of the population has some form of color blindness, most commonly red-green color blindness. Using color palettes that are color-blind friendly, such as those provided by design tools like ColorBrewer, can ensure your plot remains accessible to a broader audience.

> **⚠ Warning**
>
> Avoid red-green combinations in your line plots without additional cues like line styles or markers. These are not distinguishable for many users with color vision deficiencies.

For single-line plots, the color should be chosen to maximize contrast with the background and other elements in the plot. A simple black or dark blue line on a white background works well in most cases, but depending on the context and audience, a bolder or more distinctive color might be appropriate. In scientific and technical contexts, blue, green, and red are often preferred due to their high contrast and ease of distinction.

> **Design Tip: Use Color with Purpose**
>
> Avoid rainbow color schemes. Choose high-contrast, colorblind-safe palettes. Use color to highlight important data series or differentiate between groups – not just for decoration.

Highlighting specific data points or sections of the line with different colors can also be an effective way to draw attention to key areas of the plot. For example, using a contrasting color to highlight a peak, trend reversal, or outlier can help the viewer focus on important details without overwhelming the overall plot.

2.4.2   Line Styles: Solid, Dashed, and Dotted

Line style is another powerful aesthetic tool in line plot design. The choice of line style can convey additional information about the data, such as distinguishing between different types of data series or indicating uncertainty or forecasts. The most common line styles include solid, dashed, and dotted lines, each of which has its own use case and visual implications (Table 2.4).

Solid lines are the most frequently used line style and are ideal for representing continuous, primary data. They provide a strong visual connection between data points and are well-suited for displaying the main trends or relationships in the data. Solid lines are often the default choice for line plots, and they work well when there is only one or a few data series to represent.

Dashed lines are useful for secondary data series or for indicating a break in continuity. They can be used to represent forecasted or estimated data, differentiating it from actual measurements. For instance, if you are plotting historical data alongside projections for future performance, using a solid line for the actual data and a dashed line for the projected data provides a clear visual distinction between the two.

Dotted lines are typically used when the data points are more sporadic or when you want to de-emphasize a particular series. They can also be used in combination with solid lines to indicate uncertainty intervals or margins of error around the primary trend. Dotted lines are often lighter and less prominent than solid or dashed lines, making them a good choice for representing less critical data or for overlaying supplementary information.

Table 2.4: Comparison of Line Styles and Their Common Use Cases

Line Style	Visual Example	Common Use Case
Solid (—)	-----	Primary data series or trends
Dashed (– –)	-- -- --	Forecasts or secondary data
Dotted (⋯)	· · · · · ·	Uncertainty, less important series
Dash-dot (–·–)	-·-·-·	Alternating patterns or mixed signals

Using a combination of line styles within the same plot can be an effective way to differentiate between multiple data series or to highlight specific aspects of the data. However, it is important to ensure that the use of different styles remains intuitive and does not confuse the viewer. Consistency is key: once a line style is chosen to represent a particular type of data, it should be used consistently throughout the plot.

2.4.3   Adjusting Line Thickness for Emphasis

The thickness of the lines in a plot can significantly influence its visual hierarchy. Adjusting line thickness allows you to emphasize certain trends or data series while de-emphasizing others. In general, thicker lines draw the viewer's attention, while thinner lines are more subdued.

For single-line plots, the default thickness is often sufficient, but if you want to make a particular line stand out, increasing its thickness can help. For example, when presenting a line plot with multiple data series, increasing the thickness of the most important series can guide the viewer's attention to that specific trend, while keeping the other lines thinner ensures they remain in the background.

However, overly thick lines can obscure detail and make the plot look cluttered, particularly in plots with a dense concentration of data points or multiple overlapping lines. Therefore, it is essential to strike a balance between visibility and clarity. Thicker lines work best when the plot contains relatively few data points or when you want to emphasize a major trend, while thinner lines are more appropriate for detailed, dense plots.

Adjusting line thickness can also be useful in highlighting specific data points within the plot. For example, you might choose to make a line thicker as it approaches a peak or turning point, indicating the significance of that particular section of the data. Alternatively, you could use thicker lines for historical data and thinner lines for projections or secondary data series.

2.4.4   Backgrounds: White vs. Colored

The background color of a line plot plays a subtle yet important role in its overall design. The choice between a white or colored background can influence how the plot is perceived and how easy it is to read.

White backgrounds are the most common and are typically the default choice for line plots. They offer a clean, neutral canvas that allows the data lines and other elements (such as gridlines and annotations) to stand out clearly. White backgrounds also ensure that the plot is easily printable and suitable for a variety of presentation formats, including slides, reports, and scientific papers. In most cases, a white background is the safest and most professional option for line plots.

Colored backgrounds, on the other hand, can be used to create a more distinctive or visually striking plot. They are often used in dashboards, presentations, or marketing materials where aesthetic appeal is a higher priority. For example, a dark or gradient background can create a dramatic contrast with brightly colored lines, making the plot visually engaging. However, colored backgrounds should be used carefully to ensure that they do not detract from the clarity of the data.

When using a colored background, it is important to consider the contrast between the background and the plot elements. High-contrast combinations, such as light lines on a dark background or dark lines on a light background, are generally easier to read.

Colored backgrounds should be chosen with accessibility in mind, ensuring that they do not interfere with the legibility of the lines, text, or other plot elements (Table 2.5). In scientific and technical contexts, colored backgrounds should be used sparingly and thoughtfully. While they can add visual interest, they can also introduce visual noise that distracts from the data itself. For most professional and technical plots, a simple white background remains the best option.

Table 2.5: Accessibility Considerations: Foreground and Background Color Combinations for Line Plots

Background Color	Line Color (Example)	Accessibility Evaluation	Recommendation / Use Case
White	Dark Blue	High Contrast	Ideal for reports and publications; ensures legibility on print and screens
Light Gray	Crimson Red	Good Contrast	Suitable for presentations; provides a soft contrast with visual balance
Cream	Dark Gray	Accessible	Elegant look for slides or dashboards; works well with earthy tones
Dark Gray		Readable if bold	Use bold lines and increased font size; suitable for dark mode dashboards
Mustard Yellow	Red	Low Contrast	Avoid this pairing; red on yellow is difficult for many viewers, especially with color blindness
	Red, Green	Not Colorblind-Friendly	Avoid red-green pairs on black; consider blue-yellow for improved accessibility

Aesthetic considerations such as color selection, line styles, line thickness, and background choice are integral to creating effective and visually appealing line plots. By making thoughtful design choices, you can enhance both the clarity and impact of your visualizations, ensuring that they communicate the underlying data effectively to your audience. In the next section, we will discuss how to further improve the readability of line plots through the use of smart labels and annotations.

2.5 Enhancing Readability with Smart Labels

Labels play a critical role in making a line plot understandable by providing essential information about the axes, data points, and trends. Smart labeling techniques can enhance readability and ensure that the viewer quickly grasps the key insights from the data. However, improperly positioned or excessive labels can clutter the plot, leading to confusion. In this section, we explore techniques for positioning labels, using dynamic and interactive labeling methods, and incorporating callouts to highlight key data points effectively.

2.5.1   Positioning Labels for Maximum Clarity

Positioning labels correctly is crucial for maximizing clarity and avoiding overlap with other elements of the plot, such as lines, gridlines, or data points. Proper label positioning ensures that the plot remains clean and easy to interpret, even when multiple data series or complex trends are involved.

The most basic labels in a line plot are the axis labels, which provide context for the x-axis and y-axis. These labels should be clear, concise, and informative, typically specifying the variable being plotted and its units of measurement. Axis labels should be positioned close enough to the axes to be easily readable but far enough to avoid overlapping with the data points or axis ticks. The default placement of axis labels (below the x-axis and next to the y-axis) generally works well, but in cases where space is limited or where the plot has an unusual layout, alternative placements may be needed.

For labeling specific data points, placing the labels directly on or near the data point is the most straightforward approach. However, when multiple data points are densely packed or when several lines intersect, label placement becomes more challenging. In such cases, it is essential to avoid visual clutter by staggering labels or using leader lines (short lines connecting the label to the data point). This technique allows the viewer to associate the label with the correct point without obstructing the view of the data.

Another useful technique is to position labels at the end of the line rather than along the line itself. This method works particularly well when comparing multiple data series. By placing labels at the end of each line, the viewer can easily distinguish between different series without needing to reference a separate legend. This approach also eliminates the need to repeatedly look back and forth between the plot and a legend, making the plot more intuitive.

2.5.2   Dynamic and Interactive Labeling Techniques

In the digital age, interactive visualizations are becoming increasingly popular, particularly in dashboards, web applications, and data-driven presentations. Dynamic labeling techniques, where labels change or appear in response to user interaction, can significantly enhance the readability and usability of line plots.

One common dynamic technique is tooltips, which display additional information when the user hovers over a data point with their cursor. Tooltips allow you to keep the plot visually clean by avoiding the need for excessive static labels, while still providing the viewer with detailed information when needed. For example, in a time series plot, a tooltip might show the exact value of a data point, the time at which it was recorded, and any other relevant metadata. This approach provides an interactive experience while maintaining the simplicity of the plot.

Another interactive technique is zoomable plots, where users can zoom in on specific sections of the plot to reveal more detailed labels. As the viewer zooms in, additional labels can appear for data points that were previously too densely packed to be labeled without cluttering the plot. This dynamic adjustment of labels ensures that the plot remains legible at any zoom level, making it easier to explore large datasets or long time series.

In addition, interactive legends can be used to toggle the visibility of different data series, automatically updating the labels on the plot. This technique is especially useful in plots with multiple data series, where displaying labels for all series at once could be overwhelming. By allowing the viewer to interactively control which data series are displayed, the plot remains clean and readable, while still offering full access to the data.

2.5.3   Using Callouts to Highlight Key Data Points

Callouts are an effective way to draw attention to specific data points or trends within a line plot. Unlike standard labels, which provide general information about the axes or data series, callouts are used to emphasize key findings, anomalies, or areas of particular interest. They often include additional text or graphical elements, such as arrows or boxes, to ensure the highlighted data stands out.

Callouts are particularly useful in situations where you want to guide the viewer's attention to a specific point or pattern. For example, in a plot showing sales growth over time, a callout might highlight the point where sales peaked, along with an explanatory note such as "Highest sales in Q4 2022." By providing this additional context, callouts can help tell a more complete story about the data, going beyond mere trends to explain why certain points are significant.

When using callouts, it's important to balance visibility with clarity. The callout should be prominent enough to attract attention, but not so large or intrusive that it distracts from the rest of the plot. A good practice is to use subtle graphical elements, such as light-colored boxes, arrows, or circles, that complement the overall design of the plot without overwhelming it.

The positioning of callouts is also critical. Placing them too close to the data points may result in overlap with other plot elements, while placing them too far away may confuse the viewer as to which data point the callout refers to. Leader lines or arrows are useful tools for connecting the callout to the correct data point while maintaining a clean layout.

Callouts can also be dynamic in interactive visualizations. For example, in a web-based dashboard, a callout might appear only when the user hovers over a particular data point, providing additional information in a non-intrusive way. This method ensures that the plot remains clean and focused while still offering detailed insights when needed.

Smart labeling techniques, such as careful label positioning, dynamic and interactive labels, and the use of callouts, are essential tools for enhancing the readability and interpretability of line plots. These techniques help guide the viewer's attention, reduce visual clutter, and provide additional context without overwhelming the plot.

In the next section, we will explore how to plot multiple series and overlays, ensuring that even complex data sets remain clear and easy to interpret.

2.6   Plotting Multiple Series and Overlays

Line plots often go beyond displaying a single data series. In many scenarios, it is necessary to compare multiple datasets on the same plot to reveal trends, relationships, or differences across different categories or variables. Plotting multiple series within the same chart, however, presents unique challenges, such as ensuring that the data remains distinguishable, managing overlaps, and making the plot easy to interpret. In this section, we explore effective techniques for comparing multiple datasets, using color and line styles for differentiation, handling overlapping data, and best practices for using legends in complex plots.

2.6.1   Comparing Multiple Data Sets on a Single Plot

One of the primary reasons for plotting multiple series on a single plot is to enable direct comparison between different datasets. Whether you are comparing trends across different time periods, analyzing different categories of data, or evaluating several variables simultaneously, plotting multiple series allows you to visualize relationships and contrasts more clearly than separate plots would.

When comparing multiple datasets, it's crucial to ensure that each series is displayed clearly and that the viewer can easily distinguish between them. Overlapping or closely packed data series can make it difficult to interpret the plot, so it's essential to use visual techniques that highlight each series without creating confusion.

A common strategy is to plot each dataset using distinct line colors or line styles (such as solid, dashed, or dotted). This allows the viewer to visually separate each series while still understanding the overall trends. However, it's important to avoid plotting too many series at once, as this can lead to clutter and make it difficult for the viewer to extract meaningful information. For line plots with more than five or six series, alternative approaches, such as splitting the data across multiple plots or using interactive features, may be more effective.

> **Design Tip: Avoid Overcrowding**
>
> Don't cram more than 5–6 data series into a single plot. Use small multiples or interactive toggles in dashboards to maintain clarity.

Another effective technique for comparing datasets is to use dual y-axes. This approach allows you to plot two different datasets with different ranges on the same chart, each corresponding to its own y-axis. For example, you might plot temperature on one axis and rainfall on the other, enabling the viewer to compare both variables without distorting the scale of either dataset. However, this technique should be used cautiously, as dual y-axes can sometimes lead to confusion or misinterpretation if not properly labeled and explained.

2.6.2 Using Color and Line Style to Differentiate Data

Color and line style are two of the most important visual tools for differentiating between multiple datasets on a single plot. When used thoughtfully, they can enhance the readability of the plot, allowing the viewer to quickly and easily distinguish between different data series.

Choosing distinct colors for each dataset is the most straightforward method of differentiation. When selecting colors, it's important to ensure that they are easily distinguishable, even when the lines cross or overlap. Colors should also be chosen with accessibility in mind, using color palettes that are color-blind friendly. For example, tools like ColorBrewer offer pre-designed color schemes that work well for visualizing multiple data series without causing confusion for viewers with color vision deficiencies.

In addition to color, line style (solid, dashed, dotted) can be used to differentiate between data series. This is particularly useful in black-and-white prints or in situations where color may not be available, such as in presentations with limited color options. Using a combination of color and line style is often the best approach, as it provides multiple visual cues to help the viewer distinguish between datasets. For example, you might use solid lines for one category of data, dashed lines for another, and dotted lines for a third category, with each category using a distinct color.

Line thickness can also be adjusted to emphasize certain datasets over others. Thicker lines draw attention and can be used for the most important or primary data series, while thinner lines can be used for secondary or less critical data.

2.6.3 Handling Overlapping Data Points

One of the biggest challenges in plotting multiple series is dealing with overlapping data points. When data series overlap or intersect frequently, it can be difficult for the viewer to separate them visually and interpret the plot accurately. Fortunately, several techniques can help mitigate this issue.

One approach is to use transparency , also known as Alpha Blending. By making the lines partially transparent, you allow overlapping data to be visible without completely obscuring the other lines. This technique is particularly effective when plotting many data series on the same chart, as it helps reduce visual clutter and ensures that all datasets remain visible.

Another technique for handling overlapping data points is to stagger the data by introducing slight offsets. While you should be cautious not to distort the data, small shifts in the x or y values can help separate closely packed lines, making them easier to distinguish. For example, in a time series plot, you might slightly offset the lines for different categories to ensure they don't all converge at the same points.

Using different line styles or colors for overlapping series is another way to reduce confusion. For instance, if two lines cross frequently, assigning them different styles (such as solid and dashed) or distinctly contrasting colors will help the viewer follow each line independently, even at the points of overlap.

In cases where too much overlap makes the plot unreadable, consider using small multiples, where each dataset is plotted in a separate but similarly scaled subplot. This method avoids overlap entirely and allows for a more detailed comparison between datasets.

2.6.4 Best Practices for Legends in Complex Plots

When plotting multiple series, the legend becomes a critical tool for helping viewers understand the data. A well-designed legend ensures that the viewer can quickly identify which line corresponds to which dataset, reducing confusion and enhancing the overall clarity of the plot.

The most important rule for designing legends is to make sure they are easily accessible and understandable. In most cases, the legend is placed in a corner of the plot, but it's also important to ensure that it doesn't obscure any critical data points or overlap with the lines themselves. Alternatively, you can place the legend outside the plot area, giving the plot itself more room for the data and reducing the risk of visual clutter.

Another best practice is to use direct labeling of lines instead of, or in addition to, a traditional legend. In this approach, each line is labeled directly at its endpoint, eliminating the need for the viewer to cross-reference the legend. This method works especially well in plots with a limited number of data series and is often preferred for its simplicity and ease of interpretation.

In complex plots with many data series, consider making the legend interactive, particularly in digital or web-based visualizations. An interactive legend allows the viewer to toggle the visibility of individual data series, making it easier to focus on specific trends without being overwhelmed by too much information at once. For example, a viewer might click on a category in the legend to highlight or hide that data series, dynamically adjusting the plot to suit their analysis.

Legends should also use symbols or markers that correspond directly to the line styles, colors, and thicknesses used in the plot. Ensure that the symbols are large enough to be easily readable and that there is sufficient space between each entry in the legend to avoid confusion.

Plotting multiple series and overlays requires careful attention to color, line style, and handling of overlapping data points. By using these techniques, you can create clear and effective visualizations that allow viewers to compare and analyze different datasets simultaneously. Thoughtful use of legends, particularly in complex plots, ensures that viewers can easily interpret the data and extract meaningful insights. In the next section, we will explore advanced techniques for line plot design, including the use of shading to show uncertainty and combining line plots with other plot types.

2.7 Advanced Techniques for Line Plot Design

Beyond the basic design principles of line plots, advanced techniques can significantly enhance the depth, insight, and interactivity of visualizations. These methods are especially useful when working with more complex datasets or when you need to communicate additional layers of information, such as uncertainty or correlations between multiple variables. In this section, we will explore advanced techniques, including the use of shading to represent uncertainty, combining line plots with other plot types like bar and area plots, and designing interactive line plots for web and dashboard applications.

2.7.1 Using Shading to Show Uncertainty

In many datasets, particularly in scientific research and forecasting, it is important to communicate not only the trend in the data but also the uncertainty or variability around that trend. One of the most effective ways to represent uncertainty in a line plot is through the use of shaded areas.

Shading can be used to represent confidence intervals, prediction ranges, or variability in the data. For example, in a line plot showing the projected growth of a population, shading around the line could represent the 95% confidence interval, indicating the range within which the true values are likely to fall. The width of the shaded area conveys the degree of uncertainty–narrow shading suggests more confidence in the data, while wide shading indicates greater uncertainty.

```python
import numpy as np
import matplotlib.pyplot as plt

x = np.linspace(0, 10, 100)
y = np.sin(x)
y_lower = y - 0.2
y_upper = y + 0.2

plt.plot(x, y, color='blue', label='Mean Line')
plt.fill_between(x, y_lower, y_upper, color='blue', alpha=0.3,
    label='95% CI')
plt.title('Line Plot with Uncertainty Shading')
plt.legend()
plt.grid(True)
plt.show()
```

To use shading effectively, it's important to choose colors and opacities that complement the plot without overwhelming it (Figure 2.3). A common approach is to use a semi-transparent shade in a color that contrasts with the background but is lighter than the line itself. This ensures that the shading does not obscure the main data line but still clearly indicates the uncertainty range. For example, if the line is a dark blue, the shaded area could be a light, semi-transparent blue.

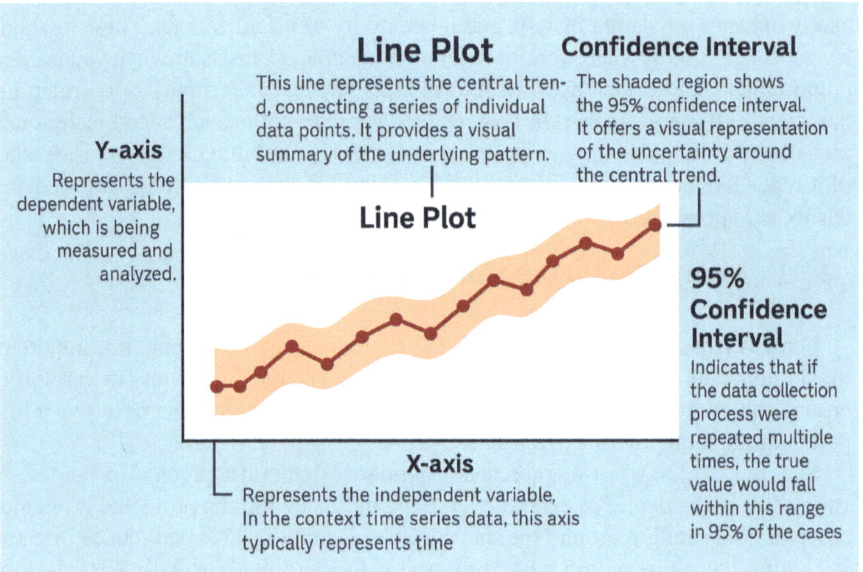

Fig. 2.3: Line plot with 95% confidence interval shading. The shaded region visually communicates uncertainty around the central trend.

In addition to showing uncertainty, shading can also be used to highlight regions of interest within the data. For instance, you might use shading to emphasize a time period of particular significance, such as the onset of a financial crisis or a peak in sales. In this case, the shading acts as a visual guide, helping the viewer focus on the most important parts of the plot.

2.7.2 Combining Line Plots with Other Plots

Line plots are powerful tools for showing trends, but they can be even more informative when combined with other plot types that offer complementary insights. By integrating line plots with bar or area plots, you can represent different aspects of the data within a single visualization, making it easier for viewers to understand complex relationships.

One common combination is the use of line plots with bar charts. This is particularly useful when you want to compare a continuous variable (represented by the line) with a discrete or categorical variable (represented by the bars). For example, in a financial plot, you might use a line to represent the stock price over time and bars to show the trading volume for each period. This combination allows the viewer to see both the trend in the stock price and how it relates to changes in trading volume, offering a more comprehensive view of the data.

Another popular combination is the use of line plots with area plots. In an area plot, the space beneath the line is filled, creating a visual emphasis on the cumulative total or magnitude of the variable. This technique works well for showing the relative proportions of different categories over time. For example, in a plot of sales data, you could use an area plot to show the total sales volume, with different segments of the area corresponding to different product categories, and overlay a line plot to show the trend in overall sales.

When combining plot types, it's important to ensure that the different visual elements remain distinguishable and that the plot doesn't become overly complex or cluttered. Using consistent color schemes and line styles helps keep the plot visually coherent. It is crucial to ensure that the y-axes are aligned or that dual axes are clearly labeled to prevent misinterpretation of the data.

2.7.3 Interactive Line Plots for Web and Dashboard

In the era of web-based dashboards and interactive visualizations, static plots are often insufficient to meet the needs of users who want to explore and analyze data in real-time. Interactive line plots offer a dynamic solution, allowing users to manipulate the plot, drill down into specific data points, and customize the view based on their needs.

One of the key features of interactive line plots is the ability to display tooltips. As the user hovers over the line or specific data points, additional information (such as the exact value, date, or related metadata) can be displayed in a small tooltip box. This interaction enhances the user's understanding of the plot without cluttering the main visualization with too many static labels or annotations.

Another useful feature in interactive plots is zooming and panning . When working with large datasets, it can be challenging to display all the data at once without losing clarity. Interactive zooming allows users to focus on what they want.

2.8 Avoiding Common Pitfalls in Line Plot Design

While line plots are powerful tools for visualizing data, poor design choices can lead to confusion, misinterpretation, or even misinformation. To ensure that your line plots remain effective and accurate, it's essential to be aware of common pitfalls in line plot design and how to avoid them. In this section, we will discuss the dangers of misleading axis scaling, the risks of visual overload when using too many lines in a single plot, and best practices for handling data gaps and breaks (Table 2.6).

Table 2.6: Common Pitfalls and How to Avoid Them

Pitfall	Solution
Misleading Axis Scaling	Use meaningful zero references or justify scale breaks
Overcrowded Lines	Limit to 5–6 lines or use small multiples
Unlabeled Axes	Always label axes with variable name and unit
Too Many Colors	Use a limited, accessible palette; reinforce with styles
Ignored Data Gaps	Show gaps transparently or annotate missing data

---

Design Tip: Highlight with Purpose

Use annotations, shaded regions, or arrows to spotlight insights – not everything. Strategic emphasis helps guide interpretation without overwhelming the viewer.

2.8.1 Misleading Axis Scaling and its Consequences

One of the most significant pitfalls in line plot design is the misuse of axis scaling. The way you scale the axes can dramatically affect how the data is perceived, and poor choices can unintentionally or intentionally mislead the viewer. Misleading axis scaling can cause viewers to overestimate trends, underplay the significance of certain data points, or misinterpret the overall message of the plot.

A common issue arises when the y-axis does not start at zero, especially when displaying data that fluctuates over a small range. For example, if the y-axis is truncated to emphasize small changes, a minor fluctuation in the data might appear as a significant trend. This can exaggerate differences and give the impression that changes are more drastic than they actually are. Starting the y-axis at zero, when appropriate, ensures that the viewer gets a more accurate sense of the data's scale.

⚠ Warning

Never truncate the y-axis to exaggerate small differences unless it is clearly justified and annotated. A misleading scale can distort perception and damage trust in your visualization.

On the other hand, starting the y-axis at zero may not always be the best choice. For datasets with small variations over a narrow range, such as stock prices or temperature differences, starting the axis at zero can compress the data, making important trends harder to discern. In these cases, it's important to carefully consider how much of the axis should be shown and ensure that any adjustments are clearly labeled and explained.

Logarithmic scales can also be used in certain cases, particularly for data that grows exponentially or spans several orders of magnitude. However, logarithmic scales can confuse viewers if they are not familiar with them, so it's important to clearly indicate when a logarithmic scale is being used and why it is appropriate for the data.

Ensuring that your axis scaling accurately reflects the nature of the data is crucial for maintaining the integrity of your line plot. When making adjustments to the axes, transparency and clear labeling are key to preventing misinterpretation.

2.8.2 Visual Overload: Too Many Lines in One Plot

Line plots are excellent for comparing multiple datasets, but there is a limit to how much information can be effectively conveyed in a single plot. Including too many lines in a single plot can lead to visual overload, making it difficult for viewers to distinguish between data series or extract meaningful insights.

When a line plot becomes cluttered with too many lines, the result is often visual noise: lines overlap, intersect, or blend together, causing confusion and reducing the overall clarity of the plot. To avoid this, it's important to strike a balance between including enough data to make the plot informative while ensuring that the visualization remains readable (Figure 2.4).

One way to reduce visual overload is by limiting the number of lines in the plot. If the dataset contains a large number of series, consider splitting the data across multiple plots or using small multiples (a series of smaller plots with consistent scales) to allow for better comparison. This approach maintains the clarity of each individual plot while still providing an overview of all the data.

Another strategy is to use color and line style thoughtfully. Assign distinct colors and line styles (such as solid, dashed, or dotted lines) to each series, but avoid using too many similar colors or styles that may be hard to differentiate. For particularly dense plots, using transparency (Alpha Blending) can also help prevent overlap and ensure that all data series remain visible.

Interactive plots offer another solution to visual overload, allowing users to toggle individual data series on or off. This approach enables the viewer to focus on specific data without being overwhelmed by too many lines at once. By giving users control over the data they wish to view, interactive plots can significantly enhance the usability of complex visualizations.

```python
import plotly.graph_objs as go
import plotly.io as pio

x = [1, 2, 3, 4, 5]
y = [2, 1, 3, 5, 4]

fig = go.Fig.()
fig.add_trace(go.Scatter(x=x, y=y, mode='lines+markers',
    name='Interactive Line'))
fig.update_layout(title='Interactive Line Plot',
                xaxis_title='X Axis',
```

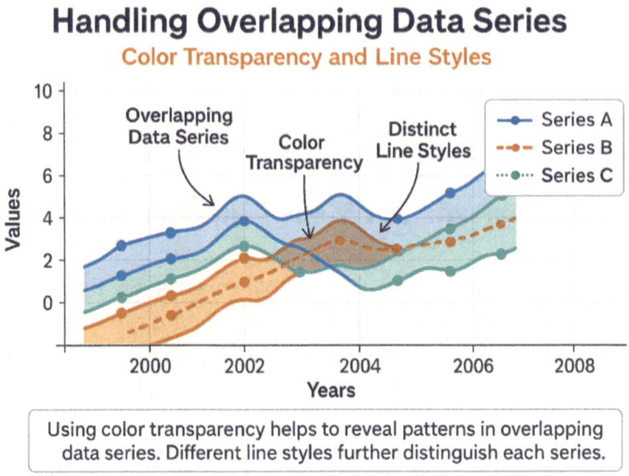

Fig. 2.4: Handling overlapping data series with color transparency and distinct line styles to maintain readability.

```
11              yaxis_title='Y Axis')
12   pio.show(fig)
```

2.8.3 Proper Handling of Data Gaps and Breaks

In real-world datasets, missing data is a common occurrence. Whether due to equipment failures, incomplete surveys, or data collection interruptions, gaps in the data can disrupt the continuity of a line plot. How these gaps are handled can significantly impact the plot's interpretation.

One of the most straightforward approaches to dealing with missing data is to leave gaps in the line where the data is unavailable. This method preserves the integrity of the plot by clearly indicating that no data was collected during that period. It also prevents the introduction of potentially misleading data through interpolation. However, this approach can make the plot look incomplete or fragmented, particularly if there are frequent data gaps.

Another option is to connect the data points before and after the missing data using interpolation. Linear interpolation draws a straight line between the two points, filling in the gap, while other forms of interpolation, such as spline interpolation, can create smoother curves. However, interpolation should be used with caution, as it introduces an assumption about the behavior of the data in the missing interval. If the data is known to vary significantly, interpolation can be misleading. When using interpolation, it is important to clearly indicate in the plot or its caption that interpolation has been applied and to provide a rational explanation.

> **⚠ Warning**
>
> Interpolating across missing data without labeling or justification can mislead users into thinking the values are known (Figure 2.5). Always disclose when and how interpolation is applied.

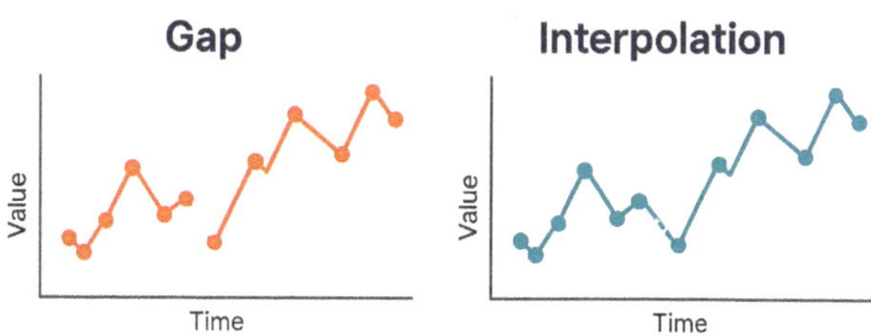

Fig. 2.5: Proper handling of data gaps and breaks in line plots. The infographic compares three techniques: (1) leaving gaps to transparently reflect missing data, (2) applying linear interpolation to bridge gaps with straight segments, and (3) using spline interpolation to generate smooth curves. Each method affects interpretability and should be clearly indicated in captions or legends to maintain data integrity.

> **Design Tip: Visualize Gaps Transparently**
>
> Never connect missing data without clearly signaling interpolation. Use dashed lines, annotations, or lighter transparency to indicate estimates or unknown values.

2.9 Case Studies and Real-World Examples

To further illustrate the concepts discussed throughout this chapter, we will explore several case studies that demonstrate the practical application of line plots in real-world contexts. These examples highlight the versatility of line plots in visualizing diverse types of data, from climate trends to financial markets and sensor monitoring. Each case study showcases how thoughtful design choices can enhance the clarity and effectiveness of a line plot while avoiding common pitfalls.

2.9.1　Line Plot for Climate Data Visualization

Climate data, which often spans decades or centuries, is a prime example of how line plots can be used to reveal long-term trends and changes in the environment. In this case study, we examine the use of line plots to visualize global temperature changes over time, a critical component of climate science communication.

The line plot in this example shows the global average surface temperature anomalies from the 1880s to the present day. The x-axis represents the years, while the y-axis displays temperature anomalies in degrees Celsius relative to a baseline period. The goal of the plot is to highlight the overall trend of increasing global temperatures, as well as the variability within individual years.

A critical design decision in this case is the use of a linear y-axis starting from a neutral baseline (0°C anomaly). This choice emphasizes both the magnitude of temperature deviations and the overall trend of warming. The line is colored in shades of red to symbolize heat, reinforcing the message of global warming. To avoid visual overload, the plot focuses on the global average rather than displaying multiple data series for individual regions, which could complicate the visualization.

In addition to the line plot itself, the design includes shading to indicate the 95% confidence intervals, which represents the uncertainty in the measurements. This shaded area provides a clear visual indication of the reliability of the data, making the viewer aware of the potential variability while maintaining focus on the overall trend (Figure 2.6).

```python
import matplotlib.pyplot as plt

# --- Visualization ---
plt.figure(figsize=(12, 6))
plt.plot(years, anomalies, color='darkred', linewidth=2.5,
    label='Temperature Anomaly')
plt.fill_between(years, lower_ci, upper_ci, color='salmon', alpha=0.3,
    label='95% Confidence Interval')

# Baseline line at 0°C
plt.axhline(0, color='gray', linestyle='--', linewidth=1)

# --- Labels and design ---
plt.title('Global Average Surface Temperature Anomalies (1880{2020)',
    fontsize=14, weight='bold')
plt.xlabel('Year', fontsize=12)
plt.ylabel('Temperature Anomaly (°C)', fontsize=12)
plt.grid(True, linestyle=':', linewidth=0.7, alpha=0.7)
plt.legend(loc='upper left', frameon=False)
plt.tight_layout()

# --- Accessibility enhancements ---
plt.xticks(fontsize=10)
plt.yticks(fontsize=10)
```

```
25  plt.gca().set_facecolor('#f9f9f9')   # Light background
26  plt.show()
```

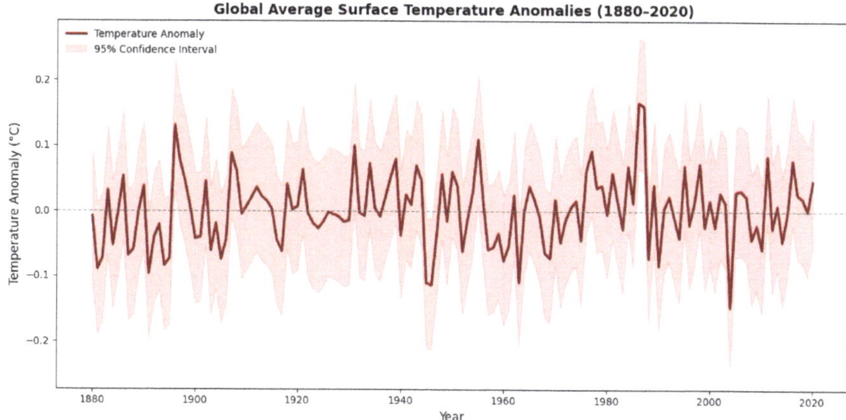

Fig. 2.6: Line plot showing global temperature anomalies from 1880 to present. Shaded area represents uncertainty; annotations highlight key events.

2.9.2   Visualizing Stock Market Trends with Line Plots

The financial sector often relies on line plots to visualize stock market trends over time, allowing investors and analysts to track price movements, identify patterns, and make informed decisions. In this case study, we examine the use of line plots to visualize stock prices for a publicly traded company over a five-year period.

The line plot displays daily closing prices on the y-axis and dates on the x-axis, providing a detailed view of the stock's performance over time (Figure 2.7). To avoid visual overload, only key events (such as quarterly earnings reports, stock splits, or major news announcements) are annotated, using callouts to draw attention to moments where the stock price experienced significant changes. A dual-axis approach is employed to compare the stock price with the company's trading volume. The left y-axis represents the stock price, while the right y-axis shows the volume of shares traded. This combination allows viewers to explore the relationship between price changes and trading activity. By observing spikes in volume alongside price fluctuations, users can gain insights into investor behavior during critical periods, such as during earnings announcements or market downturns.

To further enhance clarity, the plot uses a combination of color and line style. The stock price is represented with a solid black line, while the trading volume is shown with bars in a semi-transparent color to avoid dominating the plot. This contrast ensures that the price trend remains the focal point of the visualization, while the volume data serves as a complementary layer.

```python
import matplotlib.pyplot as plt

# Create figure and axis objects
fig, ax1 = plt.subplots(figsize=(12, 6))

# Plot closing price
ax1.plot(dates, closing_prices, color='black', linewidth=2,
    ↪ label='Closing Price')
ax1.set_xlabel('Date')
ax1.set_ylabel('Stock Price ($)', color='black')
ax1.tick_params(axis='y', labelcolor='black')
ax1.grid(True, linestyle=':', alpha=0.6)

# Plot key event annotations
for idx, label in key_events.items():
    ax1.annotate(label,
                 xy=(dates[idx], closing_prices[idx]),
                 xytext=(dates[idx], closing_prices[idx] + 5),
                 textcoords='data',
                 arrowprops=dict(arrowstyle='->', color='gray'),
                 fontsize=9,
                 color='darkred')

# Secondary y-axis for volume
ax2 = ax1.twinx()
ax2.bar(dates, volumes, width=1.0, color='skyblue', alpha=0.4,
    ↪ label='Volume')
ax2.set_ylabel('Volume (Shares)', color='skyblue')
ax2.tick_params(axis='y', labelcolor='skyblue')

# Title and layout
plt.title('Stock Price and Trading Volume Over Time', fontsize=14,
    ↪ weight='bold')
fig.tight_layout()
plt.show()
```

Interactive features, such as zooming and tooltips, are often added in digital versions of stock market line plots. These allow users to explore specific time periods or points of interest in greater detail, providing additional flexibility in how the data is viewed. By hovering over data points, users can see the exact stock price and volume for each day, enhancing the plot's utility as a tool for in-depth analysis.

2.9.3 Monitoring Sensor Data in Real-Time

In industries such as manufacturing, healthcare, and environmental monitoring, real-time sensor data is essential for maintaining systems and ensuring safety. In this case study, we explore how line plots can be used to monitor sensor data from a

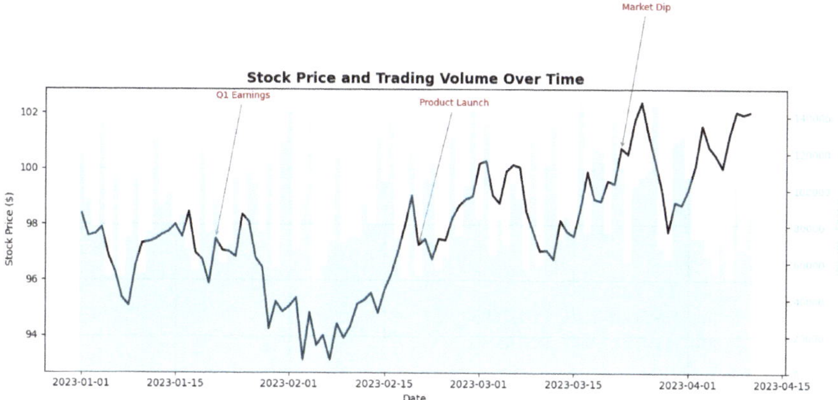

Fig. 2.7: Dual-axis line and bar plot combining stock price (line) and trading volume (bars). Vertical callouts highlight earnings reports and other key events.

manufacturing process, providing real-time insights into performance and potential issues.

The line plot displays temperature readings from a sensor installed in a critical piece of machinery (Figure 2.8). The x-axis represents time, with data points added continuously as new readings are received, while the y-axis shows the temperature in degrees Celsius. The primary objective of the plot is to detect abnormal fluctuations in temperature that could indicate malfunctions or the need for maintenance.

To make real-time monitoring effective, the plot includes several key features. First, the temperature line is updated in real-time, with each new data point being appended to the right side of the plot. This ensures that operators have an up-to-date view of the system's performance at all times. The line is color-coded to provide quick visual feedback: when the temperature remains within acceptable limits, the line is green, but if the temperature exceeds the defined safety threshold, the line turns red. This immediate visual cue allows operators to take action quickly when issues arise.

In addition to the real-time data line, the plot includes historical data, shown in a lighter color to provide context for the current readings. This helps operators compare the current temperature to previous performance and identify trends that might indicate gradual deterioration or emerging issues.

Finally, the plot includes an alarm threshold, represented by a dashed horizontal line across the y-axis. If the temperature crosses this threshold, a callout or visual alert is triggered, notifying the operator of the potential problem. This combination of real-time updates, historical context, and visual alerts ensures that the plot is not only informative but also actionable, providing the necessary information to maintain safe and efficient operations.

```python
import matplotlib.pyplot as plt
import matplotlib.animation as animation
import numpy as np
from collections import deque
import datetime

# --- Simulation Settings ---
MAX_POINTS = 100
TEMP_THRESHOLD = 75   # Celsius

# --- Data Initialization ---
x_data = deque(maxlen=MAX_POINTS)
y_data = deque(maxlen=MAX_POINTS)

# Start with historical baseline
current_time = datetime.datetime.now()
for i in range(MAX_POINTS):
    x_data.append(current_time + datetime.timedelta(seconds=i))
    y_data.append(60 + np.random.normal(0, 1))

# --- Plot Setup ---
fig, ax = plt.subplots(figsize=(12, 6))
ax.set_title("Real-Time Temperature Monitoring", fontsize=14,
    weight='bold')
ax.set_xlabel("Time")
ax.set_ylabel("Temperature (°C)")
ax.axhline(TEMP_THRESHOLD, color='red', linestyle='--', linewidth=1.5,
    label='Safety Threshold')

line_history, = ax.plot([], [], color='lightgray', linewidth=1.0,
    label='Historical')
line_live, = ax.plot([], [], color='green', linewidth=2.0,
    label='Live Temperature')

ax.legend(loc='upper left')
plt.xticks(rotation=45)
plt.tight_layout()

# --- Update Function for Animation ---
def update(frame):
    now = x_data[-1] + datetime.timedelta(seconds=1)
    new_temp = 60 + np.random.normal(0, 1.5)

    # Simulate a spike
    if frame % 60 == 0:
        new_temp += 20

    x_data.append(now)
    y_data.append(new_temp)

    # Update colors based on threshold
    line_live.set_color('green' if new_temp < TEMP_THRESHOLD else 'red')
```

```python
    line_history.set_data(x_data, y_data)
    line_live.set_data(x_data, y_data)

    ax.set_xlim(x_data[0], x_data[-1])
    ax.set_ylim(min(y_data) - 5, max(y_data) + 5)

    return line_live, line_history

# --- Animate ---
ani = animation\index{Animation}.FuncAnimation(fig, update, interval=1000)
plt.show()
```

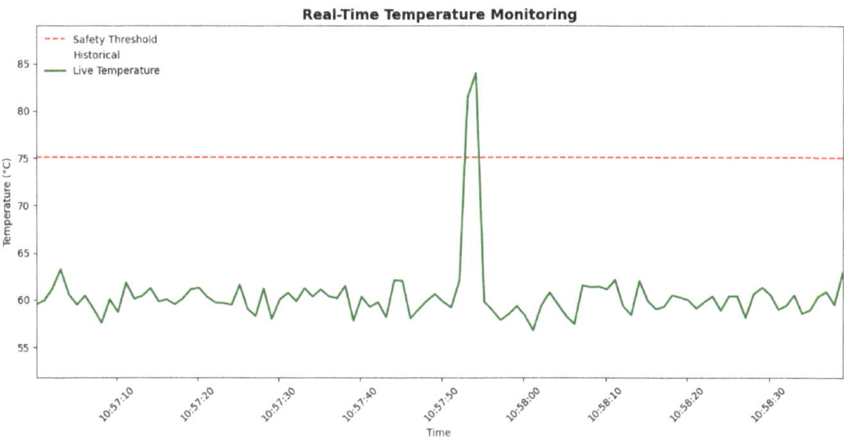

Fig. 2.8: Live-updating sensor data line plot. Green indicates normal operation; red segments show temperature exceeding thresholds. Horizontal dashed line shows the critical limit.

These case studies highlight the versatility of line plots across a range of real-world applications, from visualizing long-term climate trends to monitoring real-time sensor data (Table 2.7). By applying the principles discussed throughout this chapter–careful axis scaling, thoughtful use of color and line style, and smart labeling techniques–you can create line plots that not only convey critical information but also engage and inform the viewer.

2.10   Design Tips and Tools for Creating Line Plots

Creating effective and visually appealing line plots requires both technical skill and an understanding of design principles. Fortunately, there are numerous tools and libraries available that simplify the process of building professional-quality line plots. In this section, we will explore some of the most commonly recommended tools

Table 2.7: Expert Design Features Reflected in Case Study Visualizations

Design Feature	Global Temperature Anomalies	Stock Market Trends	Real-Time Sensor Monitoring
Use of color for semantic meaning	Shades of red to indicate warming	Black for price, blue for volume	Green for safe, red for overheat
Uncertainty visualization	95% confidence interval with shaded area	Not applicable	Not shown, but threshold used as a proxy
Axis scaling and baselines	Linear scale with 0°C baseline	Dual y-axes for price and volume	Fixed threshold with adaptive y-limits
Annotations and callouts	Highlight key climate events (e.g., volcanic eruptions)	Annotate earnings, splits, crashes	Alert callout when temperature spikes
Visual hierarchy and clarity	Focus on one global trend, subtle uncertainty shading	Line dominates over semi-transparent bars	Bright live line over faded historical trace
Interaction or animation potential	Suitable for static or animated change detection	Enables interactive zoom or tooltips	Designed for animated real-time display
Accessibility	Colorblind-friendly palettes and contrast	High contrast elements; avoids clutter	Simple cues and real-time interpretability

for designing line plots, offer specific tips for using Python libraries like Matplotlib and Plotly, and discuss the importance of using templates and themes for consistent, polished design.

2.10.1   Recommended Tools for Designing Line Plots

There are several powerful tools and libraries available for creating line plots, each with its own strengths and ideal use cases. Choosing the right tool depends on the complexity of the data, the level of customization required, and the intended audience for the plot. Here are a few of the most widely used tools for line plot creation:

- ➤ **Matplotlib (Python):** Matplotlib is a highly versatile plotting library for Python that is ideal for creating static, publication-quality line plots. It offers a wide range of customization options, from controlling axes and scales to styling lines and adding annotations. Matplotlib is well-suited for scientists and engineers who need precise control over every aspect of the plot.

- ➤ **Plotly (Python, JavaScript):** Plotly is a library that excels in creating interactive plots. It is available for both Python and JavaScript, making it a great choice for web applications and dashboards. Plotly's interactivity allows users to zoom, hover over data points for more information, and dynamically update plots based on user input.

➢ **Excel:** For basic line plots, Excel remains a popular choice, especially for business and non-technical users. Excel offers easy-to-use charting tools, making it accessible to users with little programming experience. It's ideal for simple data visualizations in reports and presentations.

➢ **Tableau:** Tableau is a business intelligence tool that enables users to create dynamic, interactive visualizations, including line plots. Its drag-and-drop interface and powerful data integration capabilities make it a popular choice for non-technical users who need to analyze large datasets.

➢ **D3.js (JavaScript):** D3.js is a JavaScript library used for creating sophisticated, interactive data visualizations on the web. It offers extensive customization and is widely used for building web-based dashboards and interactive plots. D3.js requires more technical expertise than some other tools but provides unmatched flexibility and control over plot design.

Each of these tools has its strengths, and the right choice will depend on your specific needs. For static, highly customizable plots, Matplotlib is often the best option, while Plotly and D3.js excel in interactivity and web integration (Table 2.8).

2.10.2 Tips for Designing Line Plots in Python

Python is one of the most widely used languages for data visualization, thanks to its powerful plotting libraries like Matplotlib and Plotly. Here are some specific tips for designing effective line plots using these libraries:
Matplotlib Tips:

➢ **Use Subplots for Complex Data:** When dealing with multiple data series or different types of plots, consider using Matplotlib's subplot feature. This allows you to create multiple line plots in a single figure, each with its own axes and labels, making complex comparisons easier to understand.

➢ **Customize Axes and Labels:** Matplotlib provides fine control over axis scales, ticks, and labels. Be sure to customize your axes to reflect the nature of the data accurately. For time series data, use '$plt.gca().xaxis.set_{major formatter}()$' to format dates appropriately and avoid overlapping labels.

➢ **Use Color Maps for Differentiation:** Matplotlib's built-in color maps ('plt.cm') can help you assign distinct colors to multiple lines. Consider using perceptually uniform color maps (e.g., 'viridis', 'plasma') to ensure color distinctions are clear, even for those with color vision deficiencies.

➢ **Add Error Bars and Confidence Intervals:** For scientific data, it's often important to display error bars or confidence intervals. Matplotlib makes this easy with the 'plt.errorbar()' function, allowing you to add vertical or horizontal error bars to your plot.

➢ **Annotate Key Points:** Use 'plt.annotate()' to call out important data points or trends. Annotations can be positioned precisely and connected to the corresponding data point with arrows or lines to ensure clarity.

Table 2.8: Comparison of Visualization Tools for Line Plot Design

Tool	Strengths	Interaction Level	Customization	Ease of Use	Recommended Use Cases
Matplotlib (Python)	High-quality static plots; publication-ready	Low (static only)	**Very High** – full control over all plot elements	Intermediate to advanced – scripting required	Scientific research, academic publications, reproducible analysis with full control
Plotly (Python, JS)	Interactive plots; browser and dashboard ready	**High** – zoom, pan, tooltips, toggle series	High – supports callbacks, custom themes	Intermediate – scripting knowledge helpful	Interactive dashboards, web apps, exploratory analysis, business analytics
Excel	Ubiquitous and user-friendly; quick setup	Moderate – basic tooltips and interactivity in newer versions	**Low** – limited styling and logic control	**Very High** – GUI-based, no code needed	Basic business plots, internal presentations, accessible to non-technical users
Tableau	Dynamic BI dashboards; strong data connectors	**High** – filters, tooltips, real-time updates	Moderate – GUI-based styling, no scripting	**High** – intuitive drag-and-drop interface	Business intelligence reporting, executive dashboards, multi-source analytics
D3.js (JavaScript)	Unmatched flexibility; custom web visualizations	**Very High** – complete interactivity control, animations	**Very High** – fine-grained DOM + SVG control	**Low** – steep learning curve, requires JS/HTML/CSS expertise	Advanced web-based data storytelling, interactive infographics, high-end applications

Plotly Tips:

➤ **Use Interactivity:** One of Plotly's strengths is its interactivity. Make use of hover tooltips to display additional information about each data point. This allows you to keep the plot clean while still providing detailed information when needed.

➤ **Use Legends and Filtering:** Plotly supports interactive legends, which allow users to toggle data series on and off. This is especially useful when plotting multiple lines, as users can focus on the most relevant data without cluttering the view.

➤ **Take Advantage of Built-in Themes:** Plotly offers several built-in themes ('plotly', 'ggplot2', 'seaborn', etc.) that provide pre-configured styles and colors.

Using these themes can help ensure that your plots are visually appealing and consistent.

> **Include Annotations and Shapes:** Plotly allows you to add dynamic annotations and shapes (such as shaded areas or lines) that update when users interact with the plot. These features are useful for highlighting thresholds, trends, or important events.

Both Matplotlib and Plotly offer extensive customization options, and by mastering the capabilities of each library, you can create highly effective line plots tailored to your specific data visualization needs.

2.10.3   Using Templates and Themes for Consistent Design

Consistency is key when creating multiple plots, especially in reports, presentations, or dashboards where a uniform style is important for maintaining visual coherence. Using templates and themes can help ensure that your line plots are stylistically consistent and polished, regardless of the number of plots or the tools being used.

Themes in Python (Matplotlib and Plotly):

> **Matplotlib:** Matplotlib offers style sheets ('plt.style.use()') that allow you to apply predefined themes across your plots. Popular themes include 'ggplot', 'seaborn', and 'bmh', each of which comes with its own set of colors, gridlines, and fonts. You can also create your own custom style sheets by modifying the 'rcParams' settings in Matplotlib.

> **Plotly:** In Plotly, you can apply themes such as 'plotly_dark' or 'presentation' to quickly change the look and feel of your plot. Plotly also allows for more detailed customization, including background color, font style, and gridline settings, making it easy to create consistent designs across multiple plots.

Using Templates in Other Tools:

> **Excel and Tableau:** Both Excel and Tableau offer built-in templates that allow you to apply consistent design elements to all of your visualizations. In Excel, you can save chart styles as templates and apply them to new charts with a single click, ensuring that fonts, colors, and other elements remain consistent. Tableau provides theme options and formatting templates to ensure that visualizations in dashboards maintain a cohesive design.

> **Custom Templates:** For large projects or repeated reports, consider creating your own template that includes your preferred color scheme, font choices, and line styles. This will not only save time but also ensure that all visualizations meet your design standards.

By using templates and themes, you can ensure that your line plots are consistently styled and professional, which is especially important when producing visualizations for external audiences or publication.

By choosing the right tools, mastering design techniques in Python libraries like Matplotlib and Plotly, and using templates and themes to maintain a consistent design, you can create highly effective and visually engaging line plots. These strategies help streamline the design process and ensure that your visualizations are both clear and aesthetically pleasing, making them more impactful for your audience. To ensure that these strategies are followed in practice, the designers may use the checklist presented in Table 8.5.

Table 2.9: Line Plot Design Checklist

Element	Checked?
Clear axis labels and units	☐
Appropriate scale and range	☐
Color-blind friendly palette	☐
Consistent line styles and thicknesses	☐
Annotations or tooltips included	☐
Interactive features (if needed)	☐
Gaps in data handled transparently	☐
Legend or direct labels used clearly	☐
Minimal visual clutter	☐
Template/theme applied	☐

2.11   Conclusion and Future Trends

Line plots are a fundamental tool in data visualization, offering an intuitive way to represent trends and relationships over time or across continuous variables. Throughout this chapter, we have explored both the basic and advanced techniques for designing clear, effective, and aesthetically pleasing line plots. In this concluding section, we will summarize the key takeaways from the chapter, look ahead to upcoming trends in line plot design, and reflect on the importance of a human-centered approach to visualization.

2.11.1   Summary of Key Takeaways

This chapter has provided a comprehensive guide to designing line plots, covering everything from the essential components of a line plot to advanced techniques that enhance readability and user interaction. Here are the key takeaways:

➤ **Core Components:** A well-designed line plot includes clearly labeled axes, appropriate scales, legible data points, and informative annotations or callouts to guide the viewer. Legends are crucial when displaying multiple data series, and careful attention should be paid to axis scaling to avoid misleading interpretations.

➤ **Aesthetic Choices:** Thoughtful use of color, line style, and line thickness is essential for differentiating between data series and ensuring the plot is visually engaging without becoming cluttered. Background choices, such as white or colored backgrounds, should complement the data without detracting from its clarity.

➤ **Handling Multiple Series and Overlays:** When plotting multiple datasets on the same plot, it's important to use distinct visual cues such as color, line style, and transparency to differentiate between them. Interactive features, such as legends and tooltips, can help manage complexity and prevent visual overload.

➤ **Advanced Techniques:** Shading can be used to indicate uncertainty or confidence intervals, while combining line plots with other plot types (e.g., bar or area plots) can reveal additional insights. For interactive applications, dynamic line plots in dashboards or web applications offer users greater flexibility to explore data.

➤ **Avoiding Common Pitfalls:** Misleading axis scaling, overcrowding the plot with too many lines, and mishandling data gaps are common pitfalls that can undermine the effectiveness of a line plot. Thoughtful design and transparency in handling missing data ensure that the plot remains accurate and trustworthy.

➤ **Tools and Techniques:** Tools such as Matplotlib, Plotly, and Tableau are widely used for creating line plots, each offering unique advantages. Python libraries, in particular, provide extensive customization options, allowing for both static and interactive visualizations. Consistency across multiple plots can be achieved using templates and themes.

By applying these principles, you can create line plots that not only convey data clearly but also engage and inform the viewer, making your visualizations powerful tools for communication.

2.11.2 Trends in Line Plot Design

As technology and data visualization practices continue to evolve, several emerging trends are shaping the future of line plot design. Some of the most significant upcoming trends include:

➤ **Interactive Visualizations:** As data becomes more complex, there is a growing demand for interactive visualizations that allow users to explore data dynamically. In line plots, this means offering features such as zooming, filtering, and tooltips that provide more detailed information on demand. Libraries like Plotly, D3.js, and other web-based tools are pushing the boundaries of what is possible with interactive line plots.

➤ **Real-Time Data Visualization:** With the rise of Internet of Things (IoT) devices, sensors, and real-time data streams, line plots that update in real-time are becoming increasingly important. Monitoring dashboards for industries like manufacturing, healthcare, and finance rely on real-time line plots to display critical information and alert users to anomalies as they happen.

➤ **Increased Focus on Accessibility:** Ensuring that line plots are accessible to a broad audience, including people with color vision deficiencies, is becoming a major consideration in design. Color-blind friendly palettes and increased contrast ratios are being incorporated into design tools and libraries to make visualizations more inclusive.

➤ **AI-Enhanced Data Visualization:** Artificial intelligence and machine learning are starting to influence data visualization. Automated insights, anomaly detection, and predictive analytics are being integrated into line plots to provide users with more actionable insights. These technologies will likely lead to smarter, more informative visualizations that help users uncover patterns they may not have spotted on their own.

➤ **Minimalist Design with Greater Emphasis on Clarity:** As visualizations become more ubiquitous in communication, there is a trend toward minimalist design that emphasizes clarity and simplicity. Designers are increasingly removing unnecessary elements (like excessive gridlines or cluttered legends) to focus on the key message of the data.

These trends are shaping the way line plots are used and designed, ensuring that they remain relevant and effective in an increasingly data-driven world.

2.11.3 Closing Thoughts on Human-Centered Line Plot Design

At the heart of any good line plot is a focus on the needs of the user. Human-Centered Design ensures that visualizations are not only accurate and informative but also accessible, engaging, and easy to understand. As designers and analysts, we must always consider the perspective of the viewer, making sure that our choices in color, scale, and layout serve to enhance comprehension rather than hinder it.

The best line plots are those that strike a balance between clarity and detail. They simplify complex data into an intuitive visual narrative while maintaining the richness of the underlying information. Whether you're visualizing climate data, stock market trends, or sensor readings, a human-centered approach ensures that your audience can connect with the data and derive meaningful insights.

Looking ahead, as the tools for data visualization continue to evolve and as new techniques emerge, the principles of good line plot design will remain the same: clarity, accuracy, and user-centricity. By keeping these principles at the forefront, you can create line plots that resonate with your audience and make a lasting impact.

As we conclude this chapter, remember that line plots, when designed well, are more than just data visualizations–they are tools for storytelling, enabling people to understand and engage with complex information in a way that is both intuitive and impactful. The future of line plot visualization promises even more exciting possibilities, and by staying informed of emerging trends, you can continue to create powerful, human-centered visualizations that inform and inspire.

Suggested Readings

1. Aigner, W., Kainz, C., Ma, R., Miksch, S.: Bertin was right: An empirical evaluation of indexing to compare multivariate time-series data using line plots. Computer Graphics Forum **30**(1), 215–228 (2011). DOI 10.1111/j.1467-8659.2010.01845.x
2. Lombardi, F., Goncu, C., Marinai, S.: Line recognition for generating accessible line plots. pp. 14–19 (2019). DOI 10.1109/ICDARW.2019.00008
3. Muigg, P., Hadwiger, M., Doleisch, H., Gröller, E.: Visual coherence for large-scale line-plot visualizations. Computer Graphics Forum **30**(3), 643–652 (2011). DOI 10.1111/j.1467-8659.2011.01913.x
4. Reese, R.: Line plots. Significance **3**(2), 85–87 (2006). DOI 10.1111/j.1740-9713.2006.00169.x

3 Designing Stunning Bar Charts

Abstract

Bar charts and box plots are foundational tools in data visualization, offering powerful ways to present categorical and distributional data. This chapter explores the art and science of designing stunning bar charts and box plots that go beyond mere aesthetics to achieve clarity, precision, and impact. It begins with an introduction to the purpose, importance, and use cases of these chart types, followed by detailed design principles for creating effective visualizations. Readers will learn how to choose the right chart type, optimize visual elements such as colors and labels, and avoid common pitfalls that can distort data interpretation. The chapter also analyzes the anatomy of bar charts and box plots, including advanced topics such as combining these chart types for richer insights and exploring alternative visualizations like violin plot. Practical case studies demonstrate real-world applications, highlighting the versatility and storytelling potential of these visualizations. The chapter concludes by providing best practices, a review of popular visualization tools, and guidance on next steps for advancing skills in data storytelling and design.

Aims

After reading this chapter you should be able to:

- ➢ Understand fundamental components and uses of bar charts and box plots.
- ➢ Choose the appropriate type of bar chart based on your data and communication goals.
- ➢ Apply design principles to create visually appealing and effective bar charts.
- ➢ Avoid common pitfalls that lead to misleading or unclear visualizations.
- ➢ Customize bar charts for better data storytelling, including optimizing colors, labels, and interactivity.
- ➢ Utilize popular tools and libraries to design and implement bar charts and box plots efficiently.

3.1 Introduction to Bar Charts and Box Plots

3.1.1 Purpose and Importance

Bar charts and box plots are among the most widely used visualization techniques in data analysis and communication. They serve distinct yet complementary purposes. Bar charts are effective for comparing discrete categories or groups and visualizing their frequencies, counts, or other quantitative measures. On the other hand, box plots excel in displaying the distribution and variability of continuous data, particularly in terms of highlighting central tendencies, spread, and potential outliers.

The primary objective of using these visualization tools is to convert complex datasets into visually comprehensible forms, allowing audiences to grasp key insights quickly. Well-designed bar charts and box plots help to uncover hidden trends, outliers , or anomalies within the data that may otherwise go unnoticed in textual or tabular representations. These charts are especially important in decision-making processes where clarity and precision are critical, whether in business, science, or policy making.

Given their widespread utility, understanding how to design effective and visually appealing bar charts and box plots is essential for anyone who works with data. In this chapter, we will explore the fundamentals of both types of charts, delving into the nuances of when and how to use each, and discussing best practices for optimizing their design to ensure clarity and impact.

3.1.2 Understanding the Basics

Bar charts are one of the simplest and most effective ways to present categorical data. They consist of bars, either horizontal or vertical, where the length of each bar is proportional to the value it represents (Table 3.1). The categories being compared are typically placed along the axis (either x or y, depending on orientation), and the numeric values are represented by the length of the bars (Figure 3.1). Bar charts can be used for a variety of purposes such as comparing the sales of products, the distribution of demographic groups, or the frequency of different outcomes in an experiment.

Box plots, also known as box-and-whisker plots, provide a graphical summary of a distribution of data based on five key summary statistics: the minimum, first quartile (Q1), median, third quartile (Q3), and maximum. The "box" represents the interquartile range (IQR), which contains the middle 50% of the data, while the "whiskers" extend to the minimum and maximum values within 1.5 times the IQR. Points outside of this range are plotted individually as potential outliers. Box plots are particularly useful when comparing the spread and central tendency of different datasets or groups. For example, they are often employed in scientific research to compare experimental results across different conditions or treatments.

Table 3.1: Anatomy of a Bar Plot

Component	Description
Bar	A rectangular shape whose height (vertical bar) or length (horizontal bar) represents the magnitude of the data value.
X-Axis (Category Axis)	Displays the categorical variables or labels corresponding to each bar.
Y-Axis (Value Axis)	Displays the scale of numerical values representing the bar heights or lengths.
X-Axis Label	A descriptive text explaining what the categories on the horizontal axis represent.
Y-Axis Label	A descriptive text explaining the numerical measurement represented on the vertical axis.
Gridlines	Optional horizontal or vertical reference lines that aid in estimating bar values more accurately.
Ticks	Small marks on the axes that indicate intervals for the categories (x-axis) or values (y-axis).
Tick Labels	Text annotations for ticks that identify specific values (y-axis) or categories (x-axis).
Color	Used to differentiate categories or data groups; may encode additional variables or draw attention.
Legend (if applicable)	Identifies the meaning of colors, bar groups, or patterns in grouped or stacked bar charts.
Title	A concise and informative heading that describes the purpose or content of the chart.
Error Bars (optional)	Lines extending from bars to indicate variability, uncertainty, or confidence intervals.
Annotations (optional)	Text or markers added to emphasize specific data points or trends.

While both bar charts and box plots are relatively simple to construct, there are important details to consider in their design to ensure they accurately convey the intended message. In the case of bar charts, it is crucial to avoid visual distortions such as unequal bar widths or inappropriate scaling that could mislead viewers. Similarly, box plots require careful attention to the scale and range of the axis, as well as the correct identification and treatment of outliers , to maintain the integrity of the data being represented.

3.1.3 Common Use Cases

Bar charts are highly versatile and can be used across a wide range of applications. One of the most common uses is in business analytics, where bar charts are frequently used to compare sales performance across different products, regions, or time periods. They are also heavily employed in survey research to depict the distribution of responses across different demographic groups or answer categories. In educational

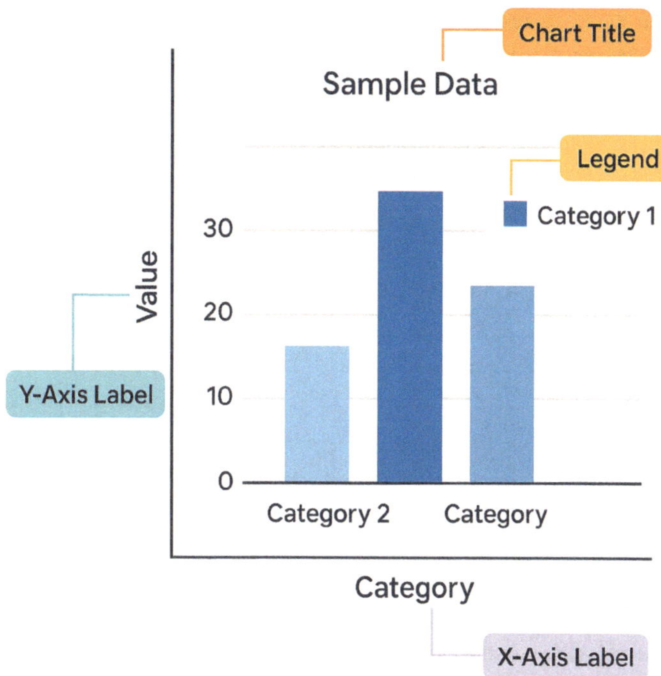

Fig. 3.1: Anatomy of a basic Bar Plot illustrating key components of a bar plot, including bars, axes, gridlines, tick marks, axis labels, and title. It helps users understand how categorical data is represented visually through bar length or height.

contexts, bar charts are useful for visualizing student performance across different subjects or groups, enabling educators to identify areas of improvement.

Stacked and grouped bar charts provide additional layers of information by allowing comparisons not only between categories but also within subcategories. For instance, a grouped bar chart might show the sales performance of multiple product categories over several years, while a stacked bar chart could reveal the proportion of male and female respondents in different age groups in a survey. These variations are invaluable when presenting multidimensional data in a compact and intuitive format.

Box plots, on the other hand, are most commonly used in statistical analysis to visualize the distribution of a dataset and to compare multiple datasets. They are especially useful in exploratory data analysis (EDA), where the goal is to understand the underlying structure and key characteristics of the data before conducting more formal analysis. In scientific research, box plots are often used to compare the variability and central tendencies of experimental results across different conditions, such as comparing the test scores of students who received different types of instruction, or the effect of different treatments in a clinical trial.

In financial analysis, box plots can help visualize the volatility of different stocks or the distribution of returns in an investment portfolio. They are also widely used in environmental science to display the range of temperatures or pollution levels across different locations or time periods. In summary, the box plot's ability to provide a concise summary of data distribution makes it a go-to tool for analyzing variability, outliers, and trends across multiple datasets.

When selecting between bar charts and box plots, the guidelines presented in Table 3.2 may be followed.

Table 3.2: Choosing Between Bar Charts and Box Plots

Use Case	Recommended Chart
Comparing categories (e.g., product sales)	Bar Chart
Summarizing distributions	Box Plot
Detecting variability across groups	Box Plot
Highlighting exact values for categories	Bar Chart
Exploring multimodal distributions	Violin Plot

In the next sections, we will focus deeper into the design principles and best practices for creating stunning bar charts and box plots that not only look good but also effectively communicate key insights.

3.2 Design Principles for Bar Charts

Bar charts are highly effective tools for visualizing categorical data, but their utility hinges on thoughtful design choices. An improperly designed bar chart can lead to confusion or even misinterpretation of the data. In this section, we will explore key design principles that will help you craft bar charts that are both visually appealing and functionally effective. These principles cover the selection of the appropriate chart type, optimizing the number of bars, choosing colors and labels, avoiding misleading visuals, and incorporating interactivity where necessary.

3.2.1 Choosing the Right Type of Bar Chart

One of the first and most important decisions to make when creating a bar chart is selecting the correct type. The most commonly used forms include vertical bar charts, horizontal bar charts, stacked bar charts, and grouped bar charts. Each serves a distinct purpose and is suited to different kinds of data and comparisons.

Vertical bar charts, sometimes called column charts, are the most frequently used type. In this format, the categories are displayed along the x-axis and the values are represented by the height of the bars along the y-axis. Vertical bar charts are ideal when the number of categories is limited and the focus is on comparing the magnitude of these categories. For example, they work well when comparing the sales figures for different products in a store or the exam scores of students in different subjects. The vertical orientation is often intuitive for readers because it aligns naturally with how we perceive increases and decreases in values.

Horizontal bar charts are similar to vertical bar charts but with the axes swapped; the categories are listed along the y-axis and the values extend horizontally from the x-axis. Horizontal bar charts are particularly useful when you have long category labels or when comparing many categories because they allow more space for the category labels to be clearly read. They are often used in survey data, where response categories might include long phrases, or in cases where the chart is presenting a large number of categories. Horizontal bars can also make it easier to compare the relative lengths of the bars, especially when the differences between categories are small.

Stacked bar charts take the basic structure of the vertical or horizontal bar chart but allow for additional data within each category by stacking multiple sub-categories on top of each other. This type of chart is useful when you want to show both the total value for each category and the breakdown of that total into sub-components. For instance, in a sales report, you might use a stacked bar chart to show the total sales for each region and stack the bars to represent the sales of different product types within those regions. However, stacked bar charts can become difficult to interpret if there are too many sub-components, as it can be hard to compare the individual segments between different bars.

Grouped bar charts, also known as clustered bar charts, are another way to compare multiple sub-categories within a category. Rather than stacking the sub-categories, they are placed side-by-side within the same category. This format makes it easier to compare the values of the sub-categories directly, as each is represented by a separate bar. Grouped bar charts are particularly useful for comparing two or more datasets across the same categories, such as comparing the sales figures for two different years across the same set of products. However, as with stacked bar charts, grouped bar charts can become cluttered if too many sub-categories are included.

3.2.2 Optimizing the Number of Bars

One of the most important design principles in bar chart creation is to avoid overwhelming the reader with too many bars. Too many categories or bars can make the chart cluttered and difficult to interpret. A key rule of thumb is to limit the number of bars to a manageable number–typically fewer than 10–so that the comparisons between categories remain clear. If your dataset contains many categories, consider aggregating them into broader groups or splitting the chart into multiple visualizations. Alternatively, you can use filtering or interactivity to allow users to focus on a subset of the data at a time.

Another strategy for managing large datasets is to use a horizontal bar chart, as the additional space for labeling categories can improve legibility. However, even with horizontal bar charts, too many bars can still obscure meaningful patterns and comparisons. As a general guideline, strive for simplicity and clarity by focusing on the most important categories in your data.

3.2.3 Choosing Appropriate Colors and Labels

Colors and labels are essential elements in the design of a bar chart, and their selection has a significant impact on the clarity and interpretability of the visualization. When choosing colors, it is important to ensure that they are both distinguishable and consistent with the message you are trying to convey. For example, use contrasting colors for different categories to make them easily identifiable, but avoid overloading the chart with too many colors, as this can cause visual confusion.

It is also important to be mindful of colorblind-friendly palettes, ensuring that your chart can be interpreted accurately by individuals with color vision deficiencies. Tools such as colorblind simulators or predefined colorblind-friendly palettes can help in selecting appropriate colors.

Labels play a crucial role in making your chart understandable. Ensure that category labels are concise yet descriptive enough for the reader to grasp the meaning quickly. Axis labels should clearly state what is being measured, and units of measurement should be included where relevant. Adding direct value labels to the bars themselves can also be helpful, especially when the exact value of each bar is important for interpretation.

3.2.4 Avoiding Misleading Visuals in Bar Charts

Bar charts are powerful tools for communication, but if not designed carefully, they can easily become misleading. One of the most common issues is manipulating the scale of the axes. For example, starting the y-axis at a value higher than zero can exaggerate differences between categories, leading to a misinterpretation of the data. It is almost always advisable to start the y-axis at zero, unless there is a compelling reason to do otherwise. Using uneven intervals on the axes or inconsistent bar widths can distort the visual representation of the data. Such design flaws might lead viewers to draw incorrect conclusions. It is essential to maintain consistency and proportionality in your design to ensure that your chart remains an honest representation of the data.

3.2.5 Adding Interactivity for Enhanced Understanding

In today's digital world, static bar charts are often replaced by interactive visualizations that allow users to explore the data in more depth. Adding interactivity to bar charts can enhance the user's understanding by allowing them to filter data, zoom in on specific sections, or hover over elements for additional information. Interactive charts are particularly useful when dealing with complex datasets or when presenting information in an online or digital format. Tools like Plotly, D3.js, and other JavaScript-based libraries enable the creation of interactive bar charts that can adapt dynamically to user input. This added layer of interactivity can make your bar charts more engaging and informative, allowing users to customize their view and extract the insights that are most relevant to them. By following these design principles, you can ensure that your bar charts are not only aesthetically pleasing but also effective in conveying meaningful insights to your audience.

3.3   Creating Stunning Box Plots

Box plots, also known as box-and-whisker plots, are valuable tools for displaying the distribution, spread, and variability of datasets. While bar charts focus on comparing categorical data, box plots excel in showcasing how data is distributed within categories, providing insights into the central tendencies, variability, and potential outliers. In this section, we will dive into the anatomy of a box plot, discuss when to use them, and explore design principles to make them visually appealing and effective for communicating data insights.

3.3.1   Introduction to Box Plot Anatomy

A box plot (Figure 3.3) provides a summary of a dataset using five key statistical measures: the minimum, the first quartile (Q1), the median (Q2), the third quartile (Q3), and the maximum. These measures are visualized in the form of a "box" that represents the interquartile range (IQR), and "whiskers" that extend to the minimum and maximum values within a defined range. Box plots are particularly powerful because they not only summarize the distribution of the data but also highlight any potential outliers that deviate significantly from the main distribution (Figure 3.2).

Table 3.3: Anatomy of a Box Plot

Component	Description
Minimum	Lowest value within 1.5 IQR below Q1
Q1 (First Quartile)	25th percentile; lower edge of the box
Median	50th percentile; line inside the box
Q3 (Third Quartile)	75th percentile; upper edge of the box
Maximum	Highest value within 1.5 IQR above Q3
Whiskers	Lines extending to min and max within range
Outliers	Individual points outside whisker range

At the heart of the box plot are the quartiles, which divide the data into four equal parts. The first quartile (Q1), located at the bottom of the box, marks the 25th percentile, meaning that 25% of the data points fall below this value. The third quartile (Q3), positioned at the top of the box, represents the 75th percentile, meaning that 75% of the data points fall below this value. The distance between Q1 and Q3 is known as the interquartile range (IQR), which captures the middle 50% of the data.

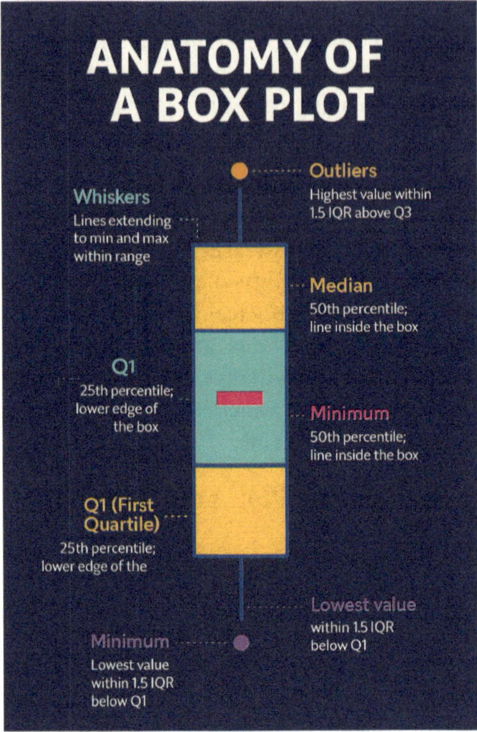

Fig. 3.2: Anatomy of a box plot showing the minimum, first quartile (Q1), median, third quartile (Q3), and maximum, with outliers plotted as individual points.

The median, represented by a line within the box, marks the 50th percentile and divides the dataset into two equal halves. The position of the median within the box provides immediate insight into the skewness of the data. If the median is closer to Q1, the data is skewed toward higher values, whereas if it is closer to Q3, the data is skewed toward lower values. This quick visualization helps viewers assess the balance of the data distribution at a glance.

The "whiskers" of the box plot extend from the top and bottom of the box to capture the range of values within 1.5 times the IQR. These whiskers effectively represent the minimum and maximum values within this range, while any data points that fall outside this range are considered outliers and are plotted individually as dots. Outliers are particularly important in box plots because they highlight unusual data points that may require further investigation or could represent meaningful anomalies.

The length of the whiskers provides insights into the variability of the dataset. Long whiskers indicate that the data is more spread out, while shorter whiskers suggest that the data is more tightly clustered around the median. Outliers, shown as individual points, can suggest either extreme values or data errors, depending on the context.

3.3.2 When to Use Box Plots

Box plots are most effective when you want to visualize the distribution of a dataset or compare distributions across multiple categories. They are particularly useful when dealing with continuous data that contains variability, and where outliers or extreme values may have a significant impact. Common use cases for box plots include:

➢ Comparing the performance of different groups in an experiment, such as the test scores of students from different schools.

➢ Visualizing the variability of stock prices over time or across different companies.

➢ Summarizing the distribution of sensor data in environmental monitoring, where there may be a wide range of measurements with potential outliers.

➢ Comparing the variability in machine learning model performance across different validation sets.

Box plots excel in highlighting differences between groups, making them ideal for exploratory data analysis (EDA). They are particularly valuable when comparing multiple datasets, as they provide a concise summary of central tendencies, spread, and outliers in a single visualization.

3.3.3 Customizing Box Plots for Impact

While box plots provide a clear and concise summary of data, their visual impact can be enhanced through thoughtful customization. Color and style choices, as well as highlighting specific data points, can help to emphasize key aspects of the dataset and make the visualization more engaging and easier to interpret.

Choosing the right color scheme is essential for making your box plots visually appealing while maintaining clarity. The following guidelines presented in Table 3.4 may be followed. It is important to select a palette that ensures high contrast between different box plots, especially when comparing multiple groups. Using a muted, consistent color palette can help the audience focus on the differences in the data rather than being distracted by overly vibrant colors.

In addition to color, consider customizing the style of the plot. You can adjust the thickness of the whiskers and the box lines, or use different line styles (such as dashed or dotted) to differentiate between categories. Transparency (alpha settings) can also be used to overlay multiple box plots, allowing for the comparison of distributions without obscuring the underlying data.

In some cases, certain data points or features of the box plot may require additional emphasis. For example, if a particular outlier represents a significant event or anomaly, you can use a different color or marker style to highlight it. You can annotate specific

Table 3.4: Color Usage Guidelines in Data Visualization

Design Element	Best Practice
Contrast	Use high-contrast colors to distinguish categories
Consistency	Use the same colors across related charts and categories
Highlighting	Use bold/saturated colors to emphasize key data points
Colorblind Accessibility	Use palettes like ColorBrewer that are colorblind-safe
Avoid Overuse	Limit the palette to 5–6 distinct colors to avoid visual clutter
Semantic Color Mapping	Use culturally intuitive colors (e.g., red for decrease) where appropriate

parts of the box plot, such as marking the median value or labeling important outliers , to provide additional context for the viewer.

By selectively highlighting key data points, you can direct the viewer's attention to the most important aspects of the data and enhance their understanding of the narrative you are trying to convey.

3.3.4   Dealing with Multiple Groups in Box Plots

Box plots are especially useful for comparing multiple groups or categories within a dataset. When dealing with multiple groups, it is important to ensure that the comparison remains clear and that the visualization does not become cluttered. One approach is to arrange the box plots side by side, with each group clearly labeled. If there are many groups, consider organizing them in a logical order, such as by median value or by the number of observations.

To avoid overwhelming the viewer, it is also important to maintain a consistent scale across all box plots in a comparison. A common axis makes it easier for the viewer to compare the spread and central tendencies of each group. In some cases, it may be helpful to include additional visual aids, such as a reference line indicating the overall median, to provide further context for comparison.

3.3.5   Handling Large Datasets with Box Plots

When working with large datasets, box plots remain a useful tool for summarizing data distributions without overwhelming the viewer. However, as the dataset grows in size, it may become necessary to adjust certain aspects of the visualization. For instance, if the dataset contains a large number of outliers , the box plot can become cluttered with dots representing those points. In such cases, you might consider using alternative methods, such as jittering the outliers to prevent overlap or summarizing extreme values in a different manner. Interactive box plots can be used to allow viewers to explore large datasets more effectively. Interactive features such as zooming, filtering, or hovering over individual data points to reveal additional details can help viewers navigate the data and focus on the aspects most relevant to them.

By carefully managing the design and presentation of box plots, you can ensure that even large and complex datasets are presented clearly and effectively, allowing viewers to derive meaningful insights without becoming overwhelmed by the amount of data.

3.4 Data Storytelling with Bar Charts and Box Plots

Data storytelling is an essential aspect of scientific visualization, where the goal is not just to present data but to convey a compelling narrative that informs, persuades, or motivates action. Both bar charts and box plots are powerful tools for communicating data-driven stories. When used effectively, these visualizations can highlight key trends, comparisons, and insights that might otherwise be buried in raw numbers or complex statistics. In this section, we explore how to align your visuals with your message, how to combine bar charts and box plots for deeper insights, and how to present comparative data in a way that strengthens your data narrative.

3.4.1 Aligning Visuals with Your Message

The most important principle in data storytelling is ensuring that the visual representation of the data aligns with the message you intend to communicate. Before designing a bar chart or box plot, it's crucial to define the story you want to tell and identify the key takeaways that you want your audience to grasp. Whether you're trying to showcase trends, highlight disparities, or point out anomalies, your choice of chart type, scale, and emphasis should be guided by your narrative.

For instance, if your message revolves around the central tendency and variability of different groups, a box plot would be the most appropriate choice, as it provides an effective summary of the distribution and any outliers. On the other hand, if the focus is on comparing the magnitude of categorical data, such as sales figures for different products, a bar chart would likely be more effective.

Another important aspect of aligning visuals with your message is maintaining consistency in design. Use colors, fonts, and formatting that emphasize the most important aspects of the data. For example, if one group in your dataset is of particular interest, consider using a bolder color or adding annotations to draw attention to that group. Always ensure that your design choices reinforce the narrative you are presenting, rather than distracting from it. Make sure your axis labels, legends, and data points are clear and concise. The goal is to provide enough context for the viewer to understand the data without overwhelming them with unnecessary details. By keeping your design aligned with your core message, you can ensure that your data visualizations are not just informative, but impactful.

3.4.2 Combining Bar Charts and Box Plots for Deeper Insights

While bar charts and box plots are each valuable on their own, combining them in a single visualization or dashboard can provide a deeper level of insight. This combination is particularly useful when you need to compare both categorical data and the distribution of values within those categories.

For instance, imagine you are analyzing sales performance across multiple regions. A bar chart could be used to show the total sales in each region, providing a straightforward comparison of the magnitude of sales. Meanwhile, a box plot could complement this by showing the variability of individual store performances within each region. This dual approach allows you to convey both the total outcome and the underlying distribution, offering a richer understanding of the data.

Another common scenario where combining these visualizations is effective is in survey analysis. A bar chart might display the average rating for each question in a survey, while a box plot could reveal the distribution of responses, highlighting the range of opinions and any outliers. This combination helps tell a more nuanced story, as it shows not only the central tendency but also the spread of opinions or results.

When designing combined visualizations, be mindful of the scale and ensure that the charts complement each other rather than compete for attention. Consistency in colors, fonts, and axis labels can help create a cohesive visual story. By presenting bar charts and box plots together, you can use the strengths of both formats to communicate a richer, more detailed narrative.

3.4.3   Presenting Comparative Data

One of the most common uses of both bar charts and box plots is for comparing data across different groups or categories. Whether you are comparing performance metrics, experimental results, or demographic data, these charts allow you to present clear, compelling comparisons that highlight key differences and similarities (Table 3.5).

When using bar charts for comparative analysis, consider grouping or stacking bars to provide an intuitive comparison between categories. Grouped bar charts work well when you have multiple series to compare side-by-side, such as comparing the performance of different product lines across multiple regions. Stacked bar charts, on the other hand, allow you to show both the total value and the contribution of subcategories within each group, which can be useful for showing the breakdown of sales by product type within each region.

Box plots are particularly effective for comparing the distribution and variability of different groups. When comparing multiple datasets, arrange the box plots side by side with a consistent axis to ensure easy comparison. For example, if you are comparing test scores across different schools, a series of box plots can reveal differences in the median, spread, and outliers for each school, allowing you to identify which schools have higher variability or more extreme scores.

In presenting comparative data, it is essential to maintain consistency across all visual elements to facilitate easy comparison. Use the same scale for all charts to prevent visual distortions, and make sure that all category labels and data points are clearly defined. Annotations can also be used to draw attention to key differences or trends, ensuring that the viewer understands the significance of the comparisons being made.

Ultimately, whether you are using bar charts or box plots, the goal is to present the data in a way that supports your narrative and makes comparisons clear and meaningful. By carefully selecting and designing your charts, you can guide the audience through the data story and help them draw the right conclusions.

Table 3.5: Comparison of Chart Types for Visualizing Data

Chart Type	**Best Use Case**	**Key Features**
Bar Chart	Comparing values across categories	Simple to interpret, uses height/length to encode value
Box Plot	Showing data distribution and outliers	Median, quartiles, IQR, whiskers, and outliers
Violin Plot	Understanding data density and distribution shape	Combines KDE with box plot for richer distribution insight
Rainbow Plot	Showing individual points, summary statistics, and density	Combines jittered data points, box plot, and density curve

3.5 Going Beyond Bar Charts and Box Plots

While bar charts and box plots are fundamental tools for visualizing data, there are other advanced visualization techniques that provide even deeper insights, especially when dealing with complex datasets. In this section, we will explore violin plots. These visualizations go beyond the basic summaries provided by bar charts and box plots, offering a more detailed view of data distributions, and they are particularly useful when you need to convey richer information about variability, density, and comparisons across multiple groups.

3.5.1 Violin Plots

Violin plots are an advanced visualization tool that combines elements of both box plots and kernel density plots. They provide a detailed view of the distribution of data, including the density of different values, making them ideal for comparing multiple datasets or groups. Violin plots are particularly useful when you want to understand the underlying distribution of data, rather than just the summary statistics provided by a box plot.

A violin plot is similar in structure to a box plot but with an added element: it includes a kernel density estimate on each side of the box plot to visualize the distribution of the data (Figure 3.3). This density curve shows the frequency of values, giving the plot a "violin" shape. The width of the violin at any given point corresponds to the density of the data at that value, with wider sections indicating more frequent values.

Violin plots are effective when you want to compare distributions across multiple categories or datasets while also showing central tendencies and variability. The inclusion of the density plot allows you to see nuances in the data that would be hidden in a traditional box plot, such as multimodal distributions or skewed data.

The anatomy of a violin plot consists of several key components (Table 3.6):

> **Kernel Density Estimate (KDE):** The smooth curve on either side of the "violin" that represents the distribution of the data.

> **Box Plot:** The inner part of the violin plot often includes a traditional box plot, which shows the median, interquartile range (IQR), and potential outliers.

> **Median Line:** A horizontal line inside the violin that indicates the median value of the data.

> **Whiskers:** Lines extending from the box plot that represent the minimum and maximum values within the 1.5 IQR range, similar to traditional box plots.

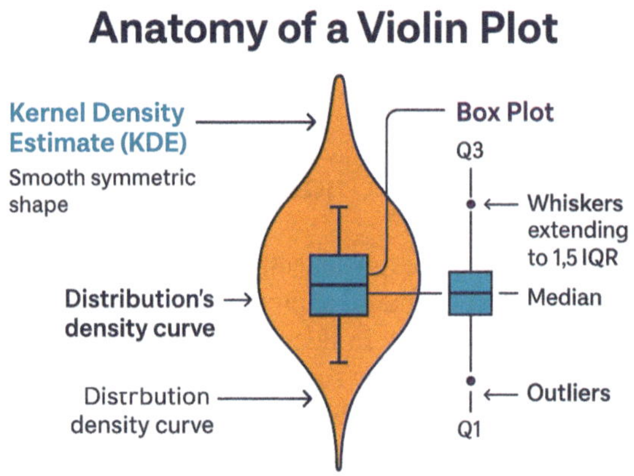

Anatomy of a Violin Plot

Kernel Density Estimate (KDE)
Smooth symmetric shape
Box Plot
Q3
Whiskers extending to 1,5 IQR
Distribution's density curve
Median
Distrbution density curve
Outliers
Q1

Fig. 3.3: Anatomy of a Violin Plot illustrating the structure and interpretation of a violin plot, combining a smooth kernel density estimate (KDE) with a traditional box plot to show distribution, central tendency, interquartile range (IQR), and potential outliers.

When designing violin plots, it is important to consider the following best practices (Table 7.8) to ensure that the plot is both informative and easy to interpret:

> **Use appropriate smoothing for the KDE:** The kernel density estimate should be smooth enough to show meaningful trends but not so smooth that it obscures important details in the distribution. Adjust the bandwidth of the KDE to strike the right balance.

> **Include the box plot:** Although the density plot provides rich information about the distribution, it is still helpful to include a traditional box plot within the violin to summarize key statistics such as the median and IQR.

Table 3.6: Anatomy of a Violin Plot

Component	Description
Kernel Density Estimate (KDE)	A smoothed, symmetrical curve that shows the probability density of the data distribution. The width at each vertical position reflects the relative frequency of data points at that value.
Density Curve Shape	The outer shape of the violin plot represents the distribution's density. Wider regions indicate more frequent values; narrower regions indicate less common values.
Box Plot (Embedded)	A standard box plot is embedded within the violin plot to provide summary statistics: median, quartiles, whiskers, and outliers.
Median	A horizontal line inside the box that represents the 50th percentile of the data. It divides the data into two equal halves.
Q1 (First Quartile)	The lower boundary of the box, indicating the 25th percentile of the data.
Q3 (Third Quartile)	The upper boundary of the box, indicating the 75th percentile of the data.
Interquartile Range (IQR)	The range between Q1 and Q3, containing the middle 50% of the data. Used to compute the whisker extent.
Whiskers	Lines that extend from the box to the smallest and largest data points within 1.5 times the IQR from the quartiles. They represent the typical range of the data.
Outliers	Individual data points plotted outside the whiskers. These points are considered extreme or unusual values in the dataset.
Symmetry	Violin plots are typically symmetrical about the central vertical axis, helping to visualize skewness or modality in the distribution.

➤ **Maintain consistency in scale:** When comparing multiple violin plots, ensure that the scales are consistent across all plots to make valid comparisons between distributions.

➤ **Use color effectively:** Violin plots can become visually complex, so choose colors that enhance readability without overwhelming the viewer. Different categories or groups should be clearly distinguished with contrasting yet complementary colors.

Interpreting a violin plot involves examining both the density and the box plot. The width of the violin at any given point represents the density of data at that value–wider sections indicate more frequent values, while narrow sections indicate less frequent values. The box plot inside the violin provides a summary of the central tendency and spread, similar to a traditional box plot, and any outliers are shown as individual points. By combining both density and summary statistics, the viewer can quickly identify patterns in the data, such as bimodal distributions or skewness, that would be less apparent in other types of plots.

Table 3.7: Best Practices in Chart Design

Principle	Recommendation
Limit the number of elements	Use fewer than 10 bars or groups to maintain clarity
Maintain consistent scales	Use a common axis scale when comparing groups
Choose accessible colors	Use contrasting and colorblind-friendly palettes
Avoid axis truncation	Start y-axis at zero unless otherwise justified
Label clearly	Include axis titles, units, legends, and annotations
Emphasize key insights	Highlight significant values or trends visually
Use interactivity when helpful	Allow exploration via filters, zoom, or tooltips for large datasets

Note: Outliers Are Not Always Errors

Outliers in box plots indicate unusual values, but they are not necessarily incorrect. Always investigate them contextually before deciding to exclude or annotate.

Violin plots are particularly useful in the following scenarios:

- **Comparing exam scores across different schools or classes:** A violin plot can show the full distribution of scores, highlighting whether the distribution is skewed or whether there are multiple peaks.

- **Visualizing distribution of salaries across different industries:** A violin plot provides insights into the variability and distribution of salaries, making it easier to see whether certain industries have more concentrated or diverse salary ranges.

- **Displaying the distribution of experimental results in scientific research:** Violin plots can help visualize the spread of results across different experimental conditions, showing not only central tendencies but also how varied the data is.

```python
import seaborn as sns
import matplotlib.pyplot as plt

# Reuse the same DataFrame from the box plot\index{Box Plot} example
sns.violinplot(x='Group', y='Score', data=df, inner='box',
    palette='Pastel1')
plt.title('Violin Plot of Scores by Group')
plt.tight_layout()
plt.show()
```

3.6 Tools and Best Practices for Bar Chart Design

The effectiveness of bar charts and box plots depends not only on choosing the right type of chart but also on how they are implemented and designed. With the right tools and techniques, you can create visualizations that are both visually stunning and easy to interpret. This section will cover the most popular visualization libraries, provide tips for achieving clarity and precision, and highlight common pitfalls to avoid in bar chart and box plot design.

3.6.1 Popular Visualization Libraries

There are numerous tools and libraries available for creating bar charts and box plots, ranging from simple spreadsheet software to advanced programming libraries. Choosing the right tool depends on your level of expertise, the complexity of your data, and the level of customization required. Below are some of the most popular libraries for creating these types of charts (Table 3.8).

➤ **Matplotlib (Python):** Matplotlib is one of the most widely used libraries for creating static, interactive, and animated visualizations in Python. It provides a high level of customization, allowing users to design bar charts, box plots, and a variety of other chart types with precision. Matplotlib's versatility makes it a favorite among data scientists and researchers for publication-quality visualizations.

➤ **Seaborn (Python):** Built on top of Matplotlib, Seaborn simplifies the creation of statistical plots and provides a more aesthetically pleasing default style. It's particularly useful for creating box plots, violin plots, and other statistical visualizations with minimal code. Seaborn's integration with Pandas makes it an excellent choice for quickly visualizing data from dataframes.

➤ **ggplot2 (R):** For R users, ggplot2 is one of the most popular libraries for data visualization. It follows the principles of the "grammar of graphics," allowing users to build complex plots in a structured and logical way. ggplot2 excels in creating both bar charts and box plots with customizable features and strong support for layered visualizations.

➤ **Plotly (Python, R, JavaScript):** Plotly is a powerful library for creating interactive visualizations. It allows users to build bar charts, box plots, and other chart types that include interactive elements such as tooltips, zooming, and filtering. Plotly is ideal for web-based visualizations or presentations where user interaction can enhance the data exploration process.

➤ **Tableau (Software):** Tableau is a popular tool for creating interactive dashboards and visualizations without requiring extensive coding knowledge. It is widely used in business analytics for its ease of use, drag-and-drop interface, and ability to handle large datasets. Tableau supports a wide range of visualizations, including bar charts and box plots, and is often employed in business intelligence reporting.

Table 3.8: Comparison of Tools for Creating Bar Charts and Box Plots

Tool	Strengths	Limitations	Best For
Matplotlib	Precise control over visuals; publication-ready	Requires more code	Scientific visualizations
Seaborn	Elegant defaults; quick summaries (violin, box, bar)	Less flexible for interactive use	Statistical exploration
ggplot2	Grammar-based layering; strong faceting support	Steeper learning curve for some users	R-based analytics
Plotly	Interactive charts with ease	Can be heavy for simple plots	Dashboards, web publishing
Tableau	No coding needed; fast dashboard building	Limited custom logic	Business intelligence and reports

Each of these libraries and tools has its strengths, so the choice depends on the level of customization and interactivity you need for your visualizations. For quick and simple designs, Excel or Tableau may suffice. For more control over the design and the ability to integrate with other programming tasks, libraries like Matplotlib, Seaborn, and ggplot2 are preferred.

3.6.2 Tips for Achieving Clarity and Precision

Achieving clarity and precision in your bar charts and box plots is essential to ensure that your audience can easily interpret the data. Here are some best practices to help you design visualizations that communicate effectively:

- ➤ **Limit the number of categories:** Overloading a bar chart with too many categories or bars can make the visualization cluttered and difficult to read. Aim to include no more than 10 categories in a single chart, and if necessary, group smaller categories together or use filtering options for better clarity.

- ➤ **Use a consistent scale:** Ensure that your axes are appropriately scaled. Starting the y-axis at zero is generally recommended for bar charts to prevent exaggeration of differences between categories. For box plots, make sure that the same scale is used across all groups to enable accurate comparison.

- ➤ **Choose color wisely:** Use contrasting colors to differentiate between categories but avoid using too many colors. Keep your color scheme consistent and consider using colorblind-friendly palettes to make your visualizations accessible to a broader audience. If possible, use color to reinforce the key message of your chart–highlighting important data points or categories with bolder colors.

> **Label everything clearly:** Your chart should be self-explanatory. Include clear axis labels, titles, and legends to ensure that the viewer understands what the chart represents. In box plots, it may also be helpful to add annotations for outliers or notable data points.

> **Simplify where possible:** Avoid unnecessary elements such as gridlines or overly complex backgrounds that detract from the data. The goal is to focus the viewer's attention on the key insights, so aim for simplicity in design.

> **Highlight important information:** If there are key findings in the data, such as significant outliers or trends, use visual cues like bold colors or annotations to highlight them. This ensures that the most important aspects of the data are immediately clear to the viewer.

> **Use interactivity wisely:** If you are creating interactive charts, ensure that the interactive elements enhance the user's understanding of the data. Interactive features such as tooltips, zooming, and filtering should provide additional context without overwhelming the viewer.

By following these best practices, you can ensure that your bar charts and box plots are not only visually appealing but also communicate your data with clarity and precision.

3.6.3 Avoiding Common Design Pitfalls

Even with the best tools and intentions, certain design pitfalls can reduce the effectiveness of your visualizations. Below are some common mistakes to avoid when designing bar charts and box plots:

> **Inconsistent scales:** One of the most frequent errors is using different scales for different parts of a chart, especially when comparing categories or groups. This can lead to misinterpretation by exaggerating or minimizing differences. Always use a consistent scale across comparisons.

> **Misleading axis truncation:** For bar charts, starting the y-axis at a value higher than zero can artificially inflate differences between bars, misleading the viewer. Ensure that your axis starts at zero unless there is a compelling reason to do otherwise, and always communicate any deviations clearly.

> **Too many bars or categories:** Including too many bars in a bar chart or too many groups in a box plot can make the chart cluttered and difficult to read. If your data contains a large number of categories, consider aggregating or splitting the data into multiple charts.

> **Caution: Visual Overload**
>
> Bar charts with too many categories become cluttered and unreadable. Limit to 10 categories or split data into multiple visualizations to retain clarity.

➢ **Inconsistent bar widths or spacing:** In bar charts, ensure that all bars are of equal width and that spacing between bars is consistent. Uneven spacing or bar widths can distort the visual representation of the data and lead to incorrect conclusions.

➢ **Overcomplicated color schemes:** Using too many colors can make your chart visually overwhelming and confusing. Stick to a simple color scheme with meaningful contrasts. If you are using color to represent different groups, make sure the colors are easily distinguishable, even for viewers with color vision deficiencies.

> **Design Tip: Use Color Consistently**
>
> Ensure color schemes are consistent across related plots (e.g., same group has the same color in bar and box plots). Inconsistencies can confuse and mislead viewers.

➢ **Ignoring outliers :** In box plots, outliers are a critical part of the data, as they can indicate unusual or interesting points. Ignoring or removing outliers without explanation can skew the interpretation of the data. If outliers are present, highlight them clearly and provide context if necessary.

➢ **Lack of context:** Providing insufficient context for the viewer is a common mistake. Make sure that your chart includes titles, labels, and legends that explain what the data represents. Without this information, viewers may misinterpret the chart or overlook important details.

Avoiding these common pitfalls (Table 3.9) will help you create bar charts and box plots that are both accurate and easy to interpret, ensuring that your visualizations effectively convey your message to your audience.

Table 3.9: Common Pitfalls in Bar and Box Plot Design

Pitfall	Why It's a Problem
Truncated y-axis in bar chart	Exaggerates differences between categories
Too many bars or groups	Leads to clutter and cognitive overload
Inconsistent scales across charts	Makes visual comparison misleading or invalid
Overuse of colors or effects	Distracts from the main message
Unlabeled axes or missing units	Reduces interpretability and trust
Ignoring outliers in box plots	Misses important anomalies or variation

3.7 Case Studies and Examples

To bring the concepts of bar chart and box plot design to life, we will explore several case studies and examples that illustrate how these visualizations can be applied to real-world data. These examples demonstrate how bar charts and box plots can be used individually and in combination to analyze demographic data, performance variability, and experimental results, offering actionable insights through effective visualization techniques.

3.7.1 Bar Chart Example: Demographic Distribution Analysis

Bar charts are ideal for comparing categorical data, such as demographic variables, across different groups or regions. In this case study, we will use a bar chart to analyze the demographic distribution of a population across several regions. The goal is to compare the percentage of different age groups within each region, revealing key trends or disparities in the age structure of the population.

Problem Statement: The local government is interested in understanding the demographic breakdown of its population across five regions to inform resource allocation and public policy planning. The dataset includes population counts for five age groups (0-18, 19-35, 36-50, 51-65, and 66+) in each region.

Visualization Design:

➤ We choose a grouped bar chart, where each region is represented by a set of bars. Each bar within the group corresponds to one of the five age groups.

➤ A consistent color palette is applied across all age groups for ease of comparison.

➤ To avoid clutter, we limit the chart to five regions, with the y-axis representing the percentage of the population in each age group.

➤ Labels for each region and age group are clearly marked, and the chart is supplemented with a legend that explains the color coding for the age groups.

Insights: By examining the chart (Figure 3.4), viewers can quickly identify regions with an aging population or regions with a higher percentage of younger people. For example, if Region 1 has a significantly larger percentage of people in the 66+ age group compared to Region 5, this insight can guide policymakers in directing healthcare and social services to where they are most needed.

Design Highlights:

➤ The use of grouped bars allows for a clear comparison across both regions and age groups.

➤ Consistent color usage makes it easy to follow trends across different age groups.

➤ Axis labels and legends ensure that the chart is immediately understandable to viewers.

3.7.2 Box Plot Example: Performance Variability in Research Groups

Box plots are invaluable for visualizing the distribution of data and comparing the variability of different groups. In this case study, we will use a box plot to analyze the performance variability across several research groups in a university setting.

Problem Statement: The university administration wants to compare the performance (measured by the number of peer-reviewed publications) of different research groups over a five-year period. The dataset includes publication counts for 10 different research groups, with each group containing data on the number of publications per year.

Visualization Design:

> A box plot is used to summarize the distribution of publication counts for each research group over the five-year period.

> Each research group is represented by a separate box plot, allowing for a comparison of the median, interquartile range (IQR), and the presence of outliers.

> Whiskers extend to 1.5 times the IQR, and any outliers (research groups with unusually high or low publication counts) are marked with individual points.

> The y-axis represents the number of publications, and the x-axis lists the research groups.

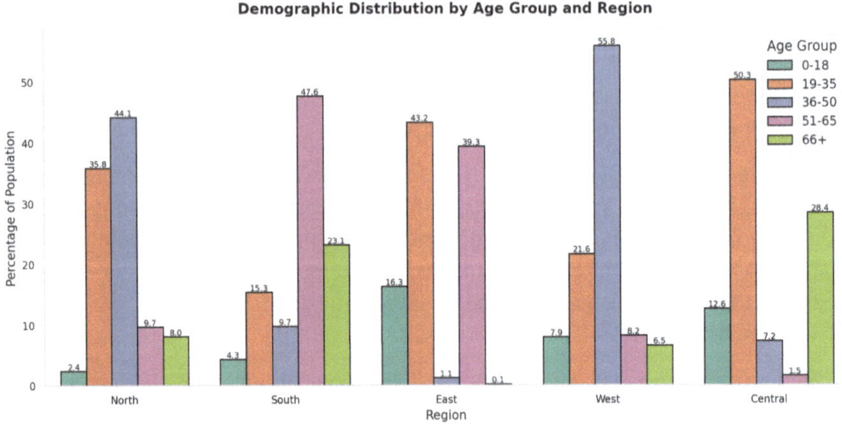

Fig. 3.4: Grouped bar chart showing the percentage distribution of population across five age groups (0–18, 19–35, 36–50, 51–65, and 66+) in each of the five regions. This visualization supports demographic analysis for regional policy planning.

Insights: The box plot reveals significant variability between research groups. Some groups have a wide IQR, indicating variability in their publication output year over year, while others have a more consistent number of publications. Outliers highlight research groups that have particularly productive or unproductive years. This can inform resource allocation or further investigation into the factors contributing to higher or lower productivity.

Design Highlights:

➤ Box plots effectively summarize the performance variability and highlight outliers , which may represent unusually productive years.

➤ Consistent y-axis scaling across all groups allows for easy comparison.

➤ The addition of outlier markers draws attention to groups with extreme values, which could prompt further analysis.

3.8 Conclusion

In this chapter, we have explored the design and application of bar charts and box plots, focusing on how to create stunning visualizations that effectively communicate data insights. By understanding the strengths and limitations of each visualization type, and applying thoughtful design principles, you can craft charts that not only look great but also provide clear and actionable insights. Whether used individually or in combination, bar charts and box plots are essential tools for data storytelling in a variety of fields.

3.8.1 Key Takeaways for Stunning Bar Charts and Box Plots

As you continue to refine your skills in data visualization, keep these key takeaways in mind for creating stunning and effective bar charts and box plots:

➤ **Choose the right chart for your data:** Bar charts are excellent for comparing categorical data, while box plots provide insights into the distribution and variability of continuous data. Select the chart that best aligns with the story you want to tell.

➤ **Keep it simple and clear:** Avoid clutter by limiting the number of bars or categories in your chart. Use a consistent and intuitive design, with clear axis labels, legends, and annotations that make the chart easy to interpret.

➤ **Pay attention to scale:** For bar charts, always start the y-axis at zero unless there's a strong reason not to. For box plots, ensure that the scales across groups are consistent to allow for meaningful comparisons.

➤ **Use color effectively:** Choose contrasting colors to distinguish between categories or groups, but avoid overloading your chart with too many colors. Consider using colorblind-friendly palettes to ensure your visualizations are accessible to all audiences.

➤ **Highlight key insights:** Use design elements such as bold colors, annotations, or outlier markers to draw attention to the most important aspects of your data. Ensure that viewers can quickly grasp the key message of the chart.

➤ **Avoid common pitfalls:** Watch out for misleading design choices such as truncated axes, inconsistent bar widths, or overly complex color schemes that can distort the interpretation of your data.

> **Warning: Misleading Y-Axis**
>
> Always start the y-axis at zero in bar charts. Truncating the y-axis can drastically exaggerate small differences between categories, leading to misleading interpretations of the data.

➤ **Combine charts for richer insights:** In some cases, combining bar charts and box plots (or other advanced visualizations) can provide a more comprehensive view of the data. Use this approach when you need to highlight both comparisons and distributions.

By keeping these principles in mind, you will be able to create bar charts and box plots that are not only visually appealing but also deliver the insights your audience needs.

3.8.2 Next Steps for Further Improvement

Mastering the basics of bar chart and box plot design is an important first step, but there is always room for growth. To continue improving your skills in data visualization, consider the following next steps:

➤ **Experiment with advanced chart types:** As discussed in the section on going beyond bar charts and box plots, consider exploring violin plots, rainbow plots, or other advanced visualization techniques that offer additional insights into your data.

➤ **Incorporate interactivity:** Interactive visualizations, enabled through tools like Plotly or Tableau, allow users to explore the data on their own terms. Consider adding interactivity to your charts when presenting them in digital formats.

➤ **Practice data storytelling:** Data visualization is not just about presenting numbers; it's about telling a story. Practice crafting narratives around your charts that lead viewers through the key insights and encourage meaningful conclusions.

➤ **Refine your design aesthetics:** Great visualizations balance form and function. Continue to experiment with design elements such as fonts, colors, and layouts to find the perfect balance between aesthetics and clarity.

➤ **Stay current with tools and techniques:** The field of data visualization is constantly evolving, with new tools and techniques emerging regularly. Stay up-to-date with the latest trends and technologies by participating in online communities, taking courses, or following experts in the field.

➤ **Gather feedback:** Share your visualizations with colleagues or users and seek feedback on clarity, impact, and usability. Incorporating feedback is one of the best ways to improve the effectiveness of your charts.

Suggested Readings

1. Krzywinski, M., Altman, N.: Visualizing samples with box plots. Nature Methods **11**(2), 119–120 (2014). DOI 10.1038/nmeth.2813
2. Larson–Hall, J.: Moving beyond the bar plot and the line graph to create informative and attractive graphics1. Modern Language Journal **101**(1), 244–270 (2017). DOI 10.1111/modl.12386
3. Lem, S., Onghena, P., Verschaffel, L., Van Dooren, W.: The heuristic interpretation of box plots. Learning and Instruction **26**, 22–35 (2013). DOI 10.1016/j.learninstruc.2013.01.001
4. Lem, S., Onghena, P., Verschaffel, L., Van Dooren, W.: On the misinterpretation of histograms and box plots. Educational Psychology **33**(2), 155–174 (2013). DOI 10.1080/01443410.2012.674006
5. Lem, S., Vandebeek, C., Onghena, P., Verschaffel, L., Van Dooren, W.: Heuristic reasoning in interpreting box plots: The influence of orientation. Studia Psychologica **56**(2), 127–136 (2014). DOI 10.21909/sp.2014.02.655
6. Nuzzo, R.: The box plots alternative for visualizing quantitative data. PM and R **8**(3), 268–272 (2016). DOI 10.1016/j.pmrj.2016.02.001
7. Prakash, Y., Nayak, A., Jayarathna, S., Lee, H.N., Ashok, V.: Understanding low vision graphical perception of bar charts (2024). DOI 10.1145/3663548.3675616
8. Skau, D., Harrison, L., Kosara, R.: An evaluation of the impact of visual embellishments in bar charts. Computer Graphics Forum **34**(3), 221–230 (2015). DOI 10.1111/cgf.12634
9. Srinivasan, A., Brehmer, M., Lee, B., Drucker, S.: What's the difference?: Evaluating variants of multi-series bar charts for visual comparison tasks (2018). DOI 10.1145/3173574.3173878
10. Streit, M., Gehlenborg, N.: Points of view: Bar charts and box plots. Nature Methods **11**(2), 117 (2014). DOI 10.1038/nmeth.2807
11. Xiong, C., Lee-Robbins, E., Zhang, I., Gaba, A., Franconeri, S.: Reasoning affordances with tables and bar charts. IEEE Transactions on Visualization and Computer Graphics **30**(7), 3487–3502 (2024). DOI 10.1109/TVCG.2022.3232959

4 Designing Amazing Scatterplots

Abstract

Scatterplots are one of the most fundamental and versatile tools in data visualization, offering an intuitive way to explore and communicate relationships between variables. This chapter provides a comprehensive guide to designing scatterplots that are not only visually appealing but also highly effective at conveying complex data to diverse audiences. Beginning with an introduction to the core principles of scatterplot design, the chapter explores advanced techniques for visualizing multivariate data, enhancing interactivity, and creating dynamic representations of temporal changes. Practical examples and case studies illustrate the application of scatterplots in scientific research, public communication, and education, highlighting both successful and unsuccessful designs. The chapter also examines tools and frameworks, including Python's Matplotlib and Plotly, R's ggplot2 and Shiny, and platforms like Tableau and D3.js, offering best practices for scatterplot creation in various contexts.

Aims

After reading this chapter you should be able to:

➢ Understand the principles of scatterplot design.

➢ Identify when scatterplots are the most appropriate visualization technique and recognize when alternative approaches should be considered.

➢ Apply design aesthetics, including color schemes, marker shapes, and gridline choices, to create clear and accessible scatterplots.

➢ Tell compelling stories with scatterplots by using annotations, focus, contrast, and visual hierarchy.

➢ Evaluate the effectiveness of a scatterplot design based on user feedback, accessibility standards, and iterative design processes.

➢ Avoid common pitfalls in scatterplot design and continuously improve your visualizations through best practices and user-centered feedback.

R. Damaševičius, *Human-Centred Scientific Data Visualisation*,
Undergraduate Topics in Computer Science,
https://doi.org/10.1007/978-3-032-01606-5_4

4.1   Introduction to Scatterplots

Scatterplots are one of the most versatile and widely used tools in data visualization. They allow us to represent the relationship between two continuous variables on a two-dimensional plane, offering an intuitive way to observe trends, patterns, and correlations in data. Whether you are dealing with scientific data, business analytics, or any other domain, scatterplots provide an efficient method for displaying the interplay between variables. This chapter will guide you through the process of designing scatterplots that are not only aesthetically pleasing but also highly functional and accessible to a broad audience.

```python
import matplotlib.pyplot as plt

# Sample data
x = [1, 2, 3, 4, 5]
y = [2, 4, 1, 3, 5]

# Basic scatterplot
plt.figure(figsize=(6, 4))
plt.scatter(x, y, color='blue', marker='o')
plt.title('Basic Scatterplot')
plt.xlabel('X-Axis Label')
plt.ylabel('Y-Axis Label')
plt.grid(True)
plt.show()
```

A well-designed scatterplot can reveal insights that would otherwise remain hidden in raw datasets. The simple act of plotting data points allows users to visually perceive patterns, such as clusters, outliers , and linear or non-linear relationships, making scatterplots invaluable in exploratory data analysis. However, creating an effective scatterplot goes beyond merely plotting points on a graph. It requires thoughtful design choices, a clear understanding of the data, and a commitment to making the information accessible to users of various backgrounds and abilities. This chapter will cover both the technical and aesthetic aspects of scatterplot design, emphasizing a human-centered approach to data visualization.

4.1.1   The Importance of Scatterplots in Data Visualization

Scatterplots play a crucial role in data visualization because they offer an immediate and intuitive way to understand relationships between variables. They help to identify trends and patterns that would be difficult to discern in numerical or tabular data. For example, scatterplots can reveal whether two variables are positively or negatively correlated, whether the relationship is linear or non-linear, and whether there are any significant outliers in the dataset.

Beyond their role in exploratory data analysis, scatterplots are also a powerful tool for communication. They enable data scientists, researchers, and analysts to convey complex relationships to non-experts in a clear and accessible manner. In fields such as healthcare, economics, and environmental science, where data-driven decisions can have significant real-world implications, scatterplots help bridge the gap between technical analysis and decision-making processes.

However, the effectiveness of a scatterplot is heavily dependent on its design. Poorly designed scatterplots can obscure relationships, mislead viewers, or overwhelm them with unnecessary complexity. Thus, designing an amazing scatterplot requires balancing simplicity and sophistication, ensuring that the most important information is highlighted while avoiding visual clutter. This section will explore the foundational aspects of scatterplot design, providing a roadmap for creating visualizations that are both informative and visually engaging.

4.1.2 Understanding the Basic Elements of a Scatterplot

At its core, a scatterplot consists of a set of data points plotted on two axes: one for each of the variables being compared. Each point represents a single observation, with its position determined by the values of the two variables. The simplicity of this basic structure allows scatterplots to accommodate a wide variety of data types, from simple two-variable datasets to more complex, multivariate cases.

The x-axis typically represents the independent variable, while the y-axis represents the dependent variable. The choice of which variable to assign to each axis is not arbitrary–it can influence how patterns are perceived and understood. Therefore, the axis assignment should be informed by the nature of the data and the questions being asked.

Beyond the axes and data points, scatterplots can also incorporate additional visual elements to enhance their functionality and interpretability. Gridlines, for example, can help viewers read exact values, while axis labels and titles provide essential context. The use of color and marker shapes can encode additional dimensions of data, making it possible to represent more than two variables in a single scatterplot. This capability is particularly valuable in cases where understanding multiple dimensions is critical, such as in multivariate scientific experiments or demographic studies.

In designing the basic elements of a scatterplot, it is important to consider the target audience. For instance, a scatterplot designed for an expert audience might include intricate details and precise scales, while a scatterplot intended for a general audience should prioritize clarity and simplicity. The balance between complexity and simplicity is key to ensuring that the scatterplot serves its intended purpose without overwhelming the viewer.

4.1.3 Principles of Human-Centered Design for Scatterplots

Human-Centered Design (HCD) is a philosophy and process that prioritizes the needs, capabilities, and experiences of users when creating products, services, or visualizations. In the context of scatterplot design, HCD requires us to think critically about how users will interact with the visualization and how it can be made more accessible and understandable. This involves considering factors such as color contrast, layout, labeling, and interactivity, all of which can significantly impact the user experience.

One of the central tenets of HCD is accessibility. Scatterplots should be designed so that they are interpretable by as many people as possible, including those with visual impairments or other disabilities. This can be achieved by using color schemes that are distinguishable by people with color vision deficiencies, providing sufficient contrast between data points and background elements, and ensuring that text labels are legible at various scales. Providing alternative means of interaction, such as tooltips or interactive filtering, can make the scatterplot more accessible to users who may have difficulty interpreting static visualizations.

Another key principle of HCD is clarity. Every design decision should serve the purpose of enhancing the user's understanding of the data. This means avoiding unnecessary embellishments or visual clutter that could distract from the main message. Instead, scatterplot design should focus on creating a clear visual hierarchy, where the most important data points are emphasized, and supporting information is presented in a way that complements the primary message.

HCD also emphasizes the importance of context. A well-designed scatterplot should not exist in isolation but should be part of a larger narrative or data story. This requires careful consideration of how the scatterplot is framed, including the use of titles, axis labels, and annotations to guide the viewer's interpretation. Annotations, in particular, can be a powerful tool for drawing attention to key data points or explaining patterns in the data, helping users to make sense of the information being presented.

4.2 Choosing the Right Data for Scatterplots

Selecting the right data for a scatterplot is a critical step in designing an effective visualization. Not all data types or relationships are well-suited for scatterplots, and understanding when and how to use this visualization method can make a significant difference in the clarity and impact of your message. Scatterplots excel at visualizing relationships between two continuous variables, but they can also accommodate more complex datasets with the help of design enhancements. In this section, we will explore the types of data that work best for scatterplots, situations where alternative visualizations may be more appropriate, and essential data preparation techniques that can improve the accuracy and interpretability of your scatterplot.

4.2.1 Data Types Suitable for Scatterplot Representation

Scatterplots are most effective when visualizing the relationship between two continuous variables. These variables can represent anything from measurements in scientific experiments to economic indicators in financial data. The primary advantage of scatterplots lies in their ability to reveal patterns, such as correlations, clusters, and outliers , in a visually intuitive manner. Continuous numerical data, where both the x and y variables have a range of values, is ideal for scatterplot representation.

While scatterplots are traditionally used for two-dimensional data, they can be adapted to visualize additional variables through the use of color, marker size, or marker shape. For example, a third variable can be represented by changing the color of the points, while a fourth variable might be encoded by altering the size of the markers. This allows scatterplots to handle multivariate data while maintaining their core functionality of visualizing relationships between variables.

Categorical data, on the other hand, is less suited for scatterplot representation unless it is used in conjunction with continuous variables. In some cases, categorical data can be encoded using different colors or shapes, allowing the scatterplot to show groupings or classifications within the continuous data. However, care must be taken to ensure that the addition of categorical variables does not overwhelm the viewer or obscure the primary relationships being explored.

Time series data can also be visualized using scatterplots, particularly when the goal is to examine the relationship between time and another continuous variable. However, line charts are often more effective for purely temporal data, as they provide a clearer representation of trends over time. Scatterplots can still be useful for visualizing specific events or intervals within a time series, especially when comparing multiple time-dependent variables.

4.2.2 When Not to Use Scatterplots: Alternatives and Pitfalls

While scatterplots are versatile, they are not always the best choice for visualizing every type of data. Understanding the limitations of scatterplots is just as important as knowing when to use them. Scatterplots are not well-suited for representing discrete or ordinal data, where the values are restricted to specific categories or ranks rather than continuous ranges. In such cases, bar charts or line charts are more appropriate because they provide a clearer representation of the distribution and order of the data.

Another common pitfall is the misuse of scatterplots when the data does not show any meaningful relationship between the variables. If the scatterplot appears to be a random cloud of points with no discernible pattern, it may not be the best way to convey the information. In such cases, it is worth considering whether a different visualization technique, such as a histogram, box plot, or heatmap, might better illustrate the insights in the data.

Scatterplots can also become cluttered or difficult to interpret when dealing with large datasets, especially when there is significant overplotting (i.e., when data points overlap). This issue can obscure important patterns or create misleading impressions of density and distribution. Techniques such as using transparency (Alpha Blending) or binning can help mitigate these issues, but for extremely large datasets, alternative visualizations such as density plots or hexbin charts may be more appropriate.

> **⚠ Warning**
>
> Scatterplots with large datasets can suffer from severe overplotting , obscuring trends and hiding data density. Use transparency (`alpha`), binning (e.g., hexbin), or jittering to mitigate this issue and reveal meaningful patterns.

> **💡 Design Tip**
>
> When visualizing large datasets, apply transparency (`alpha` between 0.3–0.6) to data points. This makes dense areas appear darker and reveals overlapping structures without clutter.

Another pitfall to avoid is the inappropriate use of scatterplots for data that is highly skewed or has extreme outliers . In such cases, the visual scale of the scatterplot may distort the relationship between variables, making it difficult to identify meaningful trends. Data transformation techniques, such as logarithmic scaling, can help address these issues, but it is important to evaluate whether the scatterplot remains the best visualization choice after such adjustments.

4.2.3 Preparing Your Data: Scaling, Normalization, and Outlier Handling

Before creating a scatterplot, it is essential to ensure that your data is properly prepared for visualization. Poorly prepared data can lead to inaccurate or misleading interpretations, so taking the time to preprocess your data is crucial to creating an effective scatterplot. Three of the most important data preparation steps for scatterplots are scaling, normalization , and handling outliers .

Scaling is necessary when the variables being compared in a scatterplot have vastly different ranges. For example, if one variable is measured in small increments (e.g., 0 to 10) and another variable is measured on a much larger scale (e.g., 1,000 to 100,000), the scatterplot may become skewed, making it difficult to discern any meaningful relationship. In such cases, scaling the data so that both variables are represented on comparable axes can help improve the readability and interpretability of the scatterplot. Common scaling techniques include min-max scaling and z-score normalization , both of which rescale the data to a common range or distribution.

Normalization is closely related to scaling and is often applied when the goal is to compare variables with different units or magnitudes. For instance, if one variable represents height in centimeters and another represents weight in kilograms, it may be useful to normalize both variables to a common scale to allow for direct comparison. Normalization can also help reduce the impact of outliers , as it adjusts the range of the data based on its overall distribution, thereby preventing extreme values from distorting the visualization.

Handling outliers is another critical step in preparing data for scatterplot visualization. Outliers can have a disproportionate effect on the appearance of a scatterplot, especially if they fall far outside the range of the majority of the data.

Depending on the context, outliers can either be a point of interest or a distraction. In some cases, it may be important to highlight outliers as they represent significant deviations from the norm. However, in other cases, outliers can obscure more important patterns in the data and may need to be removed or transformed. Techniques such as winsorizing (capping extreme values) or using robust statistical methods can help mitigate the impact of outliers on the scatterplot.

Proper data preparation is essential for creating accurate and informative scatterplots. By scaling, normalizing, and addressing outliers , you can ensure that your scatterplot provides a clear and meaningful representation of the relationships in your data. This allows users to focus on the insights revealed by the visualization, rather than being distracted by distortions caused by unprocessed data.

4.3   Design Aesthetics of Scatterplots

While scatterplots are inherently functional, their effectiveness can be significantly enhanced through thoughtful design aesthetics. Well-designed scatterplots not only communicate data clearly but also engage users visually, making complex information more accessible and easier to interpret. The aesthetic choices you make–color schemes, marker shapes, sizes, transparency, gridlines, and axis design–can either clarify or obscure the underlying patterns in the data. By following principles rooted in accessibility and visual hierarchy, you can ensure that your scatterplots are not only beautiful but also highly functional for a diverse audience. This section focuses on various aesthetic elements and how they can be used to create clear, engaging, and accessible scatterplots.

4.3.1   Anatomy of Scatter Plot

Understanding the anatomy of a scatter plot is essential for both creating effective visualizations and critically interpreting them. A scatter plot is more than just a collection of data points; it is a structured graphical representation built from several key visual components, each with a specific purpose and design considerations. Fig. 4.1 presents a colorful and annotated infographic that dissects a basic scatter plot into its constituent parts, visually highlighting the relationships between them. This visual guide serves as a foundational reference for understanding how different design elements come together to communicate data clearly and effectively.

At the core of any scatter plot are the *data points*, which represent individual observations in the dataset. These points are positioned on the plot according to their values on the x-axis (typically the independent variable) and the y-axis (typically the dependent variable). The clarity of these points can be enhanced through the thoughtful use of color, shape, and size–especially when representing additional dimensions of the data.

> **♀ Design Tip**
>
> To support colorblind accessibility, use both color and shape to distinguish categories in scatterplots. Avoid relying solely on color to encode group information.

The *x- and y-axes* are critical to interpreting the scatter plot, as they define the data space and allow viewers to map visual positions to numerical values. Axis labels should be concise and include units of measurement where applicable to ensure interpretability. Axis ticks and gridlines further assist users by providing scale references and improving value estimation. Gridlines, however, should be used with subtlety–they are guides, not the focus of the plot.

To enhance interpretability and support pattern recognition, scatter plots may include a *trend line*, such as a linear regression line, that indicates the overall direction or correlation within the data. When present, the trend line should be visually distinguishable from the data points, often through the use of dashed lines or contrasting colors. Annotations and callouts can further guide the viewer's attention to significant observations, such as outliers or key clusters.

> **♀ Design Tip**
>
> Use annotations to highlight outliers , trends, or significant data points. Callouts can provide narrative context and direct the viewer's attention to key insights.

Contextual elements such as the *title*, *subtitle*, and *source line* provide an essential narrative framework. The title should clearly convey the topic or insight the scatter plot is intended to communicate. A subtitle may include additional information, such as the time period covered or the dataset used, while a source line at the bottom of the chart provides attribution and enhances credibility.

When color, shape, or size is used to encode additional variables, a *legend* becomes necessary to explain what these encodings represent. The legend should be simple, nonintrusive, and placed logically near the plot area without overlapping data points or labels.

Table 4.1 offers a comprehensive textual overview of the components highlighted in the figure. It summarizes each element's function and provides best practice recommendations to ensure clarity, accessibility, and effectiveness. For example, it recommends the use of colorblind-safe palettes for encoding categorical variables, suggests placing annotations sparingly to avoid clutter, and emphasizes the importance of clear and legible axis labels.

Fig. 4.1 and Table 4.1 serve as a detailed reference framework for designing and evaluating scatter plots. By understanding the anatomy of a scatter plot, designers can construct visualizations that are not only aesthetically engaging but also functionally powerful in conveying multivariate relationships in a user-centered and accessible manner.

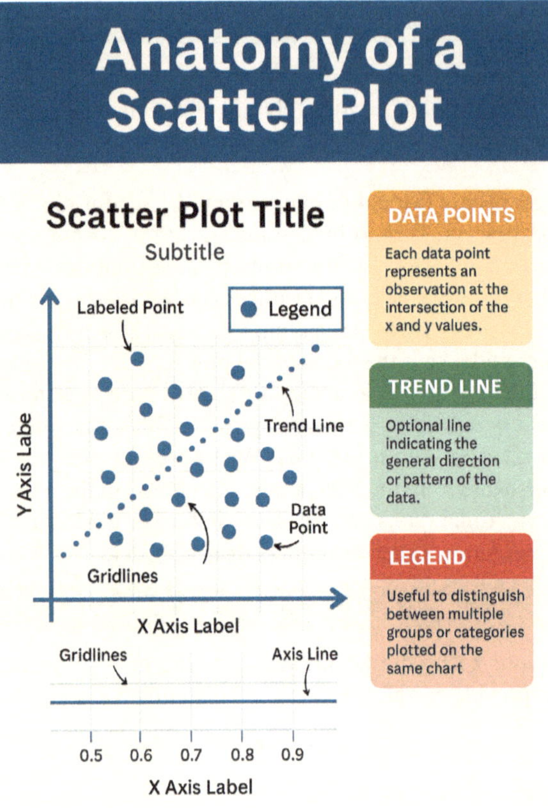

Fig. 4.1: Anatomy of a Scatter Plot. This infographic illustrates the fundamental components of a scatterplot, including data points, labeled axes, trend line, gridlines, legend, and annotations, providing a foundational understanding of scatterplot structure and function.

4.3.2   Choosing Color Schemes to Enhance Accessibility

Color is one of the most powerful tools in a designer's arsenal, and its impact on scatterplots is no exception. A well-chosen color scheme can emphasize key data points, differentiate between categories, and help viewers quickly grasp patterns. However, when selecting colors, it is essential to consider accessibility.

A popular approach for enhancing accessibility is to use color palettes designed specifically for colorblind users, such as the ColorBrewer palettes or other well-tested alternatives (Table 4.2). These palettes ensure that even those with color vision deficiencies can distinguish between different hues. Another technique is to avoid

Table 4.1: Anatomy of a Scatter Plot: Elements, Functions, and Best Practices

Component	Function	Best Practices
Title	Communicates the subject or purpose of the scatter plot	Be concise and descriptive; include context or comparison terms
Subtitle (optional)	Provides additional context such as timeframe or data source	Use to clarify scope or highlight important data limitations
X-Axis Label	Describes the independent variable	Use units if applicable; keep short and clear
Y-Axis Label	Describes the dependent variable	Ensure readability and vertical alignment; use appropriate units
Data Points	Represent individual observations plotted by their x and y values	Use distinct shapes/colors; avoid overlap; support interactivity for large sets
Gridlines	Aid in estimating values and identifying trends or alignment	Use light, unobtrusive lines; major gridlines only unless minor ones add value
Trend Line (optional)	Shows overall direction or relationship in the data (e.g., linear regression)	Use dashed or distinct color; annotate slope or equation if relevant
Legend	Identifies groups, categories, or encoded values (e.g., color, shape)	Place outside plot when possible; keep minimal and easy to read
Annotations	Highlight significant points, outliers , or explain context	Use sparingly; direct attention with arrows or callouts
Marker Color / Shape / Size	Encode categorical or quantitative variables	Use colorblind-friendly palettes; size should reflect value proportionally
Axis Ticks and Lines	Define the scale and intervals on both axes	Use logical intervals; minimize visual clutter; ensure tick labels are legible
Source Line (optional)	Cites data origin or authorship of the visualization	Place in small font below chart; include data source or link if needed

relying solely on color to convey meaning; instead, consider using additional visual encodings such as marker shapes, patterns, or sizes.

When using color to encode data in scatterplots, it is important to balance contrast and subtlety. High-contrast colors can draw attention to key data points or highlight important relationships, while more subdued tones can be used to represent background data or less significant categories. It is also crucial to ensure that the background color does not interfere with the data points. A light background with darker data points usually provides good contrast, but depending on the context and theme of the visualization, you may choose a dark background with lighter points to create a different visual effect.

Another aspect to consider is color mapping in continuous data. When representing a third variable through a color gradient, it is advisable to use perceptually uniform color scales–such as the viridis, plasma, or magma scales in Python's Matplotlib library–where differences in color correspond to equal perceptual differences in

Table 4.2: Colorblind-Friendly Color Palettes with Gradient Spectrum and Use Cases

Palette Name	Color Spectrum Preview	Use Cases and Notes
Viridis		Perceptually uniform and colorblind-safe. Ideal for continuous variables and heatmaps.
Plasma		High-contrast sequential palette. Good for intensity data and visual saliency.
Cividis		Optimized for grayscale printing and all color vision deficiencies. Widely used in scientific plots.
ColorBrewer: Set2		Soft pastel palette for categorical data. Works well for 6–8 distinct groups.
ColorBrewer: Dark2		High-contrast categories for clustering or segmentation. Colorblind accessible.
Tol Colors (Paul Tol)		Manually designed palette. Useful for colorblind-friendly categorical and diverging visualizations.

value. This ensures that the gradient is interpreted accurately by viewers, without introducing visual bias that could skew their understanding of the data.

4.3.3 Marker Shapes and Sizes: Making Data Points Stand Out

The markers in a scatterplot–the symbols representing individual data points–are essential to its design. Choosing the right shape and size for these markers can greatly enhance the clarity and interpretability of your scatterplot. In many cases, simple shapes such as circles or squares are sufficient, but other times, varying the marker shapes can convey additional information or improve accessibility.

For example, in multivariate scatterplots, you can use different marker shapes (Table 4.3) to represent different categories or groups within the data. Using a combination of circles, triangles, and squares can help viewers distinguish between groups without relying solely on color, which is particularly useful for viewers with color vision deficiencies. Similarly, varying the size of markers can encode an additional variable, such as population size or frequency, allowing you to add another dimension to the scatterplot without cluttering the design.

When choosing marker sizes, it is important to ensure that the data points remain visible without overwhelming the scatterplot. Markers that are too large can overlap and obscure important relationships, especially in dense regions of the plot, while markers that are too small may be difficult for viewers to see, particularly on high-resolution screens. A good rule of thumb is to make marker sizes proportional to the overall size of the scatterplot, adjusting them dynamically based on the number of data points and the scale of the axes.

Table 4.3: Common Marker Types Used in Scatterplots: Use Cases and Accessibility Considerations

Marker Shape	Typical Use Case	Accessibility Notes
Circle (o)	Default marker for most plots; ideal for general-purpose scatterplots	Easily distinguishable; works well at various sizes and opacities
Square (s)	Used to differentiate categories or series	Good contrast from circles; color fills can improve distinction
Triangle Up (^)	Represents a directional or ordinal category (e.g., increase)	Use sparingly; may resemble caret symbols in small sizes
Triangle Down (v)	Often paired with triangle up for dichotomies or flow direction	Ensure legibility with adequate spacing
Diamond (D)	Common for highlighting key or annotated points	Larger size recommended for clarity
Plus Sign (+)	Used when minimal visual obstruction is needed	Thin strokes may be hard to see on busy backgrounds
Cross (x)	Useful for distinguishing categories in grayscale plots	Good for black-and-white printing; prone to overplotting
Star (*)	Ideal for special points, events, or key highlights	Visually attention-grabbing; use sparingly
Pentagon / Hexagon	Used in stylized or domain-specific plots	Less common; ensure clarity at small sizes
Custom SVG/Icon	Applied in branded visualizations or UX-specific dashboards	Offers maximum flexibility; consider screen reader compatibility
Emoji / Unicode	Emerging trend in informal data journalism or mobile charts	May reduce professionalism; inconsistent rendering across systems

Marker transparency is another tool that can be used to address overplotting , which is common when large datasets are represented in a scatterplot. By reducing the opacity of the markers, you can allow overlapping data points to be seen more clearly, revealing density patterns without cluttering the plot. This technique, which we will discuss further in the next subsection, is particularly effective when combined with appropriate marker sizes and shapes.

4.3.4 Using Transparency and Overplotting Techniques

Overplotting occurs when too many data points overlap in a scatterplot, obscuring relationships and making it difficult for viewers to discern patterns. This is a common issue in scatterplots with large datasets or where data points are densely clustered. Fortunately, several design techniques can help mitigate overplotting , allowing viewers to see both the overall distribution and finer details of the data.

One of the most effective techniques for reducing overplotting is using transparency, or Alpha Blending, which involves lowering the opacity of the data points. This allows overlapping markers to be visible, revealing areas of higher density where many points are concentrated. For example, by setting the transparency level to around 50%, multiple overlapping points will appear darker, providing a visual cue to their density. This technique helps users understand how data points are distributed without obscuring the overall structure of the scatterplot.

In addition to transparency, jittering can be used to spread out overlapping points slightly, making it easier to distinguish individual data points. Jittering works by adding a small amount of random noise to the data, which helps separate points that would otherwise be plotted directly on top of each other. This technique is especially useful in cases where the data contains discrete values, leading to a significant amount of overplotting .

Another approach to dealing with overplotting is to use hexbin or 2D Density Plot, which aggregate data into bins and represent the density of points within each bin using color. While not technically scatterplots, these visualizations can be a useful alternative when overplotting becomes a major issue and you want to convey the density of data points more effectively.

4.3.5 Effective Use of Gridlines and Axis Design

Gridlines and axis design are often overlooked elements in scatterplot aesthetics, but they play a crucial role in ensuring that the data is easy to read and interpret. Well-designed gridlines can provide important reference points, helping viewers estimate the values of data points, while poorly designed gridlines can clutter the scatterplot and detract from its readability.

When incorporating gridlines, less is often more. Overly prominent or densely spaced gridlines can distract from the data points, making the scatterplot look busy and difficult to read. Instead, opt for light, subtle gridlines that provide guidance without overwhelming the visualization. Major gridlines can be emphasized slightly more, while minor gridlines should be kept faint or omitted altogether to maintain visual clarity.

> **♀ Design Tip**
>
> Include light, unobtrusive gridlines to assist value estimation–especially in static scatterplots. Avoid heavy lines that distract from the data.

Axis design is equally important for creating an effective scatterplot. The axes should provide clear, easily readable labels that help viewers understand the scale and context of the data. Avoid using overly technical or complex labels that may confuse viewers, and ensure that the axis ticks are spaced appropriately to avoid overcrowding. In some cases, it may be helpful to use logarithmic or other non-linear scales to better represent data with wide-ranging values or exponential growth patterns.

Labels on both the x and y axes should be descriptive enough to convey the nature of the variables being represented, but not so verbose that they detract from the overall simplicity of the scatterplot. The use of axis titles and captions can provide important context, explaining to viewers what the scatterplot is intended to communicate and guiding their interpretation of the data.

By thoughtfully designing gridlines and axes, you can enhance the usability of your scatterplot while maintaining its aesthetic appeal. These elements should support the primary goal of the scatterplot–communicating the relationship between variables–without distracting from the data itself (Table 4.4).

Table 4.4: Scatterplot Aesthetic Elements and Their Functions

Element	Purpose and Consideration
Color	Differentiate categories or encode numeric values; use accessible palettes like ColorBrewer or viridis
Marker Shape	Encode categories or classes; helps viewers distinguish without relying solely on color
Marker Size	Encode a quantitative variable; must be balanced to avoid distortion
Transparency (Alpha)	Reduces overplotting and clarifies dense areas
Gridlines	Aid value estimation; should be subtle to avoid distraction
Annotations	Provide context, highlight trends, and guide interpretation
Axis Design	Affects readability; label clearly and scale appropriately

4.4 Advanced Scatterplot Techniques

Scatterplots, while traditionally used to visualize relationships between two continuous variables, can be extended in a variety of ways to accommodate more complex datasets and enhance the user's understanding of the data. As data becomes more multidimensional and interactive technologies improve, scatterplots can be designed to reveal deeper insights and create more engaging experiences. In this section, we will explore advanced techniques such as multivariate scatterplots, interactive scatterplots, dynamic visualizations, and ways to combine scatterplots with other visual elements to tell a more comprehensive data story.

4.4.1 Multivariate Scatterplots: Visualizing More Dimensions

Although scatterplots are typically two-dimensional, representing one variable on the x-axis and another on the y-axis, they can be adapted to visualize additional dimensions of data. By using different visual encodings, such as color, shape, and size, scatterplots can represent more than two variables without overwhelming the viewer. This capability makes multivariate scatterplots a powerful tool for exploring complex datasets where multiple factors interact.

One common method of adding a third variable to a scatterplot is through the use of color. Each data point can be colored according to the value of a third variable, which allows viewers to discern patterns that involve more than just the relationship between the x and y variables. For instance, in a scatterplot showing income versus education level, color could represent geographic region, revealing regional trends that may not be obvious in a simple two-variable plot. It is important to choose color schemes carefully, ensuring they are intuitive and accessible, as discussed earlier in the chapter.

> **⚠ Warning**
>
> Relying solely on color to convey information can exclude colorblind users. Avoid red–green contrasts, and instead use colorblind-safe palettes like Viridis, Cividis, or ColorBrewer's Set2, combined with shape or size encodings.

Another way to represent a third or fourth variable is through the size or shape of the markers. Marker size can indicate a quantitative variable, such as population size or frequency, while different marker shapes can be used to distinguish between categories, such as different demographic groups or product types. This allows the scatterplot to convey more information in a compact and visually digestible format. However, care must be taken not to overload the scatterplot with too many variables, as this can make the plot difficult to interpret. Balancing clarity with complexity is key when designing multivariate scatterplots.

```python
import numpy as np

# Generate data
np.random.seed(0)
x = np.random.rand(50)
y = np.random.rand(50)
colors = np.random.rand(50)         # Third variable
sizes = 1000 * np.random.rand(50)   # Fourth variable

plt.figure(figsize=(8, 6))
scatter = plt.scatter(x, y, c=colors, s=sizes, alpha=0.6, cmap='viridis')
plt.title('Multivariate Scatterplot')
plt.xlabel('Variable X')
plt.ylabel('Variable Y')
plt.colorbar(scatter, label='Color Encoded Variable')
plt.show()
```

In cases where more than four variables need to be visualized, alternative methods such as parallel coordinates , radar charts, or 3D scatterplots might be considered. These approaches allow for the exploration of higher-dimensional data, but they come with their own challenges in terms of readability and user engagement.

4.4.2   Interactive Scatterplots: Enhancing User Experience

Interactive scatterplots offer a dynamic way for users to engage with data, allowing them to explore patterns, filter subsets of data, and gain deeper insights through direct interaction with the visualization. Interactive features enhance the user experience by providing tools that allow users to customize the view, zoom in on areas of interest, and access additional information about specific data points.

One of the most common interactive features in scatterplots is the tooltip, which displays detailed information when the user hovers over or clicks on a data point. This allows the scatterplot to remain uncluttered while still offering users access to more granular details as needed. For example, in a scatterplot representing sales versus time, a tooltip could display the exact sales figures, product names, and regions when hovering over specific points.

Zooming and panning functionality is another valuable interactive feature, particularly for scatterplots with large datasets. Users can zoom in to investigate clusters or outliers more closely or pan across different sections of the data without losing the overall context. This flexibility allows users to explore the data at their own pace and focus on areas that are most relevant to them.

Interactive filtering is also a powerful tool for exploring multivariate data. By allowing users to filter data based on specific criteria–such as selecting a subset of categories, time ranges, or variable thresholds–scatterplots can dynamically adjust to show only the data that meets the user's interest. This feature is especially useful in dashboards or applications where users need to slice and dice data to answer specific questions.

Creating interactive scatterplots often requires using specialized tools or libraries, such as D3.js, Plotly, or Bokeh, which provide the framework for building rich, user-responsive visualizations. These tools not only make it easy to add interactivity but also allow for seamless integration with web applications and other digital platforms, ensuring that the scatterplot remains a central part of an interactive data exploration.

```python
import plotly.express as px
import pandas as pd

# Sample DataFrame
df = px.data.iris()

fig = px.scatter(df,
                 x='sepal_width',
                 y='sepal_length',
                 color='species',
                 size='petal_length',
                 hover_data=['petal_width'],
                 title='Interactive Iris Scatterplot')
fig.show()
```

4.4.3 Visualizing Data Changes Over Time

When the data being visualized changes over time, dynamic scatterplots (also known as animated scatterplots) provide a powerful way to illustrate how relationships evolve. Dynamic scatterplots are particularly useful for showing trends, growth patterns, or fluctuations in data over time, which would be difficult to capture in a static plot.

In a dynamic scatterplot, time is typically represented through animation. The data points move as the animation progresses, showing how the relationship between variables changes across different time periods. For example, in a dynamic scatterplot visualizing economic data, points could represent countries, with the x-axis showing GDP and the y-axis showing life expectancy. As the animation plays, viewers can see how each country's position on the plot shifts over time, revealing long-term trends in development and health.

To enhance the clarity of dynamic scatterplots, it is often helpful to include time markers or a progress bar that indicates the current time period being displayed. Using color gradients or trails behind the moving data points can help viewers track individual points over time, making it easier to follow the progression of each entity throughout the animation.

Creating dynamic scatterplots typically involves using libraries or tools that support animation, such as Plotly, Highcharts, or even Python's Matplotlib with its animation functionality. These tools allow developers to build seamless animations that convey temporal data in an intuitive and engaging way. However, designers must ensure that the animation speed is appropriate and that the user has control over playback, including the ability to pause, replay, or skip through time periods for a more interactive experience.

4.4.4 Combining Scatterplots with Other Visual Elements

While scatterplots are a powerful visualization tool on their own, they can be even more effective when combined with other visual elements to provide additional context or clarity. By incorporating elements such as histograms, regression lines, or reference annotations, you can create richer visualizations that convey more information and guide the viewer toward key insights.

One common technique is to combine scatterplots with marginal histograms or density plots. These additional visual elements can be placed along the x and y axes, providing insight into the distribution of each variable. This combination allows viewers to see not only the relationship between variables but also the overall distribution of data points for each variable, helping to identify patterns such as skewness or the presence of multiple modes. Marginal plots can be especially useful for highlighting outliers or clusters that might otherwise go unnoticed in a scatterplot alone.

Another effective combination is overlaying regression lines or trendlines on a scatterplot to highlight relationships between variables. A regression line can reveal whether the relationship is linear or nonlinear, providing a clearer interpretation of the data. Including confidence intervals or error bars along the trendline can also help communicate the uncertainty or variability in the data, giving viewers a more nuanced understanding of the relationship between variables.

Annotations are another valuable tool for enhancing scatterplots. By adding text or labels to specific data points or areas of the scatterplot, you can draw attention to important details or explain patterns that might not be immediately obvious. For example, in a scatterplot showing financial performance, annotations could be used to highlight significant events, such as a market crash or a major product launch, providing important context that helps viewers interpret the data.

```python
from scipy import stats

# Generate data
x = np.linspace(0, 10, 30)
y = 2 * x + 1 + np.random.normal(size=30)

# Fit linear regression
slope, intercept, r_value, _, _ = stats.linregress(x, y)
line = slope * x + intercept

plt.figure(figsize=(8, 5))
plt.scatter(x, y, label='Data Points')
plt.plot(x, line, color='red', label=f'Trend Line (r={r_value:.2f})')
plt.annotate('Outlier', xy=(x[5], y[5]), xytext=(x[5]+1, y[5]+3),
             arrowprops=dict(facecolor='black', shrink=0.05))
plt.title('Scatterplot with Trend Line and Annotation')
plt.xlabel('X')
plt.ylabel('Y')
plt.legend()
plt.grid(True)
plt.show()
```

Finally, scatterplots can be integrated into dashboards or reports alongside other visualizations, such as bar charts, line charts, or heatmaps. By providing multiple perspectives on the same dataset, this approach helps users explore data from different angles and gain a more comprehensive understanding of the relationships within the data. When combining scatterplots with other visual elements, it is important to maintain a consistent design aesthetic and ensure that the various visualizations complement each other rather than compete for attention.

4.5   Storytelling Through Scatterplots

Scatterplots are not merely tools for visualizing data; they can also serve as powerful vehicles for storytelling. When used effectively, scatterplots can tell a compelling narrative about the relationships between variables, uncover hidden insights, and guide viewers through the data in a way that is both informative and engaging. Storytelling through scatterplots involves more than just displaying data points–it requires careful consideration of how visual elements such as annotations, contrast, and visual flow can be used to communicate key messages and craft a coherent narrative. In this section, we will explore how to transform scatterplots into effective storytelling tools by adding context, focusing attention, and creating a logical flow that helps viewers understand the story your data is telling.

4.5.1 Annotations: Adding Context and Narrative

Annotations are an essential tool for adding context and narrative to scatterplots. They allow you to highlight important data points, explain trends, and guide the viewer's interpretation of the plot. Without annotations, a scatterplot might simply present data in its raw form, leaving viewers to draw their own conclusions. Annotations, on the other hand, help tell the story of the data, providing the viewer with a deeper understanding of what the scatterplot represents and why it is important.

There are several ways to incorporate annotations into a scatterplot. The most common approach is to add text labels to specific data points. These labels can explain anomalies, emphasize outliers , or highlight key events. For instance, in a scatterplot that tracks the relationship between GDP and life expectancy over time, you might annotate significant events such as recessions or public health interventions that explain sudden shifts in the data. By doing so, you provide viewers with the necessary context to interpret the data accurately.

Another approach is to use callouts or arrows to direct attention to specific areas of the scatterplot. This can be especially useful when you want to emphasize trends or clusters that may not be immediately obvious. Callouts can include brief explanatory text or numerical values that clarify the meaning of the data points. The use of arrows or other guiding symbols helps guide the viewer's eye to the most important parts of the scatterplot, ensuring that the main narrative of the data is conveyed effectively.

It's important to strike a balance between providing enough context and overwhelming the viewer with too much information. Over-annotating a scatterplot can lead to clutter and distract from the overall message. Instead, focus on highlighting the most significant aspects of the data, using annotations sparingly to reinforce the story without compromising the visual clarity of the scatterplot.

> **⚠ Warning**
>
> Encoding too many variables in a single scatterplot (e.g., color, shape, size, animation) can overwhelm viewers and reduce clarity. Balance expressiveness with simplicity to maintain focus on the main message.

4.5.2   Highlighting Data Points: Using Focus and Contrast

One of the challenges in scatterplot design is ensuring that viewers focus on the most important data points while still appreciating the overall distribution of the data. Highlighting specific data points or areas of the scatterplot through the use of focus and contrast can help draw attention to critical information, guiding the viewer's interpretation and reinforcing the narrative.

Focus and contrast can be achieved in a number of ways. One effective technique is to use color to differentiate between important and less important data points. For instance, if a scatterplot shows the performance of various companies, you might use a bright color to highlight the top-performing companies, while using muted colors for the rest. This creates a clear visual distinction between the points you want to emphasize and the rest of the data, allowing viewers to quickly identify the key players in the narrative.

Another way to highlight important data points is by adjusting their size or shape. Larger or differently shaped markers can signify data points of particular interest, such as outliers , key trends, or anomalies. This technique is especially useful when the scatterplot contains a large number of data points, as it allows the most important points to stand out from the crowd.

transparency and opacity are also effective tools for controlling focus and contrast. By making less important data points more transparent, you can emphasize the points that are more relevant to the story. This helps to reduce visual noise, allowing the viewer to concentrate on the key messages you want to convey. Similarly, adjusting the opacity of certain areas of the scatterplot can help highlight clusters or trends, making them easier to identify.

```python
# Simulate dense data
x = np.random.normal(5, 1, 1000)
y = np.random.normal(5, 1, 1000)

plt.figure(figsize=(6, 5))
plt.scatter(x, y, alpha=0.3, color='darkgreen')
plt.title('Scatterplot with Transparency to Handle Overplotting')
plt.xlabel('X Value')
plt.ylabel('Y Value')
plt.grid(True)
plt.show()
```

It's important to use focus and contrast thoughtfully to avoid misleading the viewer. Emphasizing certain data points too heavily can create bias or misrepresent the overall dataset. The goal should be to guide the viewer's attention without distorting the true story of the data.

4.5.3   Visual Hierarchy and Flow in Scatterplots

In any visualization, including scatterplots, the concept of visual hierarchy plays a crucial role in guiding the viewer's eye through the data in a logical and meaningful way. A well-constructed visual hierarchy helps viewers prioritize the information presented, ensuring that they understand the most important points before delving into finer details. Creating a strong visual hierarchy in a scatterplot is key to making the story clear and easy to follow.

There are several ways to establish visual hierarchy in scatterplots. One method is through the use of color and size, as discussed in the previous section. Larger, more vibrant data points naturally draw more attention than smaller, muted ones, allowing you to prioritize key data in the viewer's mind. Similarly, by layering elements with varying levels of transparency, you can create a sense of depth, with the most important data points appearing in the foreground.

The layout of the scatterplot itself also contributes to visual hierarchy. For example, you can position the most important data points or clusters in prominent locations within the plot area, such as near the center or along key axes. In some cases, it may be useful to divide the scatterplot into multiple sections or panels, each focusing on a different aspect of the data. This allows you to control the flow of information, leading viewers through the narrative in a structured manner.

Flow refers to the sequence in which viewers perceive and process the elements of a scatterplot. An effective flow guides the viewer through the visualization in a logical order, from the most significant data points to supporting details. For instance, you might start with an annotation or highlighted data point that introduces the main theme of the story, then direct the viewer's attention to related clusters or trends that provide context or further evidence.

The use of guiding elements, such as arrows or lines, can help reinforce flow. These elements act as visual signposts, directing the viewer's attention from one part of the scatterplot to the next. Flow can also be enhanced by the careful placement of text labels, titles, and legends, ensuring that viewers have all the necessary information to understand the plot at each step of their journey through the data.

4.5.4   Explaining Data Through Visual Language

Crafting a narrative in a scatterplot involves more than just arranging data points; it requires a thoughtful approach to visual language that clearly communicates the relationships and trends within the data. The goal of storytelling through scatterplots is to make the data accessible and engaging, allowing viewers to understand the story at a glance while also encouraging them to explore deeper insights.

To craft a compelling story, it is important to define the key message or theme that the scatterplot is meant to convey. Is the scatterplot illustrating a correlation between variables? Highlighting outliers or anomalies? Showing trends over time? Once the central message is identified, all design elements–from color and annotations to marker size and layout–should work together to support that message.

The next step is to build a narrative structure around the scatterplot. This structure should have a clear beginning, middle, and end. The beginning might introduce the main variables and the overall trend or relationship. The middle could explore supporting evidence or additional variables, such as clusters or outliers that provide context. Finally, the end might offer conclusions, interpretations, or insights, possibly through annotations or additional visual elements like regression lines.

In crafting the story, it's important to remember that less is often more. Resist the temptation to include every detail or variable in the scatterplot. Instead, focus on the elements that contribute most to the narrative. By simplifying the plot and reducing clutter, you make it easier for viewers to grasp the main point of the story and avoid becoming overwhelmed by unnecessary information.

Ultimately, the goal is to turn raw data into a meaningful narrative that engages viewers and helps them draw conclusions. A well-crafted story can transform a simple scatterplot into a powerful tool for communication, allowing complex data to be understood, shared, and acted upon.

4.6 Practical Examples and Case Studies

Scatterplots are versatile tools used in a variety of contexts, from scientific research to public communication. By examining real-world applications and case studies, we can better understand how scatterplots are used to solve complex problems and convey important information to diverse audiences. This section will explore several case studies that illustrate the power of scatterplots, analyzing both successful and unsuccessful designs to uncover best practices for effective visual communication.

4.6.1 Case Study: Visualizing Scientific Data with Scatterplots

In the realm of scientific research, scatterplots play a critical role in analyzing and communicating relationships between variables. This case study focuses on the use of scatterplots to visualize large datasets in environmental science, specifically the relationship between atmospheric carbon dioxide (CO_2) levels and global temperatures over time.

The dataset consists of CO_2 concentrations measured in parts per million (ppm) and average global temperatures over a 100-year period. The scatterplot shows CO_2 levels on the x-axis and temperature anomalies (the deviation from the long-term average) on the y-axis. Each data point represents an annual measurement, with the color of the points indicating different decades.

In this example, the scatterplot is particularly effective in illustrating the strong positive correlation between rising CO_2 levels and increasing global temperatures. By adding a color gradient to represent time, the scatterplot also reveals how this trend has accelerated in recent decades, making the impact of human activities on climate change immediately visible.

Key features of this successful scientific scatterplot include:

> **Clear and well-labeled axes:** The x-axis is labeled with units of CO_2 (ppm), and the y-axis is labeled with temperature anomalies (°C), ensuring that viewers can easily interpret the variables.

> **Effective use of color:** The color gradient allows viewers to track changes over time without adding complexity to the plot. Using color to encode time helps to highlight how the relationship between variables has evolved.

> **Annotation of key data points:** Significant events, such as periods of major volcanic activity or industrial growth, are annotated directly on the plot, providing context for deviations in the data.

> **Regression line for trend analysis:** A regression line is added to the scatterplot to quantify the correlation between CO_2 and temperature, reinforcing the visual trend with statistical evidence.

This case study demonstrates how scatterplots can be used to communicate complex scientific data in a clear and compelling way. By focusing on the relationship between variables and using color and annotation to add context, the scatterplot effectively conveys the urgency of climate change.

4.6.2 Case Study: Scatterplots for Public Communication

In public communication and education, scatterplots are often used to make complex information accessible to a broad audience. This case study examines the use of scatterplots in a public health campaign to illustrate the relationship between physical activity and life expectancy.

The dataset for this scatterplot includes survey data from a diverse population, tracking weekly physical activity (measured in hours) on the x-axis and life expectancy (in years) on the y-axis. The plot is designed to communicate a simple but powerful message: increasing physical activity is associated with longer life expectancy.

In this example, the scatterplot is aimed at a non-expert audience, which influences several design choices:

> **Simple and intuitive markers:** The data points are represented as simple circles, with a larger size used for older age groups to emphasize their importance in the analysis.

> **Descriptive axis labels:** The x-axis is labeled "Hours of Physical Activity per Week," and the y-axis is labeled "Life Expectancy (Years)," avoiding technical jargon to ensure that the plot is understandable to a general audience.

> **Use of annotations for key messages:** Key points, such as the recommended weekly amount of physical activity (e.g., 150 minutes), are annotated on the plot to guide the viewer toward actionable insights. The annotation clearly states the public health recommendation.

> **Interactive elements for engagement:** In this digital version of the scatterplot, interactive features such as hover tooltips allow users to explore the data in more detail. Hovering over a point reveals additional information about the individuals surveyed, such as their age and health status, adding depth to the visualization.

This case study illustrates the importance of simplicity and clarity when designing scatterplots for public communication. By removing unnecessary complexity and focusing on the key message, the scatterplot effectively educates the public on the health benefits of physical activity in a visually engaging way.

4.6.3 Successful and Unsuccessful Scatterplot Designs

Not all scatterplots are created equal. In this section, we will compare examples of both successful and unsuccessful scatterplot designs, highlighting the differences in design choices and their impact on the viewer's ability to interpret the data.

Successful Design:

A successful scatterplot should prioritize clarity, accessibility, and accurate representation of data. Consider a scatterplot comparing income and education level across different regions, with points color-coded by geographic location. In this well-designed example:

> **Effective use of color:** A clearly defined color palette is used to represent geographic regions, making it easy to distinguish between data points from different areas.

> **Consistent marker size:** The size of the markers is uniform, preventing any visual distortion of the data. This ensures that viewers can easily compare the relative positions of the points without being distracted by unnecessary variations in size.

> **Well-chosen axis ranges:** The axes are scaled appropriately to include all relevant data points while avoiding excessive white space, ensuring that trends are visible.

> **Contextual information:** A legend and descriptive axis labels provide important context, helping viewers understand what the plot represents and how to interpret the data.

Unsuccessful Design:

In contrast, an unsuccessful scatterplot might suffer from several common pitfalls that hinder its effectiveness. Consider a scatterplot that attempts to show the relationship between the same variables (income and education) but introduces several design flaws:

> **Overuse of color:** Too many colors are used to represent different subcategories, leading to a cluttered and confusing plot. The color scheme lacks coherence, making it difficult for viewers to quickly identify patterns or trends.

➤ **Inconsistent marker sizes:** Marker size is used to encode a third variable (population), but the sizes vary so widely that small markers are barely visible, while large markers dominate the plot, distorting the overall interpretation.

➤ **Misleading axis scaling:** The x-axis is overly compressed, making it appear as though there is less variability in income than there actually is. This distortion misleads the viewer into underestimating the relationship between the variables.

➤ **Lack of context:** The scatterplot lacks a legend and the axis labels are ambiguous, making it difficult for viewers to understand what the data represents. Without clear labeling or annotations, the plot fails to convey its intended message.

By comparing these two examples, we can see that thoughtful design choices are crucial for creating effective scatterplots. The successful design focuses on clarity, consistency, and accessibility, while the unsuccessful design suffers from visual clutter, distortion, and a lack of context (Table 4.5). Learning from these contrasts, designers can avoid common pitfalls and ensure that their scatterplots communicate data accurately and effectively.

Table 4.5: Common Scatterplot Design Pitfalls and How to Avoid Them

Pitfall	Recommended Fix
Overplotting in dense regions	Use transparency or jitter; consider hexbin/density plots
Poor color choices	Use perceptually uniform and colorblind-safe palettes
Inconsistent or distorted axes	Scale axes appropriately; consider log scales for skewed data
Missing or unclear labels	Add descriptive axis labels, legends, and titles
Too many variables in one plot	Limit visual encodings; consider faceting or interactive filters
Ignoring accessibility	Use high-contrast design; avoid relying solely on color

4.7   Tools for Creating Scatterplots

Creating scatterplots involves not only understanding the design principles behind effective visualizations but also using the right tools to bring these concepts to life. Numerous tools and frameworks are available for building scatterplots, each offering unique features and capabilities that cater to different needs and expertise levels. From coding libraries like Python's Matplotlib and Plotly to user-friendly platforms like Tableau and Power BI, choosing the right tool depends on factors such as the complexity of your data, the level of interactivity required, and your programming skills. In this section, we will explore the best practices for using Python and R for scatterplot creation, as well as other popular tools and frameworks for data visualization.

4.7.1 Best Practices in Python: Matplotlib and Plotly

Python has become one of the most popular languages for data visualization, and it offers powerful libraries for creating scatterplots. Two of the most widely used libraries for this purpose are Matplotlib and Plotly. Both libraries provide extensive customization options, allowing users to create both static and interactive scatterplots. However, each library has its own strengths and best practices for scatterplot creation.

Matplotlib is one of the oldest and most versatile Python libraries for data visualization. It is highly customizable and provides a low-level interface for creating static scatterplots with precise control over every aspect of the plot, including axes, markers, colors, and labels. Matplotlib is particularly useful for creating publication-quality figures, and it integrates well with other scientific libraries like NumPy and Pandas.

When using Matplotlib to create scatterplots, some best practices include:

> **Start with a clean design:** By default, Matplotlib scatterplots may include unnecessary elements such as gridlines or excessive axis ticks. Customize the plot to reduce visual clutter by removing or lightening gridlines and keeping axis labels clear and concise.

> **Use the `alpha` parameter for transparency :** If your dataset contains many overlapping points, use the `alpha` parameter to introduce transparency, which helps reveal patterns in dense regions of the scatterplot.

> **Enhance the plot with annotations:** Matplotlib's `annotate()` function allows you to add text labels to specific points in the scatterplot, helping to highlight important data points or explain trends.

> **Combine scatterplots with additional elements:** Matplotlib easily supports adding elements such as trendlines, regression lines, or marginal histograms, providing deeper insight into the data.

Plotly is another popular Python library, but it specializes in creating interactive plots. It allows users to build highly interactive scatterplots with features such as tooltips, zooming, panning , and filtering. Plotly is particularly suited for web-based applications and dashboards, as its interactive plots are rendered in web browsers using JavaScript.

Best practices for creating scatterplots with Plotly include:

> **Use interactivity:** Make use of Plotly's built-in interactivity, such as hover labels (tooltips) and click events, to provide users with detailed information about specific data points without cluttering the plot with too much static text.

➢ **Customize the layout for clarity:** Plotly offers extensive customization options for adjusting the layout, axes, and legends. Ensure that your plot is clean and well-organized, with appropriate labeling and spacing between elements.

➢ **Add sliders or filters:** For multivariate or time-based data, you can add interactive sliders or dropdowns to allow users to filter the data or animate the scatterplot over time, offering a more dynamic user experience.

➢ **Export interactive plots:** Plotly allows you to export interactive scatterplots as HTML files, making them easy to share online or embed in web applications.

By using Python libraries like Matplotlib and Plotly, you can create both static and interactive scatterplots that are tailored to your specific needs, whether you're generating scientific reports or building engaging web-based dashboards.

4.7.2 Scatterplot Creation with R: ggplot2 and Shiny

For users working in R, **ggplot2** is the go-to library for creating scatterplots. Built on the principles of "The Grammar of Graphics," ggplot2 offers a high-level, declarative interface for creating complex and aesthetically pleasing visualizations with minimal code. It is widely used in academic and research settings due to its flexibility and ability to create publication-ready graphics.

Best practices for creating scatterplots with ggplot2 include:

➢ **Use `geom_point()` for scatterplots:** The `geom_point()` function is the primary method for creating scatterplots in ggplot2. You can easily customize marker size, shape, and color using this function, allowing you to tailor the scatterplot to your specific needs.

➢ **Layer additional elements to enhance the plot:** One of the strengths of ggplot2 is its ability to layer different geometric elements. For example, you can add a regression line to the scatterplot using `geom_smooth()`, or you can combine scatterplots with histograms, density plots, or boxplots to provide additional context.

➢ **Use `facet_wrap()` or `facet_grid()` for multivariate data:** When visualizing multidimensional data, ggplot2 allows you to create small multiples, or "facet" plots, which are useful for comparing relationships across different subgroups or categories.

➢ **Pay attention to themes and aesthetics:** ggplot2 comes with a variety of built-in themes, such as `theme_minimal()` and `theme_classic()`. Choosing the right theme can greatly enhance the readability and visual appeal of your scatterplot.

Shiny is another powerful tool in the R ecosystem, designed for creating interactive web applications. When paired with ggplot2, Shiny allows users to create dynamic, interactive scatterplots that respond to user inputs. This makes Shiny particularly useful for building data dashboards or exploratory data analysis tools.

These tools are compared by Table 4.6.

Table 4.6: Comparison of Popular Tools for Scatterplot Creation

Tool	Strengths	Best For	Interactivity Support
Matplotlib (Python)	Highly customizable, publication-quality plots	Static scientific plots	Low (needs extensions like mpld3)
Plotly (Python)	Built-in interactivity, web-friendly	Interactive dashboards	High
ggplot2 (R)	Elegant grammar of graphics, layered design	Academic charts	Low–Medium (higher with extensions)
Shiny (R)	Builds reactive web apps	Interactive educational dashboards	High
D3.js (JavaScript)	Maximum flexibility	Web-based storytelling	Very high
Tableau	No code, fast iteration	Business reports, dashboards	High
Power BI	Excel integration, BI context	Corporate use cases	High

Best practices for using Shiny with scatterplots include:

➤ **Provide user controls for interactivity:** Use Shiny's input widgets, such as sliders, dropdowns, and checkboxes, to allow users to filter data, adjust plot parameters, or explore different subsets of the scatterplot interactively.

➤ **Ensure responsive design:** When building Shiny apps, ensure that your scatterplots and layouts are responsive to different screen sizes and devices, so they look good on both desktop and mobile platforms.

➤ **Use reactivity wisely:** Shiny's reactive framework automatically updates the scatterplot whenever the underlying data or inputs change. However, be mindful of performance, especially when dealing with large datasets, and optimize your code to ensure smooth interactions.

➤ **Integrate ggplot2 with interactive features:** You can enhance static ggplot2 plots in Shiny with interactivity by using packages such as `plotly` or `ggiraph`, which allow ggplot2 plots to be made interactive with minimal code changes.

R's ggplot2 and Shiny provide powerful tools for both static and interactive scatterplot creation, making them ideal for a wide range of applications, from academic research to business intelligence dashboards.

4.7.3 Other Tools and Frameworks: Tableau, D3.js, Power BI

Beyond Python and R, several other tools and frameworks are widely used for creating scatterplots, particularly in business intelligence and web development contexts. These platforms provide a range of features, from user-friendly interfaces to highly customizable, code-based visualizations.

Tableau is one of the most popular data visualization tools in business intelligence, known for its user-friendly drag-and-drop interface and powerful data integration capabilities. Tableau allows users to create scatterplots without any coding, making it accessible to non-programmers. Best practices for using Tableau include:

- **Use Tableau's built-in features for customization:** Tableau allows you to easily adjust marker shapes, sizes, and colors, as well as add trendlines, reference lines, and annotations to enhance the scatterplot.

- **Use calculated fields to enrich your scatterplots:** Tableau provides powerful data manipulation capabilities through calculated fields, allowing you to create derived variables or group data in ways that enhance your scatterplot visualization.

- **Add interactivity through dashboards:** Tableau makes it easy to combine scatterplots with other visual elements in a dashboard, adding filters, parameters, and actions to create an interactive user experience.

D3.js is a JavaScript library for creating highly customizable, web-based visualizations. While D3.js requires coding knowledge, it offers unparalleled flexibility for creating dynamic and interactive scatterplots. Some best practices for using D3.js include:

- **Modularize your code for maintainability:** D3.js can get complex quickly, so it's important to write modular, reusable code to make your scatterplots easier to maintain and update.

- **Use D3's data binding for interactivity:** D3.js excels at binding data to visual elements, allowing you to create highly interactive scatterplots that respond to user actions such as hovering, clicking, or filtering.

- **Integrate with web technologies:** D3.js scatterplots can be easily integrated into websites and web applications, making them an excellent choice for data-driven storytelling on the web.

Power BI is another widely used tool for business intelligence, offering a user-friendly interface for creating scatterplots and other visualizations. It is particularly known for its integration with Microsoft Excel and its ability to handle large datasets. Some best practices for using Power BI include:

➤ **Use Power BI's built-in analytics features:** Power BI provides built-in analytics tools such as clustering, trendlines, and forecasting, which can be easily applied to scatterplots to reveal insights.

➤ **Create interactive reports:** Power BI allows you to build interactive reports that combine scatterplots with slicers, drill-downs, and other interactive elements, enhancing the exploration of your data.

➤ **Optimize for performance:** As Power BI is designed to handle large datasets, ensure that your scatterplots are optimized for performance by using measures like data aggregations and filters to prevent slow rendering.

By selecting the appropriate tool or framework based on your needs, skill level, and data complexity, you can create scatterplots that are not only visually appealing but also highly functional and interactive. Each tool brings its own strengths, from the flexibility of D3.js to the ease of use in Tableau and Power BI, making them valuable assets in the creation of compelling scatterplots.

4.8 Evaluating the Effectiveness of Your Scatterplot

Designing a scatterplot is only the first step toward effective data communication. Once the scatterplot is created, it is essential to evaluate how well it conveys the intended message and whether it is accessible to all intended users. Evaluation helps identify potential issues, refine design choices, and ensure that the scatterplot achieves its purpose. In this section, we will explore methods for evaluating the effectiveness of your scatterplot, including user testing, identifying common mistakes, improving accessibility, and adopting an iterative design process for continuous improvement.

4.8.1 User Testing and Feedback Collection

User testing is one of the most valuable ways to evaluate the effectiveness of your scatterplot. Involving real users in the testing process allows you to gather feedback on how well the visualization communicates the data and whether users can easily interpret the information presented. Testing with a diverse group of users–both experts and non-experts–can reveal insights into how different audiences perceive your scatterplot.

To conduct user testing, consider the following steps:

➤ **Define your objectives:** Clearly outline what you aim to achieve through user testing. Are you testing for clarity, ease of interpretation, or accessibility? Having specific goals helps guide the feedback collection process.

➤ **Create testing scenarios:** Develop realistic scenarios in which users interact with the scatterplot. These scenarios could involve tasks such as identifying key trends, interpreting correlations, or finding outliers . Observing how users complete these tasks will give you insight into how effectively the scatterplot communicates the data.

- **Ask targeted questions:** After the testing session, ask users open-ended questions to gather qualitative feedback. Some useful questions include: "What was the first thing you noticed?", "Were any parts of the scatterplot confusing?", and "What would you change to improve the visualization?"

- **Analyze feedback systematically:** After collecting feedback, identify recurring themes or issues. For example, if multiple users found the color scheme difficult to interpret or struggled to understand the axis labels, these areas should be prioritized for improvement.

User feedback provides direct insight into how well your scatterplot meets the needs of your audience. By incorporating this feedback into the design process, you can make adjustments that improve the overall user experience and ensure that your visualization communicates the data effectively.

4.8.2 Common Mistakes and How to Avoid Them

Even experienced designers can fall into common traps when creating scatterplots. These mistakes can range from subtle design flaws to major oversights that obscure the intended message or mislead viewers. Here are some common mistakes and how to avoid them:

1. Overcrowding the Scatterplot: One of the most frequent issues with scatterplots is overcrowding, which occurs when too many data points are plotted in a limited space. This can make it difficult for viewers to see patterns or relationships, especially in large datasets. To avoid overcrowding:

- Use transparency (alpha) to reduce overplotting , allowing viewers to see density patterns in the data.

- Consider using binning techniques such as hexbin plot or 2D Density Plot for very large datasets.

- Limit the number of variables represented on a single scatterplot to avoid visual complexity.

2. Inappropriate Use of Color: Color can be a powerful tool for conveying information, but it can also be misused. Common mistakes include using too many colors, relying on color alone to encode information, and choosing color schemes that are not accessible to colorblind users. To avoid these issues:

- Limit the number of colors used, and ensure that the color scheme is intuitive.

- Use colorblind-friendly palettes, such as ColorBrewer, to ensure accessibility.

- Avoid relying solely on color to encode data; consider adding shapes or labels for clarity.

3. Poorly Chosen Axis Scales: Incorrectly scaled axes can distort the data and lead to misinterpretation. For example, using an axis that does not start at zero or choosing an inappropriate range can exaggerate or minimize differences between data points. To avoid this:

➤ Always ensure that axes are scaled appropriately to represent the true range of the data.

➤ Consider using logarithmic scales if the data spans several orders of magnitude, but ensure that users understand the transformation.

➤ Provide clear and descriptive axis labels to guide interpretation.

> **⚠ Warning**
>
> Improper axis scaling–such as truncating the y-axis or using inconsistent intervals–can distort visual relationships and lead to misleading interpretations. Always label axes clearly and use consistent, data-appropriate scaling.

4. Inadequate Annotations or Labels: A scatterplot without adequate labels or annotations can be confusing to viewers. Key trends, outliers , or clusters may go unnoticed if not highlighted properly. To avoid this:

➤ Use annotations to call attention to important data points or trends, providing context where necessary.

➤ Ensure that axis labels are clear, concise, and informative.

➤ Add a descriptive title and legend that explain the variables and encodings used in the scatterplot.

By being aware of these common pitfalls, you can design scatterplots that avoid misinterpretation and effectively communicate the intended message.

4.8.3 Improving Accessibility and Inclusivity in Scatterplot Design

Accessibility and inclusivity are critical considerations in modern scatterplot design. A well-designed scatterplot should be interpretable by all users, including those with visual impairments or cognitive disabilities. By following best practices for accessibility, you can create visualizations that are inclusive and ensure that all users can engage with the data.

Here are several key strategies for improving accessibility and inclusivity in scatterplot design:

1. Use Colorblind-Friendly Palettes: As mentioned previously, colorblindness affects a significant portion of the population. By using color palettes that are distinguishable for individuals with different types of color vision deficiencies, such as red-green colorblindness, you can ensure that your scatterplot is accessible to a wider audience. Tools like ColorBrewer and the viridis palette in Python's Matplotlib provide colorblind-friendly options.

2. Provide Alternative Means of Encoding: Color should not be the sole means of encoding important information in a scatterplot. Consider adding alternative encodings, such as varying the shape or size of markers, to differentiate between categories or highlight specific data points. This helps ensure that users who cannot distinguish colors can still interpret the visualization effectively.

3. Ensure Sufficient Contrast: Low contrast between data points and background elements can make it difficult for users with low vision to interpret the scatterplot. Ensure that there is sufficient contrast between the data points, gridlines, and background. This can be checked using tools like WebAIM's contrast checker.

4. Provide Descriptive Text Alternatives: For users who rely on screen readers, providing descriptive text or alternative explanations for the scatterplot is essential. This could be achieved through alt text or additional explanations in accompanying documentation or captions.

5. Consider Cognitive Load: Scatterplots should be designed to minimize cognitive load, especially for users with cognitive disabilities. Simplify the design by reducing unnecessary elements and ensuring that the scatterplot has a clear focal point. Use annotations and visual hierarchy to guide users through the data without overwhelming them.

By focusing on accessibility and inclusivity, you ensure that your scatterplots are usable by the widest possible audience, helping to democratize data interpretation and analysis.

4.8.4 Continuous Improvement: Iterative Design Process for Scatterplots

Effective scatterplot design is rarely a one-time effort. Instead, it is an iterative process that involves continuous testing, refinement, and feedback collection. As data, user needs, and technology evolve, so too must your approach to scatterplot design. The iterative design process ensures that your scatterplots remain relevant, accessible, and effective over time. The iterative design process typically follows these steps:

➤ **Prototype:** Begin by creating an initial version of your scatterplot, applying best practices for data representation, aesthetics, and accessibility.

➤ **Test and Collect Feedback:** Conduct user testing and gather feedback on the initial design. Identify areas where users struggle to interpret the data or where the plot can be made more accessible.

➤ **Refine the Design:** Based on the feedback collected, make improvements to the scatterplot. This may involve adjusting color schemes, simplifying the layout, or improving annotations.

➤ **Test Again:** After refining the design, conduct another round of user testing to see if the changes have improved the user experience and data interpretation.

➤ **Implement:** Once the scatterplot design has been sufficiently refined, implement it in your report, dashboard, or publication.

➤ **Monitor and Update:** Continue to monitor user feedback after the scatterplot is in use. As new data becomes available or as user needs change, make updates to ensure that the scatterplot remains relevant and effective.

The iterative design process emphasizes the importance of flexibility and adaptability. No scatterplot is ever truly "finished"–there are always opportunities for improvement, especially as user expectations evolve and new design techniques emerge. By adopting an iterative mindset, you can ensure that your scatterplots consistently communicate data clearly and effectively.

4.9   Concluding Remarks

Evaluating the effectiveness of a scatterplot is a crucial step in ensuring that it communicates data accurately, clearly, and inclusively. By conducting user testing, avoiding common design pitfalls, improving accessibility, and adopting an iterative design process, you can create scatterplots that are both functional and engaging. The following table (Table 4.7) is recommended as a checklist when designing scatterplots.

Table 4.7: Scatterplot Evaluation Checklist

Evaluation Criterion	Pass?
Are axes clearly labeled and properly scaled?	[Yes/No]
Is the color scheme accessible (e.g., colorblind-safe)?	[Yes/No]
Are important trends or outliers clearly visible?	[Yes/No]
Is overplotting handled using transparency or binning?	[Yes/No]
Does the scatterplot support interactivity if needed?	[Yes/No]
Are annotations or tooltips used effectively?	[Yes/No]
Is the scatterplot readable on various screen sizes?	[Yes/No]
Is there a clear visual hierarchy guiding attention?	[Yes/No]

Scatterplots are a powerful tool for visualizing relationships between variables, but their effectiveness depends on thoughtful design and clear communication. Throughout this chapter, we have explored various techniques for enhancing scatterplots, from basic design principles to advanced techniques like multivariate and interactive visualizations. By focusing on human-centered design, you can create scatterplots that not only convey data accurately but also engage and inform a wide range of audiences.

As we have seen in the case studies and examples, scatterplots are used across many fields–from scientific research to public communication–each requiring different design considerations. The key to success lies in choosing the right tools, applying best practices, and always keeping the needs of your audience in mind. Whether you are creating scatterplots for a technical report or a public health campaign, the principles outlined in this chapter will help you craft visualizations that are both informative and impactful.

4.9.1 Future Trends in Scatterplot Visualization

The field of data visualization is rapidly evolving, and scatterplots are no exception. Several trends are shaping the future of scatterplot design, offering new possibilities for making complex data more understandable and engaging. Here, we explore some of the key trends that will likely influence the development of scatterplots in the coming years.

1. Increased Interactivity and Immersive Experiences: As interactive visualization tools continue to gain popularity, we can expect to see more scatterplots that allow users to manipulate and explore data in real-time. Tools like Plotly, D3.js, and Power BI already enable highly interactive scatterplots, but future developments in augmented reality (AR) and virtual reality (VR) could take interactivity to the next level. Imagine scatterplots that users can physically manipulate in 3D space or interact with through immersive environments, making data exploration a more intuitive and engaging experience.

2. Advanced Multivariate Scatterplots: While scatterplots traditionally visualize the relationship between two variables, future trends are likely to push the boundaries of multivariate visualizations. With the use of advanced visual encodings such as color, shape, size, transparency, and interactivity, multivariate scatterplots can represent many more dimensions of data. Machine learning techniques may also play a role in automatically identifying patterns in high-dimensional scatterplots and suggesting optimal ways to visualize complex datasets.

3. AI-Assisted Visualization Design: Artificial intelligence (AI) is making its way into data visualization, enabling designers to create more efficient and effective scatterplots. AI-driven design tools could help automate the process of choosing optimal axis ranges, color schemes, and marker sizes, based on the characteristics of the dataset and the target audience. By providing intelligent suggestions for improving design and accessibility, AI could significantly reduce the time and effort required to create high-quality scatterplots.

4. Personalized Visualizations: In the future, scatterplots may become more personalized, adjusting their design and content to match the specific preferences or needs of individual users. Personalization could involve tailoring the level of detail presented, offering different levels of interactivity, or even adapting the visualization to accommodate different cognitive styles and preferences. This trend aligns with the broader move toward user-centric design, ensuring that visualizations are not just informative but also tailored to the needs of each user.

5. Integration of Real-Time Data: With the increasing availability of real-time data, scatterplots are likely to become more dynamic, updating in real-time as new data streams in. This is particularly useful for applications such as financial analytics, environmental monitoring, or social media analysis, where the ability to visualize changing data in real-time can provide immediate insights. As real-time data visualization tools improve, scatterplots will become an even more powerful tool for decision-making in fast-paced environments.

4.9.2 Adopting Human-Centered Design for Visualization

Human-Centered Design (HCD) is more than just a buzzword–it is a critical approach to designing visualizations that are accessible, intuitive, and meaningful to a wide range of users. In scientific visualization, where complex datasets are often presented to diverse audiences with varying levels of expertise, HCD ensures that the visualization not only communicates the data but also resonates with users, helping them derive actionable insights.

1. Prioritizing Usability and Clarity: One of the core principles of Human-Centered Design is usability. In the context of scatterplots, this means creating visualizations that are easy to interpret, even for non-expert audiences. Usability involves clear labeling, appropriate use of color and contrast, and the inclusion of interactive elements that enhance, rather than complicate, the user experience. By prioritizing simplicity and clarity, scatterplots can become more effective tools for communication, especially when used in scientific publications or public education campaigns.

2. Focusing on User Needs: A human-centered approach starts with understanding the needs, goals, and challenges of the target audience. This might involve conducting user research, gathering feedback from stakeholders, or iterating on design based on user testing. In the context of scatterplot design, user needs can vary widely: researchers might require detailed, multi-dimensional scatterplots with statistical overlays, while a general audience might benefit from simplified scatterplots with minimal annotation. By focusing on these needs, designers can create scatterplots that effectively convey the intended message to their specific audience.

3. Enhancing Accessibility: Accessibility is a key component of human-centered design, and ensuring that scatterplots are accessible to users with disabilities is essential. This involves making conscious choices about color schemes, font sizes, and interactive features that accommodate users with visual impairments, cognitive disabilities, or motor impairments. For instance, offering alternative text for non-visual users, providing adjustable contrast levels, or allowing keyboard navigation for interactive scatterplots can make visualizations more inclusive.

4. Encouraging Exploration and Engagement: Human-centered scatterplot design also emphasizes engagement, allowing users to explore the data in ways that are meaningful to them. By incorporating interactivity, such as hover-over tooltips, filtering, and zooming capabilities, users can engage with the data more deeply, uncovering insights that may not be immediately visible in a static plot. Encouraging this exploration can help users connect with the data, making it more relevant and impactful in their decision-making processes.

5. Storytelling and Context: Incorporating storytelling into scientific visualization is another important aspect of human-centered design. A well-crafted scatterplot should not only display data but also tell a compelling story that guides the viewer through the insights. Annotations, narrative elements, and context are essential for helping viewers understand the significance of the data. This storytelling approach is particularly important when communicating complex scientific findings to a broader audience, such as policymakers, educators, or the general public.

4.9.3 Concluding Thoughts

In conclusion, scatterplots are one of the most versatile and powerful tools in data visualization, capable of revealing complex relationships between variables in an intuitive and visually engaging way. However, to maximize their effectiveness, scatterplots must be designed with the user in mind. By adopting Human-Centered Design principles–focusing on usability, accessibility, engagement, and storytelling–you can create scatterplots that not only present data accurately but also make it meaningful and actionable for a wide range of users.

Ultimately, the goal of scatterplot design–whether in scientific research, business analytics, or public communication–should always be to make complex data more accessible and understandable to all. By embracing Human-Centered Design and keeping the needs of your audience at the forefront of the design process, you can create scatterplots that bridge the gap between data and human insight.

Suggested Readings

1. Chen, H., Chen, W., Mei, H., Liu, Z., Zhou, K., Chen, W., Gu, W., Ma, K.L.: Visual abstraction and exploration of multi-class scatterplots. IEEE Transactions on Visualization and Computer Graphics **20**(12), 1683–1692 (2014). DOI 10.1109/TVCG.2014.2346594
2. Cuffe, P.: In praise of simple scatterplots. IEEE Potentials **40**(5), 36–38 (2021). DOI 10.1109/MPOT.2019.2891169
3. Giovannangeli, L., Lalanne, F., Giot, R., Bourqui, R.: Guaranteed visibility in scatterplots with tolerance. IEEE Transactions on Visualization and Computer Graphics **30**(1), 792–802 (2024). DOI 10.1109/TVCG.2023.3326596
4. Guha, T., Fertig, E., Deshpande, A.: Generating colorblind-friendly scatter plots for single-cell data. eLife **11** (2022). DOI 10.7554/ELIFE.82128
5. Itoh, T., Nakabayashi, A.: A technique for selection and drawing of scatterplots for multi-dimensional data visualization. pp. 62–67 (2019). DOI 10.1109/IV.2019.00020
6. Janetzko, H., Hao, M., Mittelstädt, S., Dayal, U., Keim, D.: Enhancing scatter plots using ellipsoid pixel placement and shading. pp. 1522–1531 (2013). DOI 10.1109/HICSS.2013.197
7. Lamberti, F., Manuri, F., Sanna, A.: Multivariate Visualization Using Scatterplots (2024)
8. Nguyen, Q., Huang, M., Simoff, S.: Using hybrid scatterplots for visualizing multi-dimensional data. Studies in Computational Intelligence **1014**, 517–538 (2022)
9. Okada, K., Itoh, T.: Scatterplot selection for dimensionality reduction in multidimensional data visualization. Journal of Visualization **28**(1), 205–221 (2025). DOI 10.1007/s12650-024-01025-6
10. Sadana, R., Stasko, J.: Designing and implementing an interactive scatterplot visualization for a tablet computer. pp. 265–272 (2014). DOI 10.1145/2598153.2598163
11. Sarikaya, A., Gleicher, M.: Scatterplots: Tasks, data, and designs. IEEE Transactions on Visualization and Computer Graphics **24**(1), 402–412 (2018). DOI 10.1109/TVCG.2017.2744184
12. Yuan, J., Xiang, S., Xia, J., Yu, L., Liu, S.: Evaluation of sampling methods for scatterplots. IEEE Transactions on Visualization and Computer Graphics **27**(2), 1720–1730 (2021). DOI 10.1109/TVCG.2020.3030432

13. Zheng, Y., Suematsu, H., Itoh, T., Fujimaki, R., Morinaga, S., Kawahara, Y.: Scatterplot layout for high-dimensional data visualization. Journal of Visualization **18**(1), 111–119 (2015). DOI 10.1007/s12650-014-0230-5

5 Designing Effective Histograms

Abstract

This chapter provides a comprehensive guide to designing histograms, one of the most fundamental and widely used tools for visualizing data distributions. The chapter begins by exploring the purpose of histograms and their applications in fields such as data analysis, machine learning, and decision-making. Key design principles are discussed, including the selection of appropriate bin sizes to balance detail and clarity, the use of color schemes and shading to enhance visual appeal, and techniques for labeling axes and data points effectively. The chapter also addresses common challenges in histogram design, such as over-clustering and under-clustering, and provides strategies for mitigating these issues. Advanced topics include handling different data distributions and customizing histograms for specific datasets. Practical examples demonstrate techniques for adjusting bin width, managing overlapping bars, and incorporating interactive elements to enhance user engagement. The chapter also highlights accessibility considerations, such as designing color-blindness friendly histograms and ensuring label readability.

Aims

After reading this chapter you should be able to:

- ➤ Understand the principles of histogram design and when to use histograms effectively.
- ➤ Choose appropriate bin sizes for accurate data representation.
- ➤ Optimize the use of color, spacing, and labeling for readability.
- ➤ Identify and avoid common pitfalls such as over-clustering or misrepresentation of data distributions.
- ➤ Implement accessibility best practices to ensure histograms are usable by individuals with visual impairments.

© The Author(s), under exclusive license to Springer Nature Switzerland AG 2026   126
R. Damaševičius, *Human-Centred Scientific Data Visualisation*,
Undergraduate Topics in Computer Science,
https://doi.org/10.1007/978-3-032-01606-5_5

5.1   Introduction to Histograms

Histograms serve as one of the fundamental tools in statistical data visualization, widely employed across various scientific and academic disciplines. They are utilized to visually represent the distribution of a dataset, allowing users to gain insights into the underlying structure of data without the need for complex numerical analysis. In essence, histograms simplify the process of understanding large datasets by condensing information into a graphical form that is both intuitive and accessible.

As a type of bar chart, histograms aggregate data into contiguous intervals, known as bins, where the height of each bar corresponds to the frequency of data points within that interval. While histograms may appear visually similar to bar charts, their underlying purpose and design principles differ in critical ways, which will be explored in this section. By grouping data into bins and graphically displaying frequency distributions, histograms allow for an immediate understanding of patterns such as skewness, central tendency, and variability within the data.

This section introduces the definition, purpose, and fundamental principles behind histograms as tools for exploratory data analysis. It aims to provide readers with a comprehensive foundation upon which more advanced design considerations will be built in subsequent sections.

5.1.1   Definition and Purpose

A histogram is defined as a graphical representation of the distribution of a dataset, wherein data points are sorted into a series of adjacent, non-overlapping intervals, referred to as bins. Each bin represents a range of values, and the height of the corresponding bar reflects the number of data points that fall within that interval. The key characteristic distinguishing histograms from other types of charts is their use of continuous intervals on the x-axis, which is appropriate when visualizing quantitative data.

The primary purpose of a histogram is to provide a visual summary of the frequency distribution of a dataset. Through this representation, histograms facilitate the identification of important features in the data, such as the presence of any patterns, trends, or anomalies. By revealing how data is distributed across different ranges of values, histograms enable users to identify aspects such as central tendency (e.g., mean, median, mode), spread (e.g., range, variance), and the shape of the distribution (e.g., skewness, kurtosis).

For instance, a dataset that exhibits a normal (or Gaussian) distribution will produce a bell-shaped histogram, where the majority of data points cluster around the central value. In contrast, datasets with skewed distributions may display a pronounced asymmetry in the histogram, with the majority of data points concentrated towards one end of the range. Histograms can also reveal the presence of multi-modal distributions, wherein multiple peaks or modes exist within the data, suggesting the presence of sub-populations or distinct groups.

Beyond merely visualizing data, histograms play a crucial role in informing further statistical analyses. The shape of a histogram provides valuable insights that can guide decisions about which statistical models or transformations may be appropriate for analyzing a dataset. For example, a histogram with a heavy tail may indicate the need for data normalization , while a multi-modal distribution might suggest the necessity of disaggregating the data into distinct subgroups.

In applied fields, the use of histograms is widespread. In scientific research, histograms are instrumental in illustrating experimental results, particularly when summarizing large datasets that would otherwise be unwieldy to interpret. Histograms also play a significant role in business analytics, where they are used to analyze sales, operational performance, and customer behavior by showing the frequency distribution of key metrics. In education, histograms are used to visualize exam scores and grade distributions, helping instructors identify trends in student performance. Similarly, in medical research, histograms are commonly used to display patient data, such as age distributions or the frequency of particular health outcomes.

5.1.2 When to Use Histograms

Histograms are primarily employed when there is a need to understand the distribution of continuous or discrete quantitative data across different ranges or intervals. As a frequency distribution tool, histograms are most useful when the dataset is large and requires summarization to reveal inherent patterns, such as central tendencies, variability, or data skewness. The suitability of histograms becomes apparent when dealing with continuous data–data that can take any value within a specified range, such as height, temperature, or age–where categorization of the data into predefined bins facilitates an immediate understanding of its overall distribution.

One of the essential use cases for histograms is in exploratory data analysis (EDA). During the initial stages of data exploration, histograms provide a quick visual overview of how data points are distributed across different ranges. This is invaluable when attempting to determine whether the data follows a normal distribution, whether there are any outliers , or if there is a skewness in the dataset that may influence subsequent statistical analyses. For instance, a researcher analyzing a dataset of patient ages in a medical study can use a histogram to quickly determine whether the sample population is concentrated within a specific age range or evenly spread across multiple age groups.

Histograms are well-suited for situations where comparisons within a single dataset are required. For example, in quality control processes, histograms allow practitioners to observe the frequency of occurrences of certain product characteristics (such as dimensions or weight), facilitating the identification of any deviations from the expected range. Such deviations can signal a need for process adjustments or further investigation into the cause of variability.

Histograms are also crucial in educational settings, where they can be used to summarize and analyze student performance across a range of test scores or grades. By grouping scores into intervals, histograms make it easy to detect trends such as whether the majority of students are performing at a particular level, or whether scores are skewed toward the lower or higher ends of the distribution.

Histograms are particularly effective when:

➢ The data is continuous or can be categorized into intervals.

➢ There is a need to understand the distribution or shape of the data.

➢ The goal is to identify central tendencies, spread, or outliers within the data.

➢ A clear visual summary is required for exploratory data analysis or quality control purposes.

Histograms provide a graphical means to encapsulate complex data patterns in an intuitive format, making them an indispensable tool for anyone engaged in data-driven research or analysis.

5.1.3 Comparison with Other Charts

While histograms are widely used for displaying the distribution of quantitative data, they are not the only tool available for visualizing data patterns. Other types of charts, such as bar charts, line charts, box plots, and scatter plots, serve different purposes and are better suited for specific types of data or analytical objectives. Understanding when to use histograms in comparison with these alternative visualization methods is crucial for effective data communication.

The most frequently compared chart to the histogram is the bar chart, as both use bars to display frequency. However, the key difference between histograms and bar charts lies in the type of data they represent and the way that data is grouped. Histograms are used for continuous data that is grouped into bins or intervals, with the bars touching each other to indicate the continuity of the data range. In contrast, bar charts are designed for categorical data, where the bars are separated from one another to indicate distinct categories. For instance, a histogram would be appropriate for visualizing the distribution of exam scores (continuous data), while a bar chart would be more suitable for showing the number of students enrolled in different courses (categorical data).

Line charts, another common visualization tool, differ from histograms in their ability to represent trends over time. While histograms excel at showing the distribution of data across a fixed range, line charts are used to track changes in data points over a period, connecting data points with a line to display the trend. For instance, in time series analysis, line charts are preferred over histograms for visualizing how a variable, such as stock prices or temperature, changes over time.

Box plots provide another alternative to histograms, particularly when the goal is to summarize key statistical properties of a dataset, such as the median, quartiles, and potential outliers . While histograms offer a detailed view of the frequency distribution across different intervals, box plots provide a more compact summary, making them ideal for comparative studies of multiple datasets. For example, when comparing the distributions of exam scores across different classes, a box plot can give a quick overview of the central tendency and variability of scores, without the level of detail provided by a histogram.

Scatter plots, unlike histograms, are used to examine the relationship between two continuous variables. Whereas histograms are concerned with the distribution of a single variable, scatter plots allow for the visualization of correlations between variables. For instance, a scatter plot could be used to show the relationship between study time and exam performance, providing insights into whether higher study time is associated with better exam outcomes. Scatter plots, however, are not suitable for revealing the distribution of a single variable, a task for which histograms are far more effective.

5.2 Histogram Design Fundamentals

The construction of an effective histogram depends not only on the data being represented but also on the careful selection of design elements that impact its readability and utility. Among these elements, the choice of bin size and the determination of the optimal number of bins are critical. These factors directly influence how data patterns, distributions, and potential outliers are perceived by the audience. Missteps in bin selection can lead to over-simplified or overly complex histograms, obscuring key insights.

5.2.1 Choosing the Right Bins

The process of choosing the right bin size in a histogram is a fundamental design consideration because it directly affects how the underlying data distribution is visualized. The bin size (or bin width) defines the range of values that each bin in the histogram represents. Larger bin sizes aggregate data points over broader intervals, potentially smoothing out variability, while smaller bin sizes capture more detail but can introduce noise or overemphasize minor fluctuations.

Let the dataset consist of n observations $\{x_1, x_2, \ldots, x_n\}$. The bin size h, which determines the range of each bin, can be expressed as:

$$h = \frac{\text{Range of Data}}{\text{Number of Bins}} \tag{5.1}$$

where the **range** of the data is the difference between the maximum and minimum values, $\text{Range} = x_{\max} - x_{\min}$. The number of bins, typically denoted as k, is influenced by various considerations, including the nature of the data, sample size, and the intended granularity of the visualization.

> **Design Tip**
>
> Avoid choosing bin widths arbitrarily. Always use statistical rules like Scott's or Freedman-Diaconis to guide initial choices.

The choice of h has practical implications: excessively large bins may obscure important data features by combining distinct values, while excessively small bins can fragment the data, making the histogram difficult to interpret. Thus, selecting an appropriate bin size requires balancing data resolution with the need for clarity.

Data resolution refers to the level of detail captured by the bins in a histogram. A high resolution, corresponding to a small bin size, provides a more granular view of the data, but this can lead to overfitting–where the histogram reflects random variations in the data rather than meaningful patterns. In contrast, a low-resolution histogram, with larger bins, smooths over these variations, potentially revealing broader trends but at the risk of oversimplifying the data.

Consider a dataset with a range R and n data points. If the bin size is too small (i.e., high resolution), the histogram may exhibit significant variation between adjacent bins, even when no meaningful differences exist in the underlying data. Conversely, if the bin size is too large, the histogram may fail to show important features such as multimodal distributions or the presence of outliers .

To balance this trade-off, data analysts often use rule-of-thumb methods, such as Sturges' formula or Scott's normal reference rule, to calculate an appropriate bin size. Scott's rule, for instance, is defined as:

$$h = 3.49 \frac{\sigma}{n^{1/3}} \tag{5.2}$$

where σ represents the standard deviation of the dataset, and n is the number of observations. This formula minimizes the integrated mean square error (IMSE) and is particularly useful when the data approximately follows a normal distribution. However, for skewed or multimodal data, more adaptive approaches may be necessary.

Selecting the optimal number of bins is another crucial aspect of histogram design. The number of bins k should reflect the balance between data summarization and detail preservation. While there are no hard rules for choosing k, several established heuristics guide this decision.

One of the most commonly used methods for determining k is Sturges' formula, which is based on the assumption of normally distributed data and works well for smaller datasets:

$$k = \lceil \log_2(n) + 1 \rceil \tag{5.3}$$

Sturges' formula is computationally simple and scales well with small to medium-sized datasets. However, for larger datasets or those that deviate significantly from normality, Sturges' formula may result in an overly simplified histogram, masking important details. In such cases, alternative methods like Scott's rule or the Freedman-Diaconis rule may be more appropriate. The Freedman-Diaconis rule, which is robust to outliers , calculates the number of bins as:

$$h = 2 \frac{IQR}{n^{1/3}} \tag{5.4}$$

where IQR is the interquartile range, a measure of statistical dispersion that is less sensitive to outliers than standard deviation. The Freedman-Diaconis rule is particularly useful for non-normal or highly skewed data, as it adapts the bin width based on the spread of the central data.

Ultimately, determining the optimal number of bins involves considering the characteristics of the data and the objectives of the analysis. In exploratory data analysis, a finer resolution may be warranted to detect subtle patterns, while for presentations intended for broader audiences, fewer bins may simplify the visual without sacrificing interpretability.

5.2.2 Handling Different Data Distributions

One of the key benefits of histograms is their ability to reveal the underlying shape of a data distribution. The characteristics of the distribution, such as symmetry, skewness, and modality, can influence the appropriate statistical methods and interpretations. This section explores how histograms handle different types of data distributions, with a focus on normal distributions, skewed distributions, and multi-modal distributions.

A normal distribution, also known as a Gaussian distribution, is one of the most common data distributions in statistics. It is characterized by its bell-shaped curve, where the mean, median, and mode are all located at the center of the distribution. The histogram of a normally distributed dataset is symmetric, with data values clustering around the central peak and tapering off towards the tails.

Mathematically, a normal distribution is defined by the probability density function (PDF):

$$f(x) = \frac{1}{\sigma\sqrt{2\pi}} \exp\left(-\frac{(x-\mu)^2}{2\sigma^2}\right) \tag{5.5}$$

where μ is the mean, σ is the standard deviation, and x represents the data points. The mean μ dictates the center of the distribution, while the standard deviation σ controls the spread. In a histogram of a normally distributed dataset, approximately 68% of the data falls within one standard deviation of the mean, and 95% falls within two standard deviations.

The key feature of a normal distribution in a histogram is its symmetry. The shape provides insight into statistical properties, allowing for assumptions in hypothesis testing and inferential statistics, where normality is often assumed.

Unlike the symmetric bell curve of a normal distribution, skewed distributions display asymmetry, where data values cluster toward one side of the distribution. A histogram representing a skewed distribution will show an elongated tail on either the right or the left.

A right-skewed (or positively skewed) distribution occurs when the tail is longer on the right side, with the bulk of the data concentrated on the left. Conversely, a left-skewed (or negatively skewed) distribution has a longer tail on the left, with data concentrated on the right. The degree of skewness can be quantified using the skewness statistic S, defined as:

$$S = \frac{n}{(n-1)(n-2)} \sum_{i=1}^{n} \left(\frac{x_i - \mu}{\sigma} \right)^3 \tag{5.6}$$

where n is the number of data points, μ is the mean, σ is the standard deviation, and x_i represents individual data points. A positive skewness value indicates right skewness, while a negative value indicates left skewness.

Histograms of skewed distributions are essential for identifying patterns where data values deviate from normality. For instance, in income data, which is often right-skewed, the majority of individuals may earn below the mean, while a small number of high-income individuals create a long tail to the right. Recognizing skewness is critical for selecting appropriate statistical methods, as many parametric tests assume normally distributed data.

A multi-modal distribution occurs when the dataset exhibits multiple peaks, or modes. Unlike unimodal distributions, which have a single prominent peak, multi-modal distributions suggest the presence of distinct subgroups within the data. In a histogram, this is visually represented by multiple peaks separated by valleys.

The number of modes m can be used to describe the distribution's complexity. A dataset with two peaks is called bimodal, and more than two peaks indicate a multi-modal distribution. Mathematically, multi-modal distributions may be described using a mixture model, where the overall distribution is a combination of multiple underlying distributions. For instance, a mixture of two normal distributions could be modeled as:

$$f(x) = p_1 \frac{1}{\sigma_1 \sqrt{2\pi}} \exp\left(-\frac{(x - \mu_1)^2}{2\sigma_1^2} \right) + p_2 \frac{1}{\sigma_2 \sqrt{2\pi}} \exp\left(-\frac{(x - \mu_2)^2}{2\sigma_2^2} \right) \tag{5.7}$$

where p_1 and p_2 are the mixing proportions, and μ_1, σ_1 and μ_2, σ_2 are the means and standard deviations of the two component distributions. Multi-modal histograms often indicate the existence of multiple populations or processes within the data. For example, a dataset of heights may show a bimodal distribution due to the presence of both male and female subjects, each with a distinct height distribution.

Recognizing multi-modality is critical for proper analysis, as it suggests that the data should not be treated as a single homogeneous group. Instead, further investigation is warranted to understand the distinct subgroups that contribute to the overall distribution.

Understanding the shape of the data distribution is fundamental to interpreting histograms. Normal distributions provide a symmetrical and well-defined structure, skewed distributions highlight asymmetries and potential biases in the data, and multi-modal distributions reveal the presence of subgroups or multiple underlying processes. The ability to identify and appropriately handle these different distributions is essential for accurate statistical analysis and data interpretation.

5.3   Enhancing Histogram Readability

The effectiveness of a histogram is determined not only by its ability to represent data but also by how easily it can be read and interpreted by its audience. Readability hinges on several factors, including the use of appropriate color schemes, clear labeling, and management of visual elements like overlapping bars. In this section, we explore key techniques to enhance the clarity and interpretability of histograms, making them more accessible to a wide audience.

5.3.1   Color Schemes and Shading

Choosing the right color scheme is essential for creating visually appealing and accessible histograms. Colors not only help to distinguish between different categories or data points but also play a role in directing attention to specific aspects of the histogram. However, poor color choices can introduce confusion, especially for viewers with visual impairments such as color blindness. Hence, the selection of colors should consider both aesthetic appeal and accessibility.

Color schemes should be designed with colorblindness in mind. A recommended approach is to use perceptually uniform color maps, such as the "Viridis" or "Plasma" palettes in Python's Matplotlib library. These palettes ensure that the changes in color intensity are perceived consistently across the entire spectrum, regardless of a viewer's ability to distinguish colors.

Example in Python:

```python
import matplotlib.pyplot as plt
import numpy as np

# Sample data
data = np.random.normal(0, 1, 1000)

# Create histogram \index{Histogram} with a colorblind-friendly colormap
plt.hist(data, bins=30, color='purple', edgecolor='black', alpha=0.7)
plt.title('Histogram with Color Scheme')
plt.xlabel('Value')
plt.ylabel('Frequency')

# Display the histogram
plt.show()
```

The resulting histogram is shown in Figure 5.1.

In this example, the use of a purple color with a black edge helps distinguish each bar clearly, and the colorblind-friendly approach ensures accessibility. Alpha

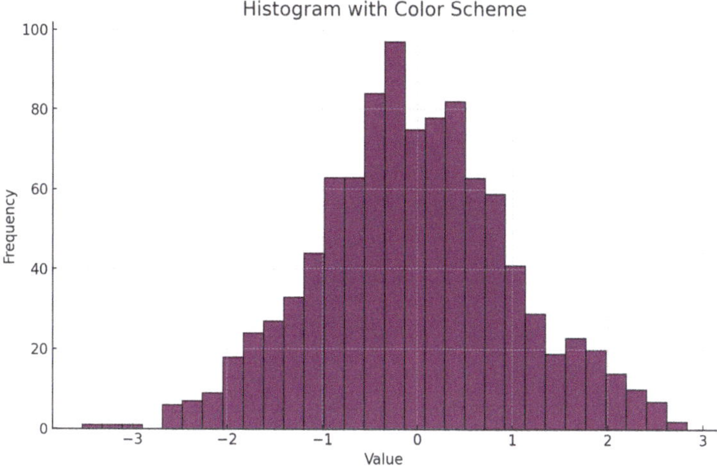

Fig. 5.1: Histogram With Color Scheme

transparency ($alpha = 0.7$) reduces the intensity of the color, making the histogram
visually balanced.

5.3.2   Labeling Axes and Data Points

Clear and informative labeling is essential for interpreting histograms accurately.
Labels should include descriptive titles for both the x- and y-axes, as well as the
overall title of the histogram. Ambiguous or missing labels may confuse the audience
and obscure the data's meaning.

In addition to labeling the axes, data point annotations can be used to provide
more detailed information, such as the exact frequency of each bin. This is especially
useful in histograms where precise values are important, such as in business analytics
or scientific research.

Example in Python:

```python
import matplotlib.pyplot as plt
import numpy as np

# Sample data
data = np.random.normal(0, 1, 1000)

# Create histogram \index{Histogram} with labels
counts, bins, patches = plt.hist(data, bins=20, color='skyblue',
    edgecolor='black', alpha=0.75)

# Labeling axes
```

```
11   plt.title('Labeled Histogram with Data Annotations')
12   plt.xlabel('Value')
13   plt.ylabel('Frequency')
14
15   # Annotate each bar with its height (frequency)
16   for count, patch in zip(counts, patches):
17       plt.text(patch.get_x() + patch.get_width() / 2, count, int(count),
18               ha='center', va='bottom')
19
20   plt.show()
```

The resulting histogram is shown in Figure 5.2.

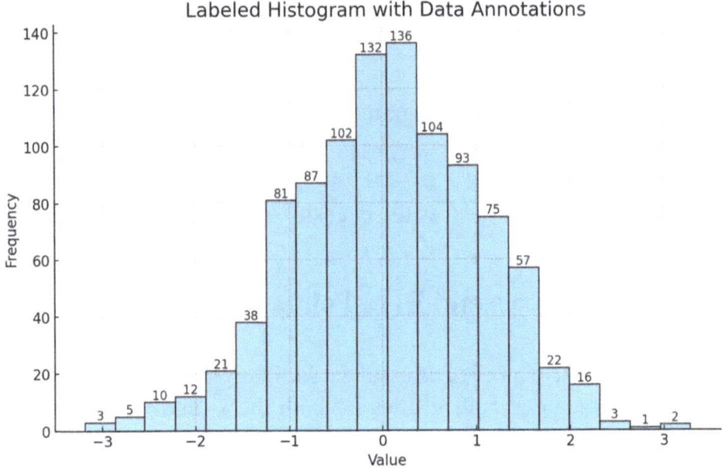

Fig. 5.2: Labeled Histogram With Data Annotations

In this example, we add labels for each bin directly on the bars using the *plt.text*() function. This provides immediate, clear insights into the distribution of the data and makes it easier for viewers to comprehend the histogram without guessing the values.

5.3.3 Managing Overlapping Bars

Overlapping bars occur frequently when multiple histograms are plotted together in a single figure. This can lead to misinterpretations or make it difficult to differentiate between datasets. One effective way to manage overlapping bars is to apply transparency (alpha) to the bars or to plot them side by side using the bar or barh functions in Matplotlib.

Another approach is to use stacked histograms, which display multiple datasets cumulatively within the same bins. Stacked histograms are particularly useful when comparing parts of a whole across categories.

Example in Python:

```python
import matplotlib.pyplot as plt
import numpy as np

# Generate two sets of data
data1 = np.random.normal(-2, 1, 1000)
data2 = np.random.normal(2, 1, 1000)

# Create overlapping histograms with transparency
plt.hist(data1, bins=30, color='blue', alpha=0.5, label='Dataset 1')
plt.hist(data2, bins=30, color='green', alpha=0.5, label='Dataset 2')

# Add labels and legend
plt.title('Overlapping Histograms with Transparency')
plt.xlabel('Value')
plt.ylabel('Frequency')
plt.legend()

plt.show()
```

The resulting histogram is shown in Figure 5.3.

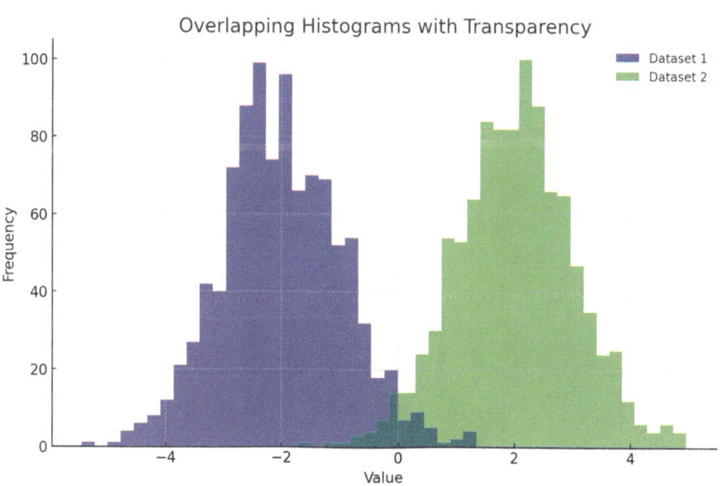

Fig. 5.3: Overlapping Histograms With Transparency

In this example, two datasets are plotted together using overlapping histograms.
By adjusting the transparency (alpha), both distributions remain visible, avoiding the

visual clutter that would result from opaque overlapping bars. The legend ensures that each dataset is clearly identified.

Alternatively, stacked histograms can be used to emphasize the cumulative contributions of different datasets:

Stacked Histogram Example:

```python
import matplotlib.pyplot as plt
import numpy as np

# Generate two sets of data
data1 = np.random.normal(0, 1, 1000)
data2 = np.random.normal(2, 1, 1000)

# Create stacked histograms
plt.hist([data1, data2], bins=30, stacked=True, color=['blue', 'green'],
    label=['Dataset 1', 'Dataset 2'])

# Add labels and legend
plt.title('Stacked Histogram')
plt.xlabel('Value')
plt.ylabel('Frequency')
plt.legend()

plt.show()
```

The resulting histogram is shown in Figure 5.7.

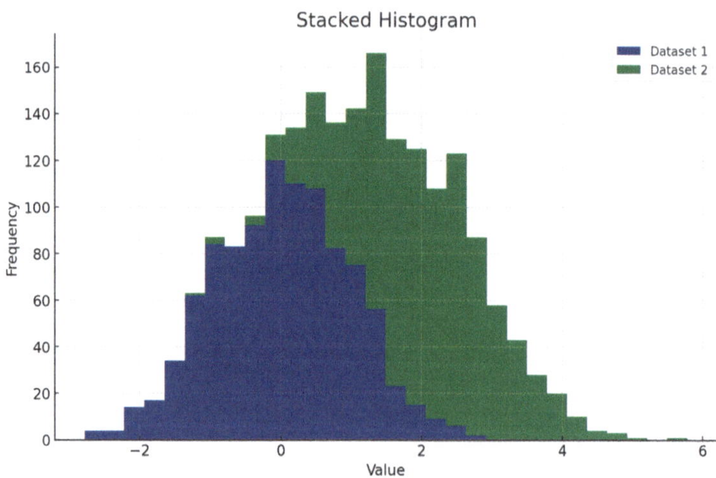

Fig. 5.4: Stacked Histogram

This Stacked Histogram enables comparison of the relative contributions of each dataset in the same bins, offering a clear and compact representation of cumulative frequency distribution.

5.4 Advanced Techniques in Histogram Design

While standard histograms effectively visualize data distributions, certain advanced techniques can provide deeper insights or accommodate more complex data scenarios. This section explores three such advanced techniques: density histograms, stacked histograms, and overlaying histograms for comparison. Each technique is illustrated with Python code examples, offering a practical guide to implementing these methods.

5.4.1 Density Histograms

A density histogram provides a smoothed estimate of the data distribution by plotting the probability density function (PDF) instead of the frequency of occurrences. Unlike a traditional histogram, which counts the number of data points per bin, a density histogram scales the y-axis to represent a probability density, ensuring the total area under the histogram sums to 1. This is particularly useful for visualizing the distribution of continuous data in a way that reflects relative probabilities.

The Python library Matplotlib makes it easy to create density histograms by setting the 'density' parameter to 'True'. This normalizes the histogram, transforming the y-axis from raw counts to probability densities.

Density Histogram example in Python:

```python
import matplotlib.pyplot as plt
import numpy as np

# Generate sample data
data = np.random.normal(0, 1, 1000)

# Create density histogram
plt.hist(data, bins=30, density=True, color='teal', edgecolor='black',
    alpha=0.7)
plt.title('Density Histogram')
plt.xlabel('Value')
plt.ylabel('Density')

# Display the plot
plt.show()
```

The resulting histogram is shown in Figure 5.5.

In this example, we generate a normal distribution of data and plot its density histogram. The histogram's y-axis represents probability density instead of absolute frequency, offering a smoothed representation of the distribution. To enhance the

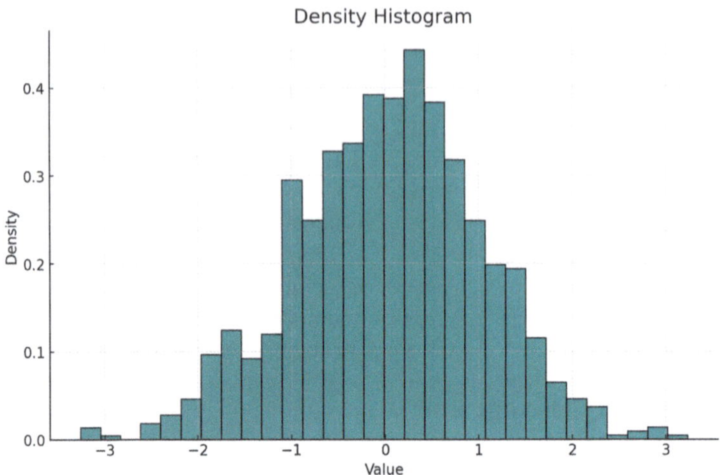

Fig. 5.5: Density Histogram

user experience (UX) and accessibility of the code for the density histogram, we can consider aspects like readability, color accessibility, and visual appeal.

Here's an improved version of the code with these principles in mind:

```python
# Generate sample data
data = np.random.normal(0, 1, 1000)

# Set up the plot with larger fonts and gridlines
plt.figure(figsize=(10, 7))  # Larger figure size for more impact

# Create density histogram\index{Density Histogram} without any color yet
n, bins, patches = plt.hist(data, bins=30, density=True, edgecolor='white',
    alpha=0.9)

# Applying a sleek, professional color palette\index{Color Palette} with a
    vertical gradient fill
for patch, height in zip(patches, n):
    color = plt.cm.cividis(height / max(n))  # Gradient based on bar
        height with professional palette
    patch.set_facecolor(color)

# Removing the title and focusing on labels with modern styling
plt.xlabel('Data Values', fontsize=16, color='#2d3436', fontweight='bold')
plt.ylabel('Density', fontsize=16, color='#2d3436', fontweight='bold')

# Adding grid with a modern, sleek style
plt.grid(True, which='both', linestyle='--', linewidth=0.7,
    color='#b2bec3', alpha=0.6)

# Customizing ticks for professional readability
```

```python
23  plt.xticks(fontsize=14, color='#636e72')
24  plt.yticks(fontsize=14, color='#636e72')
25
26  # Set background color to a light metallic grey for a professional feel
27  plt.gca().set_facecolor('#dfe6e9')
28
29  # Adding a shadow effect on the bars for uniqueness
30  plt.gca().patch.set_alpha(0.85)   # Slight transparency effect for depth
31
32  # Highlight the mean with a clean, professional vertical line
33  plt.axvline(np.mean(data), color='#d63031', linestyle='-', linewidth=2,
    ↪  label='Mean')
34  plt.legend(fontsize=12, frameon=False)   # Sleek legend without a border
35
36  # Display the plot
37  plt.show()
```

The resulting histogram is shown in Figure 5.6.

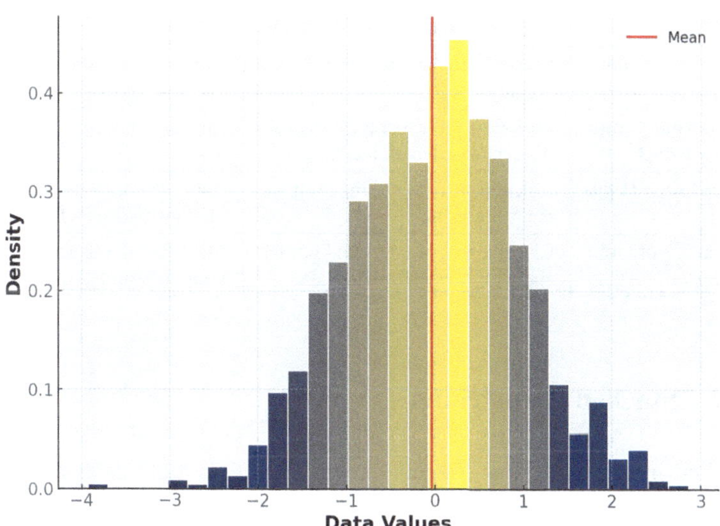

Fig. 5.6: Improved Density Histogram

The improvements are summarized as follows (also see Table 5.1):

➤ Color Accessibility: Use a color palette that is accessible for people with color blindness, such as blues and oranges. We can also provide a way to adapt colors dynamically.

➤ Font and Labeling: Use larger, readable fonts for titles and axis labels.

➤ Gridlines and transparency : Include gridlines for better readability, and make the histogram slightly transparent to create a sense of depth.

➤ Descriptive Title: Provide a more informative and engaging title for clarity.

➤ Ticks and Spacing: Improve tick marks and spacing for better precision and less clutter.

Table 5.1: Summary of Histogram Code Changes and Their Impact on UX

Change Implemented	Description	Impact on UX
Color Palette	Shifted from basic teal to a professional gradient palette with blues, greys, and yellow	Improved visual appeal and professionalism
Typography	Enhanced with modern, bold fonts for labels; title removed for minimalism	Increased readability and clarity
Gradient Effect	Introduced vertical gradient based on bar heights	Added depth and visual engagement
Gridlines	Subtle, grey dashed gridlines replace the standard black gridlines	Softer appearance, reducing visual clutter
Background Color	Changed from white to light metallic grey	Adds sophistication and contrasts the bars more effectively
Transparency and Shadow Effect	Added soft transparency and subtle shadow effect to the bars	Creates a modern, sleek look with depth
Mean Line Highlight	Added a vertical red line to mark the mean value	Enhanced information visibility, guiding the user's focus
General Layout and Spacing	Cleaner spacing, reduced clutter, and a more open layout	Improved overall aesthetic and easier to interpret

5.4.2   Stacked Histograms

Stacked histograms allow for the comparison of multiple datasets within the same bins. This technique is particularly useful when analyzing the contributions of different groups or categories to the overall distribution. Each dataset is "stacked" on top of the others, providing a clear visualization of the cumulative effect. In Python, the 'stacked' parameter in the 'hist()' function facilitates this approach, making it easy to plot multiple datasets as a single stacked histogram.

Example of overlaying histograms in Python:

```python
import matplotlib.pyplot as plt
import numpy as np

# Generate two sets of data
data1 = np.random.normal(0, 1, 1000)
```

```python
6   data2 = np.random.normal(2, 1, 1000)
7
8   # Create stacked histograms
9   plt.hist([data1, data2], bins=30, stacked=True, color=['blue', 'green'],
    ↪   label=['Dataset 1', 'Dataset 2'])
10
11  # Add labels and legend
12  plt.title('Stacked Histogram')
13  plt.xlabel('Value')
14  plt.ylabel('Frequency')
15  plt.legend()
16
17  # Display the plot
18  plt.show()
```

The resulting histogram is shown in Figure 5.7.

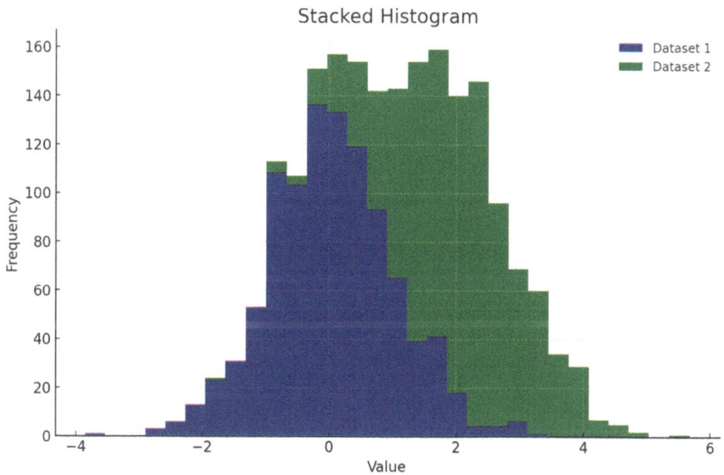

Fig. 5.7: Stacked Histogram

In this stacked histogram, two datasets are plotted within the same bin structure. Each color represents a different dataset, and the stack shows how the two datasets collectively contribute to the overall frequency for each bin.

5.4.3 Overlaying Histograms for Comparison

Overlaying histograms is another technique used for comparing the distributions of two or more datasets. Unlike stacked histograms, where data is accumulated within the same bins, overlaying histograms plots multiple datasets on the same axes with transparency applied to each dataset. This allows for the comparison of data distributions without merging them into cumulative values. In Python, this is achieved by plotting multiple histograms with the 'alpha' parameter, which controls transparency.

Example of overlaying histogram implementation in Python:

```python
import matplotlib.pyplot as plt
import numpy as np

# Generate two sets of data
data1 = np.random.normal(-2, 1, 1000)
data2 = np.random.normal(2, 1, 1000)

# Create overlapping histograms with transparency
plt.hist(data1, bins=30, color='blue', alpha=0.5, label='Dataset 1')
plt.hist(data2, bins=30, color='green', alpha=0.5, label='Dataset 2')

# Add labels and legend
plt.title('Overlaying Histograms for Comparison')
plt.xlabel('Value')
plt.ylabel('Frequency')
plt.legend()

# Display the plot
plt.show()
```

The resulting histogram is shown in Figure 5.8.

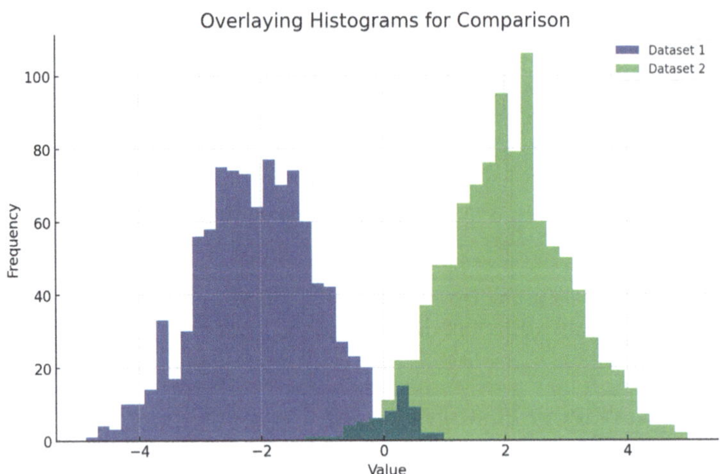

Fig. 5.8: Overlaying Histograms for Comparison

In this example, two datasets are plotted together with transparency , allowing viewers to see the overlap between them. This technique is particularly useful for identifying where datasets differ or share commonalities.

These advanced histogram design techniques–density histograms, stacked histograms, and overlaying histograms–offer more nuanced and flexible methods for visualizing data distributions. Each technique serves a specific purpose: density histograms provide smooth approximations of data distributions, stacked histograms allow for cumulative comparisons, and overlaying histograms enable side-by-side comparisons of multiple datasets. These methods, when used appropriately, can greatly enhance the interpretability of histograms, making them valuable tools for exploratory data analysis and statistical reporting.

5.5 Avoiding Common Pitfalls

Histograms are powerful tools for visualizing data distributions, but improper design choices can obscure the data or, worse, mislead the audience. Two common pitfalls in histogram design include over-clustering and under-clustering, as well as the use of misleading bin sizes and scaling. This section addresses these issues, providing practical guidance on how to avoid them, with examples implemented in Python.

5.5.1 Over-clustering and Under-clustering

One of the most significant challenges in histogram design is selecting an appropriate number of bins. Over-clustering occurs when too many bins are used, causing excessive granularity that obscures broader patterns in the data. Conversely, under-clustering results from too few bins, which can oversimplify the data, masking important details and nuances.

Over-clustering often makes a histogram difficult to interpret, as each bin contains very few data points, leading to a jagged, erratic appearance. under-clustering , on the other hand, produces a histogram that smooths over essential features of the data distribution, making it impossible to identify any real trends or outliers .

To strike a balance, choosing the number of bins based on the data's distribution and size is critical. While rules of thumb, such as Sturges' formula or Scott's normal reference rule, can provide guidance, they are not foolproof and must be adapted to the context.

Example of implementing under-clustered histogram in Python:

```python
import matplotlib.pyplot as plt
import numpy as np

# Generate sample data
data = np.random.normal(0, 1, 1000)

```

```python
 7  # Over-clustered histogram \index{Histogram} (too many bins)
 8  plt.subplot(1, 2, 1)
 9  plt.hist(data, bins=100, color='blue', edgecolor='black')
10  plt.title('Over-clustered Histogram')
11  plt.xlabel('Value')
12  plt.ylabel('Frequency')
13
14  # Under-clustered histogram \index{Histogram} (too few bins)
15  plt.subplot(1, 2, 2)
16  plt.hist(data, bins=5, color='green', edgecolor='black')
17  plt.title('Under-clustered Histogram')
18  plt.xlabel('Value')
19  plt.ylabel('Frequency')
20
21  # Adjust layout and show plot
22  plt.tight_layout()
23  plt.show()
```

The resulting histogram is shown in Figure 5.9.

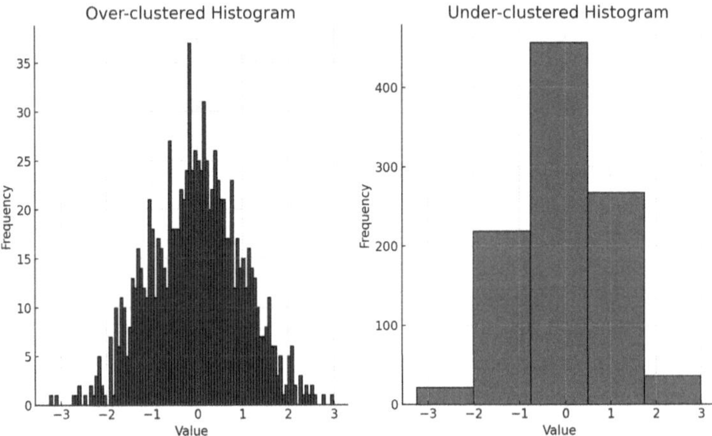

Fig. 5.9: Under-clustered Histogram

In this example, we demonstrate two common pitfalls. The first subplot illustrates over-clustering with 100 bins, which results in a highly granular and noisy histogram. The second subplot shows under-clustering with only five bins, which oversimplifies the data and obscures the underlying distribution. Finding the right balance between these two extremes is essential for accurate data representation.

5.5.2 Misleading Bin Sizes and Scaling

The choice of bin sizes and the scaling of the histogram can significantly impact the interpretation of the data. Inappropriate bin sizes can distort the appearance of the data, leading to misinterpretations. A bin size that is too large will aggregate too many data points into a single bin, concealing important patterns, while a bin size that is too small will break the data into overly specific segments, creating visual noise. Scaling issues arise when the x-axis or y-axis is manipulated in a way that changes the apparent density or distribution of the data. For example, unequal scaling on the axes can exaggerate or downplay variations in the data, leading to inaccurate conclusions.

It is important to choose bin sizes that accurately reflect the structure of the data, and to ensure that the scaling of the axes remains consistent and true to the data's distribution. In Python's Matplotlib, appropriate use of bin sizes and axis scaling can help avoid these pitfalls.

Code example in Python code:

```python
import matplotlib.pyplot as plt
import numpy as np

# Generate sample data
data = np.random.normal(0, 1, 1000)

# Misleading large bin size
plt.subplot(1, 2, 1)
plt.hist(data, bins=3, color='red', edgecolor='black')
plt.title('Misleading Large Bin Size')
plt.xlabel('Value')
plt.ylabel('Frequency')

# Proper bin size
plt.subplot(1, 2, 2)
plt.hist(data, bins=30, color='blue', edgecolor='black')
plt.title('Proper Bin Size')
plt.xlabel('Value')
plt.ylabel('Frequency')

# Adjust layout and show plot
plt.tight_layout()
plt.show()
```

The resulting histogram is shown in Figure 5.10.

In this example, we highlight how inappropriate bin sizes can mislead the audience. The first histogram uses an overly large bin size, which masks the finer details of the distribution, giving the false impression of uniformity. The second histogram employs a more appropriate bin size (30 bins), revealing the actual shape of the distribution.

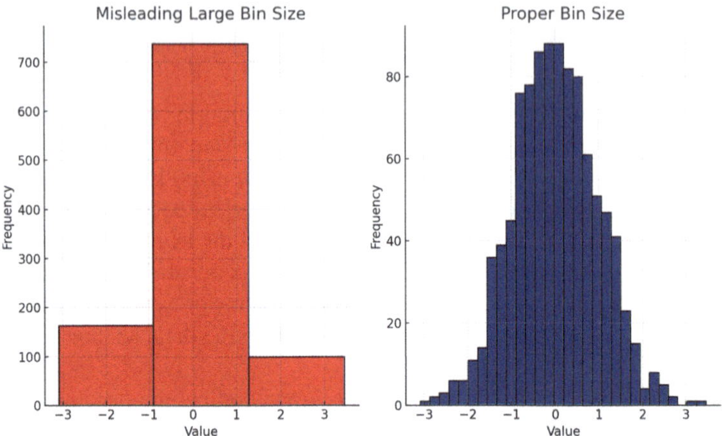

Fig. 5.10: Proper Bin Size in Histogram

Ensuring that bin sizes accurately represent the data without introducing distortion is crucial for effective data visualization.

Avoiding common pitfalls in histogram design is vital for ensuring that the data is represented clearly and truthfully. Over-clustering and under-clustering can lead to either excessive noise or oversimplification, while inappropriate bin sizes and scaling can mislead the audience. Through careful consideration of bin sizes, scaling, and balance, histograms can be designed to provide an accurate and meaningful representation of the underlying data.

5.6 Accessibility in Histogram Design

The concept of accessibility in data visualization is rooted in the principle of making visual information understandable and usable for as many people as possible, including those with disabilities. In histogram design, this involves considerations such as color accessibility, the use of patterns for visibility, and providing descriptive text. Ensuring that histograms are accessible enhances inclusivity and ensures that a wider audience can interpret the data effectively.

5.6.1 Color Accessibility

Color plays a significant role in data visualization, but reliance on color alone can be problematic, particularly for individuals with color vision deficiencies. Approximately 8% of men and 0.5% of women have some form of color blindness, the most common being red-green color blindness. To ensure that histograms are accessible, it is essential to use color schemes that are distinguishable by those with color vision deficiencies. The use of high-contrast colors and appropriate shading can further enhance visibility.

One approach to ensure color accessibility is to use colorblind-friendly palettes, such as 'Viridis', 'Plasma', or 'Cividis', which are available in the Matplotlib library in Python. These palettes are designed to be perceptually uniform, making them more accessible to individuals with different types of color blindness.

Code example in Python:

```python
import matplotlib.pyplot as plt
import numpy as np

# Generate sample data
data = np.random.normal(0, 1, 1000)

# Create histogram \index{Histogram} with colorblind-friendly colormap
plt.hist(data, bins=30, color='cividis', edgecolor='black', alpha=0.7)
plt.title('Color Accessible Histogram (Cividis)')
plt.xlabel('Value')
plt.ylabel('Frequency')

# Display the plot
plt.show()
```

The resulting histogram is shown in Figure 5.11.

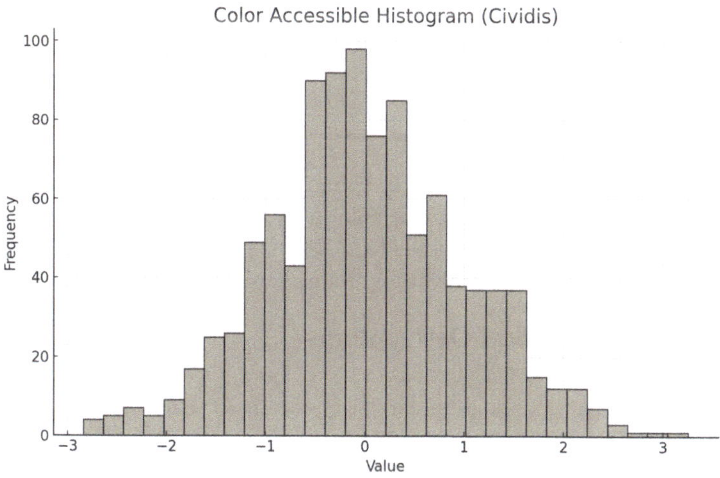

Fig. 5.11: Color Accessible Histogram (Cividis)

In this example, we use the 'Cividis' colormap, which is designed for accessibility, particularly for users with color vision deficiencies. The colorblind-friendly palette ensures that individuals can differentiate between different segments of the histogram.

5.6.2 Using Patterns for Visibility

In addition to color, patterns can be used to distinguish different sections of a histogram. Patterns are particularly useful when printing in grayscale or when color is not sufficient to differentiate between data categories. Applying textures such as stripes, dots, or hatching to different sections of the histogram enhances readability without relying solely on color.

Matplotlib allows for the use of hatching (pattern fills) in histograms, which adds visual texture to each bin.

Code example in Python:

```python
import matplotlib.pyplot as plt
import numpy as np

# Generate sample data
data = np.random.normal(0, 1, 1000)

# Create histogram \index{Histogram} with patterns for visibility
plt.hist(data, bins=30, color='lightblue', edgecolor='black', hatch='//',
    alpha=0.7)
plt.title('Histogram with Patterns (Hatching)')
plt.xlabel('Value')
plt.ylabel('Frequency')

# Display the plot
plt.show()
```

The resulting histogram is shown in Figure 5.12.

In this example, we use a light blue color with a hatching pattern ('//') to provide additional differentiation between the bars. This is particularly helpful when colors alone are insufficient or when the histogram is viewed in grayscale.

5.6.3 Providing Text Descriptions

Text descriptions serve as an essential supplement to visual elements, particularly for individuals who rely on screen readers or have visual impairments. Descriptive text, such as captions and annotations, can provide critical context and insights that are not immediately apparent from the visual itself.

In histograms, it is important to provide clear titles, axis labels, and, where necessary, annotations that describe key points in the data. Alternative text (alt-text) can be included for screen readers, summarizing the main findings and features of the histogram.

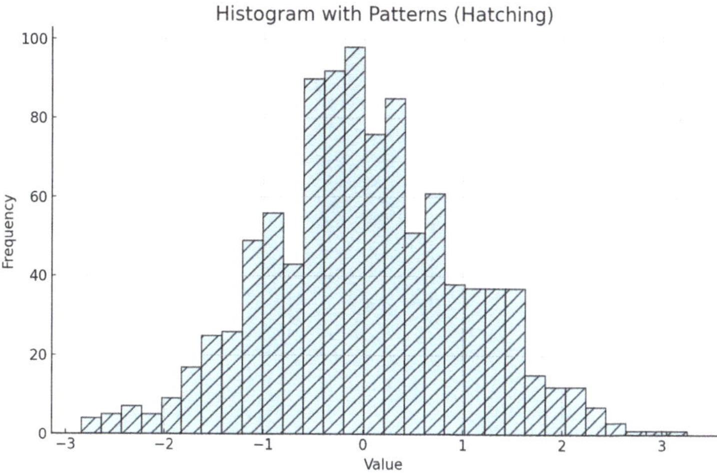

Fig. 5.12: Histogram with Patterns (Hatching)

Code example in Python:

```python
import matplotlib.pyplot as plt
import numpy as np

# Generate sample data
data = np.random.normal(0, 1, 1000)

# Create histogram
plt.hist(data, bins=30, color='skyblue', edgecolor='black', alpha=0.7)

# Add labels and annotations
plt.title('Annotated Histogram with Descriptions')
plt.xlabel('Value')
plt.ylabel('Frequency')

# Annotate key points
plt.text(-2, 50, 'Left Tail', fontsize=12, color='red')
plt.text(2, 50, 'Right Tail', fontsize=12, color='red')

# Display the plot
plt.show()
```

The resulting histogram is shown in Figure 5.13.

In this example, we add annotations to highlight specific areas of the histogram, such as the left and right tails. These annotations help explain key aspects of the

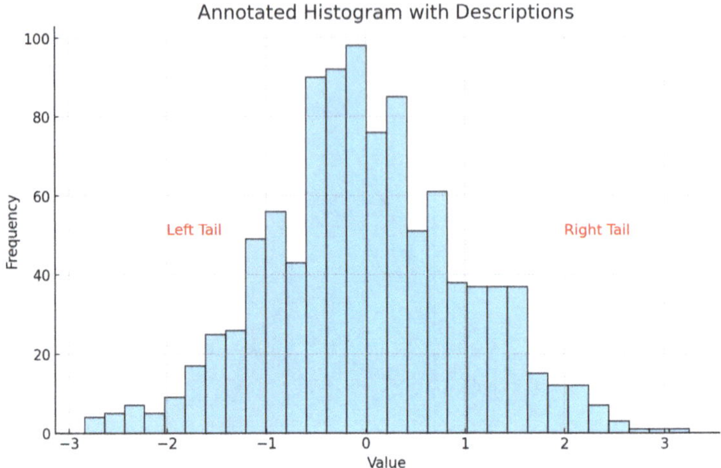

Fig. 5.13: Annotated Histogram with Descriptions

distribution and provide additional context. In practice, such descriptions would be supplemented with alt-text or captions when integrating the histogram into reports or websites.

In conclusion, designing accessible histograms involves thoughtful consideration of color accessibility, the use of patterns for enhanced visibility, and the inclusion of descriptive text. By incorporating these elements, data visualizations become more inclusive and interpretable, ensuring that individuals with visual impairments or color vision deficiencies can engage with the data effectively. The Python examples presented here provide practical methods for implementing accessibility features in histogram design, creating more equitable data visualizations.

5.7 Best UX Practices in Histogram Design

The User Experience (UX) of a histogram plays a critical role in how effectively users can interpret and interact with data. Well-designed histograms not only present data clearly but also consider accessibility, usability, and engagement. This section explores best UX practices for histogram design, focusing on consistency in axis scaling, the use of color and contrast, interactive elements, and more. Each subsection provides practical Python examples to demonstrate how these principles can be applied.

5.7.1 Ensuring Consistency in Axis Scaling and Labels

Consistent axis scaling is essential for ensuring that data is accurately represented and easily interpretable. Inconsistent scaling between histograms can lead to misleading comparisons, and poor labeling can confuse users. Axis labels must clearly describe what the data represents, and tick marks should be appropriately spaced to avoid overcrowding.

Code example in Python:

```python
import matplotlib.pyplot as plt
import numpy as np

# Generate two sets of data
data1 = np.random.normal(0, 1, 1000)
data2 = np.random.normal(2, 1, 1000)

# Create two subplots with consistent axis scaling
fig, (ax1, ax2) = plt.subplots(1, 2, figsize=(10, 4))

# First histogram
ax1.hist(data1, bins=30, color='skyblue', edgecolor='black')
ax1.set_title('Histogram 1')
ax1.set_xlabel('Value')
ax1.set_ylabel('Frequency')
ax1.set_xlim([-4, 6])
ax1.set_ylim([0, 100])

# Second histogram
ax2.hist(data2, bins=30, color='lightgreen', edgecolor='black')
ax2.set_title('Histogram 2')
ax2.set_xlabel('Value')
ax2.set_ylabel('Frequency')
ax2.set_xlim([-4, 6])
ax2.set_ylim([0, 100])

plt.tight_layout()
plt.show()
```

In this example, we ensure that both histograms have consistent x- and y-axis ranges, enabling direct comparison between the two datasets. This avoids the pitfall of misleading axis scaling, which can distort the interpretation of the data.

5.7.2 Using Color and Contrast Effectively

Color is a powerful tool in UX design, but it must be used carefully in histograms to enhance readability and accessibility. High-contrast color schemes should be used to differentiate between multiple datasets or categories. Additionally, the use of colorblind-friendly palettes ensures that the visualizations are accessible to a broader audience.

Code example in Python:

```python
import matplotlib.pyplot as plt
import numpy as np

# Generate sample data
data = np.random.normal(0, 1, 1000)

# Create histogram \index{Histogram} with high-contrast colors
plt.hist(data, bins=30, color='darkorange', edgecolor='black', alpha=0.8)
plt.title('High-Contrast Histogram')
plt.xlabel('Value')
plt.ylabel('Frequency')

# Display the plot
plt.show()
```

This example uses a high-contrast color (dark orange) for the bars, with black edges for clear delineation. The alpha transparency is set to 0.8 to ensure that the color is vibrant without being overwhelming.

5.7.3 Using Interactive Elements for Data Exploration

Interactive histograms allow users to explore data more deeply by offering dynamic interactions such as tooltips, zooming, and filtering. These elements help users engage with the data, making histograms more informative and flexible. While traditional histograms are static, libraries like Plotly in Python enable the creation of interactive visualizations.

Code example in Python:

```python
import plotly.express as px
import numpy as np

# Generate sample data
data = np.random.normal(0, 1, 1000)

# Create interactive histogram \index{Histogram} using Plotly
fig = px.histogram(data, nbins=30, title='Interactive Histogram')
fig.update_layout(xaxis_title='Value', yaxis_title='Frequency')

# Show the plot
fig.show()
```

This example uses Plotly to create an interactive histogram where users can hover over bars to view details, zoom in on specific ranges, and dynamically explore the data. This enhances user engagement and data comprehension.

5.7.4 Avoiding Overloading with Information

Overloading histograms with too much information–such as excessive data series, cluttered labels, or unnecessary embellishments–can confuse users and reduce the clarity of the visualization. To ensure simplicity, focus on the most relevant data and avoid adding elements that do not contribute to the understanding of the data.

Code example in Python:

```python
import matplotlib.pyplot as plt
import numpy as np

# Generate sample data
data = np.random.normal(0, 1, 1000)

# Create a simple histogram \index{Histogram} with minimal elements
plt.hist(data, bins=30, color='lightcoral', edgecolor='black', alpha=0.75)
plt.title('Simple Histogram')
plt.xlabel('Value')
plt.ylabel('Frequency')

# Show the plot
plt.show()
```

This histogram design focuses on simplicity by avoiding unnecessary decorations. The choice of a single dataset and clear labeling ensures that the visualization is easy to interpret.

5.7.5 Providing Clear Annotations and Callouts

Annotations and callouts can be used to highlight key points in the data, such as outliers , peaks, or specific ranges of interest. Effective use of annotations enhances the histogram's explanatory power by drawing attention to critical insights.

Code example in Python:

```python
import matplotlib.pyplot as plt
import numpy as np

# Generate sample data
data = np.random.normal(0, 1, 1000)

# Create histogram \index{Histogram} with annotations
plt.hist(data, bins=30, color='steelblue', edgecolor='black', alpha=0.7)
plt.title('Annotated Histogram')
plt.xlabel('Value')
```

```
11  plt.ylabel('Frequency')
12
13  # Annotate key points
14  plt.text(2, 60, 'Peak', fontsize=12, color='red')
15  plt.text(-2, 50, 'Left Tail', fontsize=12, color='green')
16
17  # Show the plot
18  plt.show()
```

In this example, annotations are added to highlight important features of the distribution, such as the peak and the left tail. These annotations help users quickly identify significant aspects of the data.

5.7.6 Responsive Design for Different Screen Sizes

Ensuring that histograms are responsive to different screen sizes and devices is essential for modern data visualization, especially in the context of web and mobile applications. Histograms should be designed to automatically adjust their size and layout based on the available screen real estate.

While Matplotlib does not support responsive design natively, tools like Plotly offer responsive features. For web-based histograms, using CSS and JavaScript frameworks can help make visualizations more flexible.

5.7.7 Balancing Aesthetic Appeal and Functionality

A well-designed histogram should strike a balance between aesthetic appeal and functionality. While the visual appearance of the histogram matters, it should never compromise the clarity and accuracy of the data. Clean designs, harmonious color schemes, and minimal distractions are key to achieving this balance.

Code example in Python:

```
1  import matplotlib.pyplot as plt
2  import numpy as np
3
4  # Generate sample data
5  data = np.random.normal(0, 1, 1000)
6
7  # Create a well-designed histogram \index{Histogram} with balanced
       aesthetics and functionality
8  plt.hist(data, bins=30, color='teal', edgecolor='white', alpha=0.85)
9  plt.title('Aesthetic and Functional Histogram')
10  plt.xlabel('Value')
11  plt.ylabel('Frequency')
12
13  # Show the plot
14  plt.show()
```

This example combines aesthetic appeal (with a clean color scheme and clear edges) with functionality. The histogram remains simple and readable while being visually appealing.

Table 5.2: Accessibility Checklist for Histogram Design

✓	Use colorblind-friendly palettes such as `Viridis`, `Cividis`, or `Plasma`.
✓	Ensure high contrast between bars, background, and labels for better visibility.
✓	Avoid using color as the only means of conveying information; supplement with patterns or textures (e.g., hatching).
✓	Provide clear, descriptive titles and axis labels for context and orientation.
✓	Include annotations or callouts to highlight important regions or trends.
✓	Use readable font sizes and sans-serif typefaces for screen-friendly displays.
✓	Include alt-text or caption summaries for screen readers and accessibility tools.
✓	Avoid overcrowding by minimizing unnecessary visual elements and decorations.

In conclusion, applying best UX practices in histogram design enhances both the usability and interpretability of data visualizations. Through consistent scaling, effective use of color, and minimal design elements, histograms can be made more accessible and engaging. Interactive elements and clear annotations further improve the user experience by encouraging deeper exploration of the data. These Python examples demonstrate practical ways to incorporate these best practices into histogram design.

5.8   Real-World Examples of Effective Histograms

Histograms are a powerful tool for visualizing the distribution of data points, revealing underlying patterns, clusters, and outliers . When combined with thoughtful design principles, histograms transform from raw data displays into engaging visual narratives. Below, we explore such applications through real-world examples.

5.8.1   Case Study 1: Scientific Data Visualization

In this case study, we explore the *Wine Dataset*, which contains information on various chemical properties of different wines, such as alcohol content, acidity, sugar levels, and more. The dataset can be used to understand the characteristics of wines and how they differ across wine types.

Concept: To capture the elegance and refinement that wine symbolizes, our first visualization is a sophisticated **histogram of alcohol content** across different wine types, presented in a way that resembles a vintage wine label. Using a minimalist color palette, elegant fonts, and a crisp layout, the visual serves both informative and aesthetic purposes.

Design Features:

➤ **Curved Axes:** The X-axis (alcohol percentage) and Y-axis (frequency of wines) have curved labels, mimicking the design of wine bottle labels, adding an artistic touch while maintaining clarity.

➤ **Warm Tones:** The bars representing alcohol distribution use deep wine-red and sunset-gold hues to reflect the color tones of red and white wines.

➤ **Overlayed Information:** Each bar displays the average quality rating of wines in that alcohol range, subtly annotated above each bar in a small, serif font.

Python Code for Implementation:

```python
import pandas as pd
import matplotlib.pyplot as plt

# Load wine dataset
wine_data = pd.read_csv('wine.csv')

# Create histogram \index{Histogram} for alcohol content
plt.figure(figsize=(10, 6))
plt.hist(wine_data['alcohol'], bins=15, color='#D2691E', edgecolor='white',
    alpha=0.75)

# Aesthetic design
plt.title('Alcohol Content Distribution in Wines', fontsize=20,
    fontweight='bold', color='#8B0000')
plt.xlabel('Alcohol Percentage (%)', fontsize=14, fontstyle='italic')
plt.ylabel('Number of Wines', fontsize=14, fontstyle='italic')
plt.grid(False)  # Remove gridlines \index{Grid Lines} for a cleaner look

# Add vintage label-inspired touches
plt.gca().set_facecolor('#FAF0E6')  # Soft beige background to resemble
    label paper
plt.tight_layout()
plt.show()
```

The result of this code is shown in Figure 5.14.

Concept: Wine acidity is a key factor influencing its taste profile. Our second visualization is an innovative **pH level distribution heatmap** that allows users to quickly spot the most common acidity ranges among wine samples. The gradient design, inspired by watercolor, emulates the complexity and subtlety of wine tasting.

Design Features:

➤ **Watercolor Gradient:** A color scheme shifting from light pastel hues for lower pH wines to deep, rich tones for higher pH wines, reflecting changes in wine acidity.

➤ **Interactive Tooltips:** Hovering over sections of the heatmap provides exact numbers of wines in each pH range, enhancing engagement.

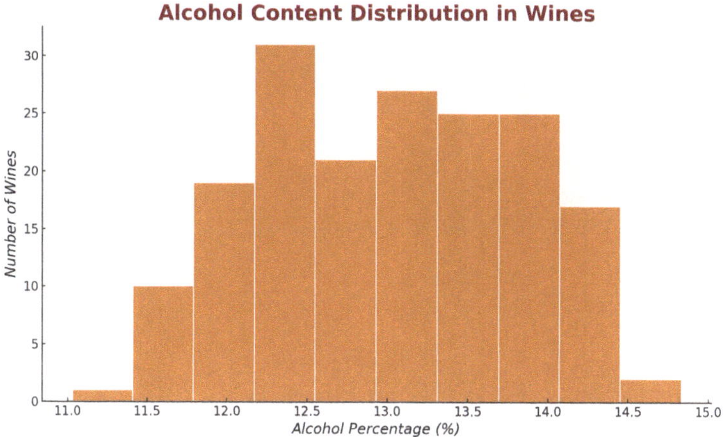

Fig. 5.14: Alcohol Content Distribution in Wines

➢ **Sleek Typography:** Clean serif fonts ensure readability and a magazine-like aesthetic.

Python Code for Implementation:

```python
import seaborn as sns

# Create a polished heatmap \index{Heatmap} for pH distribution
plt.figure(figsize=(10, 8))
pH_data = wine_data[['pH', 'quality']]

# Use seaborn to create a heatmap \index{Heatmap} with custom color
    palette
sns.histplot(pH_data['pH'], bins=20, kde=False, color='#8B0000',
    alpha=0.6)

# Design tweaks for elegance
plt.title('Distribution of pH Levels in Wine Samples', fontsize=22,
    fontweight='bold', color='#4B0082')
plt.xlabel('pH Level (Acidity)', fontsize=16, fontstyle='italic')
plt.ylabel('Count of Wines', fontsize=16, fontstyle='italic')
plt.gca().set_facecolor('#FFF5EE')  # Soft background resembling parchment
    paper
plt.grid(False)

plt.show()
```

The result of this code is shown in Figure 5.15.

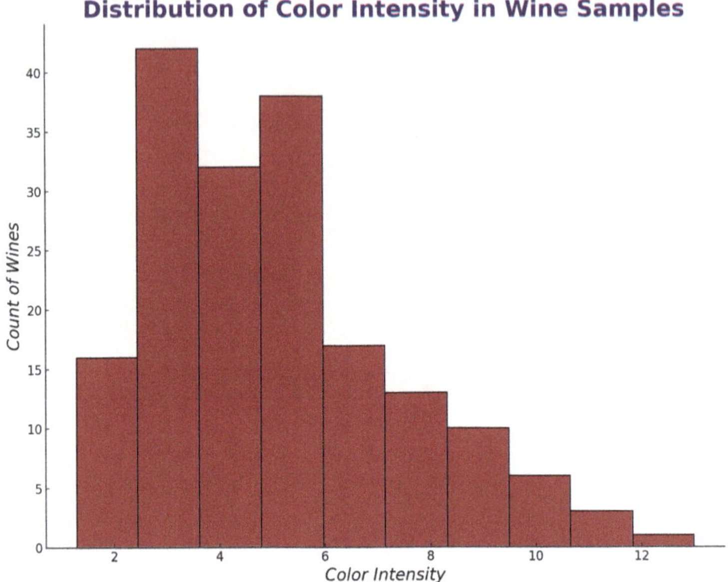

Fig. 5.15: Distribution of pH Levels in Wine Samples

5.9 Tools for Creating Visually Appealing Histograms

Histograms are an essential part of data visualization, offering insights into the distribution and shape of data. In modern data analysis, various libraries provide powerful tools for creating visually appealing histograms. This section focuses on four popular Python libraries: Matplotlib, Seaborn, Bokeh, and Plotly. Each library offers distinct advantages in terms of aesthetics, interactivity, and ease of use. We provide practical examples to demonstrate how to create effective histograms using these tools.

5.9.1 Creating Histograms in Python

Python offers a wide range of libraries that allow for the creation of highly customizable histograms. Whether you are aiming for static, publication-quality charts or interactive web-based visualizations, the right tool can significantly enhance the interpretability of your data.

Matplotlib is the foundational library for data visualization in Python. It offers granular control over every aspect of histogram design, making it the go-to choice for creating static, high-quality plots suitable for reports, papers, and presentations.

Although Matplotlib requires more coding to achieve complex designs compared to other libraries, its flexibility makes it indispensable for detailed customization.

Example of Python code:

```python
import matplotlib.pyplot as plt
import numpy as np

# Generate sample data
data = np.random.normal(0, 1, 1000)

# Create a simple histogram \index{Histogram} using Matplotlib
plt.hist(data, bins=30, color='skyblue', edgecolor='black', alpha=0.7)
plt.title('Histogram in Matplotlib')
plt.xlabel('Value')
plt.ylabel('Frequency')

# Show the plot
plt.show()
```

The result of this code is shown in Figure 5.16.

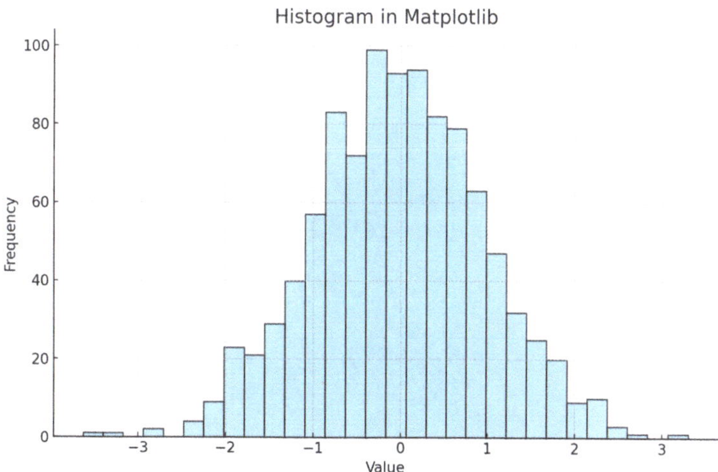

Fig. 5.16: Histogram in Matplotlib

In this example, Matplotlib is used to create a basic histogram. Its strength lies in its ability to customize every element, from colors and labels to bin size and axis limits. This control makes it ideal for professional presentations and reports.

Seaborn builds on Matplotlib but focuses on making statistical visualizations simpler to create and aesthetically appealing by default. It automatically handles many aspects of plot design, such as color schemes and axis labeling, while also providing advanced features like kernel density estimation (KDE). This makes Seaborn a great choice for users who want beautiful, easy-to-create plots with minimal effort.

Code example in Python:

```python
import seaborn as sns
import matplotlib.pyplot as plt
import numpy as np

# Generate sample data
data = np.random.normal(0, 1, 1000)

# Create a histogram \index{Histogram} with Seaborn
sns.histplot(data, bins=30, kde=True, color='teal', edgecolor='black')
plt.title('Histogram with KDE \index{Kernel Density Estimation} in Seaborn')
plt.xlabel('Value')
plt.ylabel('Frequency')

# Show the plot
plt.show()
```

The result of this code is shown in Figure 5.17.

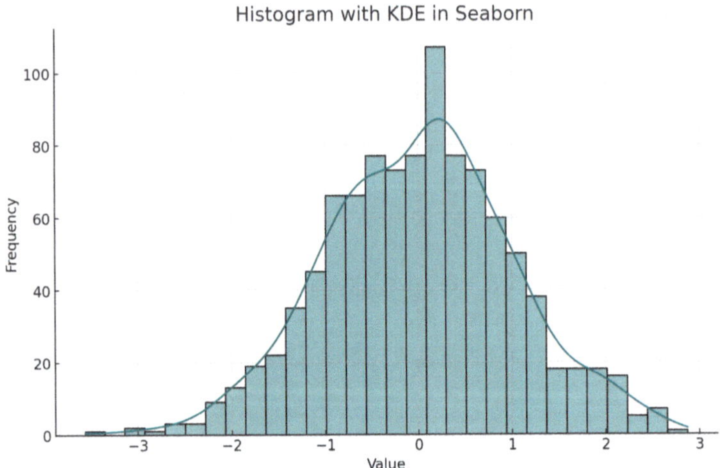

Fig. 5.17: Histogram with KDE in Seaborn

This example demonstrates a histogram with kernel density estimation (KDE) overlaid using Seaborn. Seaborn simplifies the process of adding statistical insights, making it ideal for exploratory data analysis and quick visualizations.

Bokeh is a powerful library for creating interactive, web-based visualizations. It enables users to create dynamic histograms with features like zooming, panning , and tooltips. This is particularly useful for web dashboards and applications where users need to interact with data in real-time.

Code example in Python:

```python
from bokeh.plotting import figure, show
from bokeh.io import output_notebook
import numpy as np

# Generate sample data
data = np.random.normal(0, 1, 1000)

# Create a histogram \index{Histogram} using Bokeh
hist, edges = np.histogram(data, bins=30)

# Create Bokeh plot
output_notebook()
p = figure(title="Interactive Histogram in Bokeh",
    background_fill_color="#fafafa")
p.quad(top=hist, bottom=0, left=edges[:-1], right=edges[1:],
    fill_color="navy", line_color="black", alpha=0.7)

# Show the interactive plot
show(p)
```

This Bokeh example illustrates how to create an interactive histogram with zoom and pan functionality. Bokeh is an excellent choice for users who need to embed visualizations into web-based applications or dashboards.

Plotly offers high-quality, interactive visualizations that are easy to embed in web applications and Jupyter notebooks. It is well-suited for users who want the flexibility of interactive features without sacrificing aesthetic quality. Plotly's histograms come with built-in interactivity, such as tooltips and dynamic binning.

```python
import plotly.express as px
import numpy as np

# Generate sample data
data = np.random.normal(0, 1, 1000)

# Create an interactive histogram \index{Histogram} using Plotly
fig = px.histogram(data, nbins=30, title='Interactive Histogram in Plotly',
    color_discrete_sequence=['darkorange'])
fig.update_layout(xaxis_title='Value', yaxis_title='Frequency')

# Show the plot
fig.show()
```

This Plotly example showcases how to create an interactive histogram with dynamic tooltips. Plotly's ability to create polished, interactive visualizations with minimal code makes it a popular choice for web developers and data scientists.

Python provides a diverse set of tools for creating histograms, ranging from Matplotlib's highly customizable plots to Plotly's interactive visualizations. Choosing the right tool depends on your specific needs: Matplotlib excels in static, publication-quality graphics, Seaborn offers simplicity and beauty for statistical plots, while Bokeh and Plotly are ideal for building interactive, web-based histograms. Each library brings unique strengths to the table, allowing data analysts and UX professionals to create histograms that not only represent data accurately but also engage the user effectively.

5.9.2 Creating Histograms in R

R is a powerful statistical programming language that offers robust tools for data visualization, including histograms. Two popular libraries for creating visually appealing histograms in R are `Plotly` for interactive charts and `ggplot2` for static, highly customizable visualizations. Both libraries are widely used for data analysis, and each offers unique strengths depending on the user's needs.

Similar to its Python counterpart, `Plotly` in R provides a framework for creating interactive, web-based histograms. Plotly histograms can be easily embedded in R Markdown reports, dashboards, and web applications, offering users the ability to interact with the data through tooltips, zooming, and filtering.

Code example in R:

```r
# Install the necessary library if not already installed
# install.packages("plotly")

# Load the plotly library
library(plotly)

# Generate sample data
data <- rnorm(1000)

# Create an interactive histogram
fig <- plot_ly(x = ~data, type = "histogram", nbinsx = 30)
fig <- fig %>% layout(title = "Interactive Histogram in Plotly \index{Plotly} (R)",
                      xaxis = list(title = "Value"),
                      yaxis = list(title = "Frequency"))
fig
```

In this example, `Plotly` is used to create an interactive histogram in R. The plot allows users to zoom in, pan, and interact with the chart, making it an excellent tool for presentations and web-based dashboards.

`ggplot2` is one of the most widely used visualization packages in R, known for its aesthetic quality and flexibility. With `ggplot2`, users can create publication-quality static histograms that are highly customizable in terms of colors, themes, and labeling.

Code example in R:

```r
# Install the necessary library if not already installed
# install.packages("ggplot2")

# Load the ggplot2 library
library(ggplot2)

# Generate sample data
data <- rnorm(1000)

# Create a static histogram \index{Histogram} with ggplot2
ggplot(data.frame(data), aes(x = data)) +
  geom_histogram(bins = 30, fill = "steelblue", color = "black") +
  labs(title = "Histogram in ggplot2", x = "Value", y = "Frequency") +
  theme_minimal()
```

This example demonstrates how to create a static histogram in R using `ggplot2`. The `theme_minimal()` function is applied to give the histogram a clean and modern look, suitable for academic reports or publications.

5.9.3   Creating Histograms in MATLAB

MATLAB is another powerful tool for scientific computing and data visualization. Its robust graphics engine allows users to create highly customizable histograms suitable for engineering, scientific analysis, and academic purposes. MATLAB's `histogram` function provides numerous options for adjusting bin size, color, and display properties.

Code example in MATLAB:

```matlab
% Generate sample data
data = randn(1000,1);

% Create a basic histogram \index{Histogram} in MATLAB\index{MATLAB}
histogram(data, 'NumBins', 30, 'FaceColor', 'c', 'EdgeColor', 'k');
title('Histogram in MATLAB\index{MATLAB}');
xlabel('Value');
ylabel('Frequency');
grid on;
```

In this example, MATLAB's `histogram` function is used to create a basic histogram. Users can easily modify bin sizes, color schemes, and add grid lines for better readability. MATLAB's simplicity and speed make it an excellent choice for quick prototyping and scientific analysis.

In conclusion, both R and MATLAB offer powerful tools for creating histograms, each suited for different use cases. `Plotly` and `ggplot2` in R provide robust options for interactive and static visualizations, respectively, while MATLAB excels in high-performance visualizations for scientific and engineering applications. Choosing the appropriate tool depends on the specific needs of the analysis, whether it's for web-based interactivity, static publication-quality charts, or scientific data exploration.

5.10   Histogram Visualization Tips and Tricks

Designing histograms is not only about plotting data; it's about crafting an experience that ensures clarity, accessibility, and insight for the audience. As a professional data analyst and UX expert, understanding and applying the following tips and tricks will elevate your histogram visualizations, transforming raw data into actionable and interpretable insights.

5.10.1   Select the Right Number of Bins

Choosing the correct number of bins is critical for avoiding both over-clustering and under-clustering. A good rule of thumb is to use automatic binning techniques, such as Freedman-Diaconis or Sturges' formula, to ensure that your data is neither too granular nor too oversimplified. In Python, using libraries such as `numpy.histogram` provides easy control over bin calculation.

> **♀ Design Tip**
>
> For exploratory analysis, use different bin sizes to spot patterns, but ensure final visualizations are concise and clear.

5.10.2   Consistency is Key

Consistency in axis scaling and bin size is essential when comparing multiple histograms. When visualizing multiple datasets, ensure that all histograms use the same x-axis and y-axis ranges to avoid misleading interpretations. Inconsistent scaling can make data seem more varied or uniform than it truly is, leading to biased conclusions.

> **♀ Design Tip**
>
> Always synchronize the axes when comparing distributions side-by-side to avoid misinterpretation of relative distributions.

5.10.3 Optimize Color Usage for Accessibility

Colors should serve a purpose beyond aesthetics in histograms. Consider your audience and use colorblind-friendly palettes to ensure accessibility for individuals with visual impairments. Avoid using too many colors, as this can overwhelm the viewer. Instead, use high-contrast color schemes, and apply softer hues for less important elements to maintain visual hierarchy.

> **♀ Design Tip**
>
> Use palettes like `viridis`, `plasma`, or `cividis` to make your visualizations accessible to colorblind users.

5.10.4 Use Labels and Annotations Effectively

Clear labeling of axes and values is essential for guiding users through the data. Use concise, descriptive labels for the x- and y-axes, and consider adding annotations to highlight key insights, such as outliers , peaks, or key thresholds in the data. Annotations provide context and focus the viewer's attention on the most relevant aspects of the histogram.

> **♀ Design Tip**
>
> Avoid overly complex labels–stick to straightforward titles and callouts that enhance understanding.

5.10.5 Consider Interactive Elements for Enhanced Engagement

In digital presentations, consider using interactive histogram visualizations that allow users to explore the data more deeply. Interactive features, such as tooltips that display data values on hover, or filters that allow users to adjust bin sizes dynamically, can significantly enhance user engagement.

> **♀ Design Tip**
>
> Use libraries like Plotly or Altair for building interactive histograms that invite deeper exploration of the data.

5.10.6 Avoid Visual Clutter

Overloading a histogram with too much information, such as excessive gridlines, data series, or overly intricate designs, can confuse your audience. Simplicity is crucial for clarity. Focus on the most critical aspects of the data, and avoid including unnecessary details or elements that distract from the key insights.

> **♀ Design Tip**
>
> When in doubt, simplify. Less is often more in data visualization, particularly when aiming for clarity.

5.10.7 Provide Context with Titles and Captions

Titles and captions are vital for providing the context of your histogram. A well-chosen title can summarize the main insight, while a caption can offer background information or clarify what the data represents. This is especially useful when presenting to stakeholders who may not have deep familiarity with the dataset.

> **♀ Design Tip**
>
> Write concise, action-oriented titles that guide users directly to the insight the histogram reveals.

5.10.8 Prioritize Mobile-Responsive Design

If your histogram is part of a web-based dashboard or online report, it's crucial to ensure that it is responsive across various devices. Histograms should adjust their size and layout to fit smaller screens without losing clarity. Tools such as Plotly or D3.js provide built-in features for responsive design.

> **♀ Design Tip**
>
> Always test your visualizations on mobile devices to ensure that they maintain clarity and usability in smaller formats.

5.10.9 Balance Aesthetics with Functionality

While visually appealing histograms can capture attention, aesthetics should never come at the expense of data accuracy or readability. Striking a balance between design and function is key. Use clean designs with well-chosen colors and avoid unnecessary embellishments.

Use simple, elegant designs that prioritize data integrity and user comprehension over decorative elements.

Histograms are foundational tools for data analysis, but their effectiveness is heavily influenced by design choices. By following these tips and tricks, you ensure that your histograms are not only visually appealing but also functional, accessible, and capable of conveying

Suggested Readings

1. Lindholm, S., Falk, M., Sundén, E., Bock, A., Ynnerman, A., Ropinski, T.: Hybrid data visualization based on depth complexity histogram analysis. Computer Graphics Forum **34**(1), 74–85 (2015)
2. Neuroth, T., Sauer, F., Wang, W., Ethier, S., Chang, C.S., Ma, K.L.: Scalable visualization of time-varying multi-parameter distributions using spatially organized histograms. IEEE Transactions on Visualization and Computer Graphics **23**(12), 2599–2612 (2017)

6 Designing Astonishing Pie Charts

Abstract

Pie and donut charts are among the most widely used visualization tools for representing proportional data, offering a clear and intuitive way to convey part-to-whole relationships. This chapter provides a comprehensive guide to designing astonishing pie and donut charts that are not only visually compelling but also effective in communicating data insights. We begin by exploring the historical evolution of pie charts and the rise of donut charts as their modern counterpart, highlighting their unique advantages and use cases in data visualization. The chapter focuses on essential design principles, emphasizing the importance of simplicity, clarity, and appropriate data selection. Strategies for enhancing user engagement through color, contrast, labeling, and interactive features are discussed in detail, with a focus on accessibility and inclusivity. Readers are guided on how to creatively use the donut chart's central space and incorporate animations and interactivity to transform static charts into dynamic storytelling tools.

Aims

After reading this chapter, you should be able to:

➤ Apply design principles for creating clear, effective, and aesthetically pleasing pie and donut charts.

➤ Choose suitable color schemes and contrasts to enhance user engagement and ensure readability.

➤ Properly label and annotate pie and donut charts to maximize clarity and minimize misinterpretation.

➤ Use the advantages of donut charts for storytelling and combining multiple datasets.

➤ Incorporate animation and interactivity into pie and donut charts to create dynamic and engaging visualizations.

© The Author(s), under exclusive license to Springer Nature Switzerland AG 2026
R. Damaševičius, *Human-Centred Scientific Data Visualisation*,
Undergraduate Topics in Computer Science,
https://doi.org/10.1007/978-3-032-01606-5_6

6.1   Introduction to Pie and Donut Charts

Pie and donut charts are among the most recognized and widely used types of charts in data visualization. Their primary function is to display data in a manner that communicates the proportion of parts relative to a whole, making them ideal for datasets that represent percentages or ratios. The example of pie chart can be seen in Figure 6.1.

Despite their popularity, the design and application of these charts require careful consideration to avoid misrepresenting data or overwhelming the viewer. In this chapter, we will explore the evolution of pie charts, the nuances of donut charts, and the situations where these visualizations are most effective.

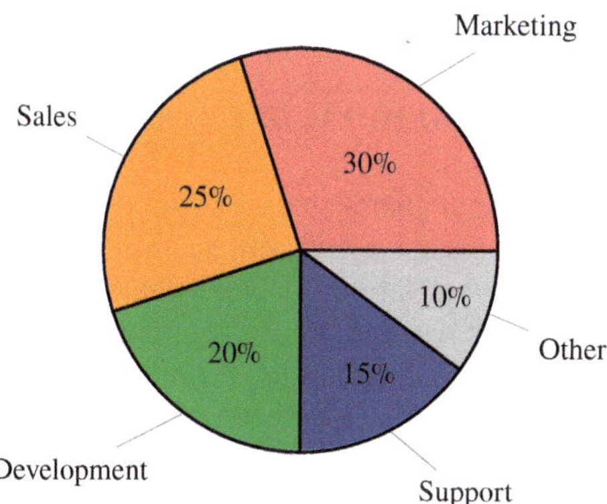

Fig. 6.1: A basic pie chart showing department budget allocation.

6.1.1   History and Evolution of Pie Charts

The pie chart is often credited to the Scottish engineer William Playfair, who is widely regarded as one of the fathers of graphical methods in statistics. In 1801, Playfair introduced the pie chart in his work *Statistical Breviary* as a tool for illustrating the proportions of the Turkish Empire's land area in Europe, Asia, and Africa. The simplicity and intuitive appeal of pie charts allowed them to gain traction in the field of statistical representation.

Through the 19th and 20th centuries, pie charts became synonymous with representing proportional data in a visually compelling manner. Their circular shape and division into slices made them a natural metaphor for understanding fractions of a whole, and they were quickly adopted in business, economics, and government statistics. However, despite their widespread use, pie charts have also faced criticism, particularly in regard to their efficacy in accurately representing complex data.

Cognitive research suggests that humans struggle with comparing angular areas as effectively as they do linear distances, making pie charts less effective when used with too many segments or for fine comparisons.

> **♀ Design Tip**
>
> Avoid using more than 5–7 slices in a pie or donut chart. Too many segments reduce clarity and make comparison difficult. Group smaller categories into an "Other" segment when necessary.

Nevertheless, pie charts have evolved. With the advent of modern data visualization tools and software, pie charts have been augmented with interactive features, animation, and additional layers of information. These enhancements have allowed pie charts to maintain their place in the data visualization toolbox, particularly for non-expert audiences who value simplicity and intuitive representations.

6.1.2 Differences Between Pie and Donut Charts

While pie charts and donut charts share a common goal–displaying parts of a whole–their visual differences are significant, and these differences can affect how data is perceived by the viewer. A traditional pie chart is a complete circle divided into slices, with each slice representing a portion of the data. The area of each slice is proportional to the corresponding data value. In contrast, a donut chart is essentially a pie chart with a central circle removed, creating a ring-like appearance. The hollow center of a donut chart opens up additional design possibilities without altering the underlying data representation.

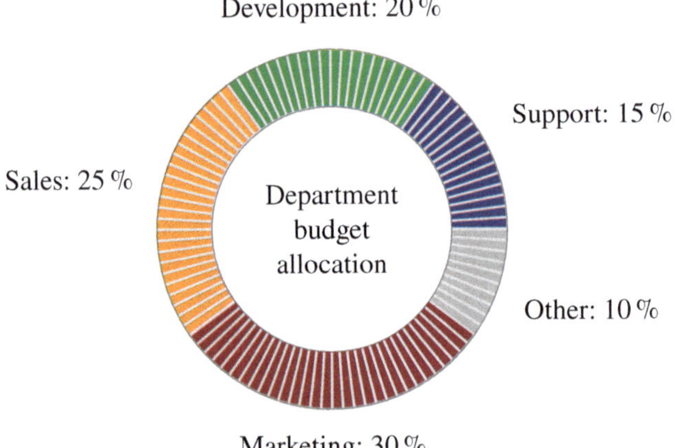

Fig. 6.2: Donut chart showing department budget allocation.

One of the most obvious differences is aesthetic (Table 6.1). The empty space in the center of a donut chart can be used to display additional information, such as total values or category labels, which is a useful feature when summarizing key takeaways.

The donut chart's design also allows for more complex visualizations where multiple rings can represent layered datasets, a feature often utilized in multi-dimensional data representations. From a perceptual standpoint, donut charts may be easier for viewers to interpret in some scenarios, particularly when fewer segments are involved. The reduced visual clutter provided by the central hole can make it easier to focus on individual segments, especially when used with well-chosen colors and spacing. However, both pie and donut charts share limitations in terms of accurately representing small differences in data values, particularly when the chart contains numerous small slices. In such cases, alternative visualizations like bar charts or stacked bar charts may be more appropriate.

Table 6.1: Comparison of Pie and Donut Charts

Feature	Pie Chart	Donut Chart
Visual Style	Solid circle divided into slices	Ring with a central hole
Space Usage	Entire area used for data	Center used for additional info
Perception Clarity	Effective with few segments	Better with fewer segments and less clutter
Use Case	Simple part-to-whole relationships	Dashboards, multiple data layers, central metrics
Customization	Limited to outer area	Supports inner annotations, logos, or metrics
Aesthetic Flexibility	Traditional and classic look	Sleek and modern design

6.1.3 When to Use Pie and Donut Charts in Data Visualization

Choosing the appropriate chart type for a dataset is a fundamental aspect of effective visualization design, and pie and donut charts are no exception. These charts are most effective when used to display data with a limited number of categories that make up a whole. They are particularly well-suited to illustrating proportional relationships, such as market shares, demographic distributions, or survey results, where the viewer needs to understand not only the individual category sizes but also how they relate to the entire dataset.

Pie charts excel when you want to emphasize a small number of key categories or when the relative sizes of these categories are easily distinguishable by the human eye. For example, when showing how a budget is divided into major categories, a pie chart can quickly convey the proportions allocated to each category in a visually intuitive manner. However, the limitations of pie charts become apparent when dealing with data that includes numerous small slices. In such cases, the viewer may struggle to accurately compare the proportions, leading to potential misinterpretations.

Donut charts, on the other hand, are often preferred in modern visual design for their aesthetic flexibility and the ability to incorporate additional information in the center space. They are particularly useful when the visualization needs to include a total value or when multiple datasets need to be visualized simultaneously in concentric rings. Donut charts are frequently used in dashboards and reports, where their sleek design and flexibility help present data in a concise yet informative way.

Pie and donut charts should be used sparingly and purposefully. They are best suited for datasets with clear proportional relationships, particularly when the number of categories is small and easily distinguishable. In cases where finer comparison is required, or when dealing with larger datasets, alternative visualizations such as bar charts, histograms, or treemaps may provide a more accurate and informative representation.

6.2 Principles of Effective Pie and Donut Design

The key to creating visually compelling and functionally effective pie and donut charts lies in adhering to fundamental design principles that prioritize simplicity and clarity. While these charts are popular for their intuitive appeal, without careful design, they can easily become overwhelming or misleading. In this section, we will explore how simplicity, appropriate data selection, and the avoidance of common design pitfalls can ensure that pie and donut charts are both aesthetically pleasing and informative.

6.2.1 Importance of Simplicity and Clarity

Simplicity and clarity are the cornerstones of good chart design, especially when dealing with pie and donut charts. These charts should be used to present data in a straightforward and easily digestible format. Overcomplicating the design with excessive elements, colors, or labels can dilute the message and confuse the audience.

A well-designed pie or donut chart should allow the viewer to quickly grasp the proportional relationships between categories. This is achieved by limiting the number of slices and ensuring the most important categories stand out. Ideally, a pie chart should contain no more than five to seven slices. Any more than this and the viewer may struggle to distinguish between segments, particularly if the proportions are close in size. When there are too many categories to display clearly, it may be better to group smaller categories into an "Other" segment or opt for a different visualization such as a bar chart.

> **⚠ Warning**
>
> Do not use pie or donut charts when you have too many categories or segments of nearly equal size. This makes visual comparison difficult and leads to misinterpretation.

Clarity is enhanced by maintaining sufficient contrast between slices. Selecting appropriate color schemes that make each slice distinct helps to ensure that the viewer can differentiate between them at a glance. Avoid overly intricate patterns or heavy gradients, which can distract from the data and make the chart visually confusing. The use of clean lines, consistent segment sizes, and careful spacing between slices will contribute to the overall readability of the chart.

Labels should be simple and informative. Rather than overcrowding the chart with lengthy text or numbers, use concise labels and provide detailed explanations in adjacent legends or tooltips. This allows the chart to remain uncluttered while still conveying essential information. The goal is to guide the viewer's eye to the most important insights without overwhelming them with excessive detail.

6.2.2 Choosing Appropriate Data for Pie and Donut Charts

Pie and donut charts are highly effective when displaying data that represents parts of a whole. However, choosing the right kind of data to visualize using these charts is critical for their success. The most suitable datasets for pie and donut charts are those that involve categorical data with clear proportional relationships. For instance, market shares, demographic distributions, budget allocations, and survey results are all examples of data that lend themselves well to this type of chart.

It is essential that the data be mutually exclusive and collectively exhaustive, meaning the categories should not overlap, and they should sum to a meaningful whole, typically 100%. Pie and donut charts are not suitable for datasets where the relationships between categories are ambiguous or where the total of the categories does not add up to a significant figure. For example, using a pie chart to display raw numbers without any reference to the whole can mislead the audience, as the proportional nature of the chart suggests a totality that may not exist.

> **⚠ Warning**
>
> Never use pie or donut charts for data that does not represent parts of a meaningful whole. These charts are inappropriate for unrelated or non-additive categories.

Another important consideration is the size of the dataset. As mentioned earlier, pie and donut charts are best used for visualizing a small number of categories. When dealing with a large dataset or when fine differences between categories are important, other visualization techniques such as bar charts or histograms may be more appropriate. Pie and donut charts excel at showing broad comparisons rather than detailed analysis, so it's important to choose data that aligns with this purpose.

Finally, ensure that the data is meaningful in a proportional context. Pie and donut charts are designed to show relative sizes of categories within a whole. If the data does not reflect these relationships–such as in the case of nominal data without an inherent ordering–then a different chart type would likely be a better choice. For example, time series data is better represented by line charts, while comparisons between nominal categories might be better suited to bar or column charts.

6.2.3 Avoiding Common Pitfalls in Design

Even with simplicity and appropriate data selection in mind, pie and donut charts can still fall victim to a number of common design mistakes (Table 6.2). Avoiding these pitfalls is essential for ensuring that the visualization remains clear, informative, and visually appealing.

Table 6.2: Common Pitfalls in Pie and Donut Charts and Recommended Solutions

Common Pitfall	Recommended Solution
Too many segments	Limit to 5–7 segments; group small ones into "Other"
Overuse of similar colors	Use high-contrast or colorblind-safe palettes
Use of 3D effects	Avoid 3D; use flat design for accurate perception
Crowded labels inside chart	Use external labels with leader lines or tooltips
No clear total or summary	Use donut center to display total or key value

One of the most frequent mistakes in pie chart design is the inclusion of too many slices. As previously mentioned, more than five to seven slices can make the chart difficult to interpret. This issue is compounded when the segments are very small or similar in size, making it hard for viewers to distinguish between them. If it is necessary to display a large number of categories, consider whether they can be consolidated into broader groups or presented in a different type of chart that better handles fine comparisons.

Another common pitfall is the use of 3D pie charts. While these may seem visually impressive, they often distort the data and make it harder for viewers to accurately judge the proportions. The perspective of 3D pie charts can exaggerate or diminish the apparent size of segments, leading to misinterpretations. Flat, two-dimensional charts are almost always preferable because they present the data in a straightforward and undistorted manner.

> **♀ Design Tip**
>
> Avoid using 3D effects in pie and donut charts. While they may seem visually appealing, they distort perception and make it harder for viewers to interpret proportions accurately.

Inconsistent or poor use of color is another frequent problem. A well-designed pie or donut chart uses color to highlight differences between segments without overwhelming the viewer. Overly bright or clashing colors can be distracting, while colors that are too similar make it difficult to distinguish between segments. It's important to select a color palette that is both aesthetically pleasing and functional, ensuring sufficient contrast between slices while maintaining harmony within the overall design.

Lastly, improper labeling can undermine the clarity of the chart. Labels that are too small, too long, or placed in awkward positions can detract from the viewer's ability to quickly interpret the data. Ensure that labels are clear, concise, and positioned in a way that makes them easy to read without obstructing the visual elements of the chart. When necessary, use legends or interactive elements like tooltips to convey additional information without cluttering the chart.

> **💡 Design Tip**
>
> When using labels inside chart segments, make sure they are large enough to be legible. For smaller slices, use leader lines and place the labels outside the chart.

By avoiding these common mistakes, designers can create pie and donut charts that effectively communicate data insights in a visually compelling way. The principles of simplicity, appropriate data selection, and thoughtful design choices ensure that pie and donut charts remain a powerful tool in the visualization toolkit.

6.3   Enhance User Engagement with Color and Contrast

Color and contrast play pivotal roles in creating pie and donut charts that not only communicate information effectively but also engage users visually. The use of color in data visualization serves two primary purposes: to differentiate between categories and to direct attention toward key insights. When applied correctly, color can make a chart easier to read and more aesthetically appealing, enhancing the overall user experience. In this section, we will explore how to select color schemes that promote readability, use contrast to distinguish between segments, and use gradients and color harmony for visual appeal.

6.3.1   Selecting Color Schemes that Promote Readability

One of the most important considerations in designing pie and donut charts is the selection of a color scheme that enhances readability. The human brain processes color quickly, making it an ideal tool for distinguishing between different data categories in a visualization. However, if colors are chosen poorly, they can confuse rather than clarify.

A well-designed color scheme should provide clear visual distinctions between segments, especially in pie and donut charts where segment comparison is essential. Start by selecting colors that are distinct and easily recognizable. Avoid using too many similar shades, as this can make it difficult for viewers to differentiate between categories. It is often advisable to use a color palette that includes both warm and cool tones, which naturally provides contrast.

> **⚠ Warning**
>
> Be cautious when using similar shades of the same color. Without enough contrast, slices may blend together, especially for users with color vision deficiencies.

Consider the overall color accessibility of the chart. Approximately 8% of men and 0.5% of women are affected by some form of color vision deficiency, meaning that charts relying too heavily on color distinctions may be unreadable to a portion of the audience. To mitigate this, designers should avoid relying solely on color to convey information. Combining color with other visual elements, such as labels, patterns, or annotations, can improve accessibility. Tools like ColorBrewer and Adobe Color can help designers create color palettes that are both visually appealing and accessible.

> **⚠ Warning**
>
> Relying solely on color to encode data is risky. Always supplement with labels or patterns to ensure accessibility for users with visual impairments.

Finally, it's important to maintain consistency in your color scheme across multiple charts and visualizations. By assigning specific colors to recurring categories (e.g., different product types or demographic groups), you help the viewer develop a mental association between the color and the data, making it easier to interpret multiple charts within a report or presentation.

6.3.2   Contrast for Differentiation Between Segments

Contrast is another key factor in ensuring that pie and donut charts are readable and effective. Adequate contrast between segments allows the viewer to quickly distinguish between categories, even when the chart contains several slices or when the values represented by the slices are close in size. Without sufficient contrast, the chart can become visually confusing, leading to misinterpretations or missed insights.

Contrast can be achieved in several ways, the most obvious being through color selection. The difference between light and dark colors, or between complementary colors (such as blue and orange, or green and red), can make segments stand out from one another. However, contrast is not just about hue–it also involves brightness and saturation. For example, using different levels of brightness within the same color family can create subtle yet effective contrast. A bright yellow and a muted mustard, for instance, can be used to differentiate two segments while maintaining a cohesive color scheme.

In situations where color contrast is limited, designers can introduce other visual elements to differentiate between segments. One method is to use borders or outlines to separate adjacent slices. A thin, neutral-colored line can provide enough separation to improve readability without overwhelming the visual design. Similarly, slight variations in texture or pattern, such as using solid colors for some segments and light shading or hatching for others, can further differentiate categories while preserving the overall design harmony.

Another way to enhance contrast is through the use of spacing between segments. While pie and donut charts are traditionally displayed as continuous shapes, slight separation between slices can help viewers distinguish between categories, especially when the chart is crowded. This technique, often referred to as a "separated pie chart" or "exploded view," can be particularly effective in emphasizing the most important segments of the chart.

6.3.3 Using Gradients and Color Harmony for Aesthetic Appeal

Beyond the functional use of color and contrast, there is also an aesthetic dimension to designing pie and donut charts. Gradients and harmonious color schemes can be used to create charts that are not only informative but also visually striking. When used thoughtfully, these design elements can engage users more deeply, encouraging them to explore the data further.

Gradients, when applied correctly, can add a sense of depth and visual interest to pie and donut charts. For instance, a gradient that moves from light to dark within each segment can subtly draw attention to certain areas of the chart without overwhelming the viewer. However, it's important to use gradients sparingly and consistently. Overuse of complex gradients can make the chart look overly busy or distract from the actual data being presented. A simple linear gradient that enhances one part of the chart (such as a highlighted segment) is usually sufficient to achieve the desired effect without sacrificing clarity.

Color harmony is another powerful tool for enhancing the aesthetic appeal of pie and donut charts. By selecting colors that work well together, designers can create a sense of balance and unity within the chart, making it more visually appealing. Color harmony can be achieved through various approaches, such as analogous color schemes (using colors that are next to each other on the color wheel) or complementary color schemes (using colors from opposite sides of the color wheel). Analogous schemes are often softer and more cohesive, making them ideal for subtle, professional visualizations, while complementary schemes provide high contrast and energy, drawing attention to key data points.

It's also worth considering cultural and psychological associations with color when designing charts. Different cultures and contexts assign varying meanings to colors, so it's important to choose colors that align with the intended message of the visualization. For example, red is often associated with urgency or danger in Western cultures but may carry different connotations in other parts of the world. Understanding these associations can help in designing charts that resonate with the intended audience.

Color and contrast are not only functional tools for differentiating between segments but also powerful aesthetic devices that can engage the viewer and enhance the overall impact of the chart. By carefully selecting color schemes, applying contrast effectively, and using gradients and harmonious color combinations, designers can create pie and donut charts that are both clear and captivating.

> **♀ Design Tip**
>
> Choose color palettes that are accessible to users with color vision deficiencies. Tools like `ColorBrewer` can help you select visually distinct and colorblind-friendly colors.

6.4 Labeling and Annotations for Maximum Clarity

Labels and annotations play a crucial role in ensuring that pie and donut charts effectively communicate data. While color and contrast help differentiate between segments, it is the labels and annotations that provide the necessary context for understanding what the data represents. Poorly placed or overly complex labels can hinder comprehension, while well-thought-out labels guide the viewer seamlessly through the data. In this section, we will explore strategies for placing labels, using callouts for complex data, and considering alternatives such as legends and interactive tooltips.

6.4.1 Strategic Placement of Labels

The placement of labels is a critical design decision in pie and donut charts. Proper labeling ensures that viewers can easily associate data values with their corresponding chart segments without visual confusion. For pie charts, labels are often placed either directly on the slices or just outside the perimeter of the chart, connected by lines or leader lines. Both approaches have advantages and disadvantages, depending on the complexity of the data and the chart's overall design.

When placing labels inside slices, it's important to ensure that the text size is large enough to be legible without overwhelming the chart. This approach works best when there are relatively few slices and each slice is large enough to accommodate both the label and the corresponding percentage or value. However, for smaller slices, placing labels inside the chart can become problematic, as cramped labels can be difficult to read. In such cases, it's often better to place the labels outside the chart.

Labels placed outside the chart are typically connected to their corresponding slices by thin lines, allowing the viewer to follow the relationship between the label and the segment it describes. This method reduces clutter within the chart itself and allows for more flexible text formatting. However, it is important to avoid crossing lines or placing labels too far from their segments, as this can introduce confusion and force the viewer to work harder to understand the chart.

Another key consideration is the ordering of labels. If the chart segments are arranged in a particular order–such as from largest to smallest–arranging the labels in the same order can help viewers quickly make sense of the data. Consistency in label placement and alignment ensures that the chart remains clean, accessible, and easy to navigate.

6.4.2 Using Callouts for Complex Segments

For charts that contain complex data or multiple small segments, standard labels may not provide enough space or clarity to convey the necessary information. In such cases, callouts are an effective solution. Callouts are a form of annotation that expands upon the label, providing additional information about specific segments or highlighting key insights.

Callouts typically consist of a short descriptive text box connected to the corresponding segment with a leader line. This format allows the designer to provide more detailed explanations or context, especially when particular data points need to be emphasized. For example, in a donut chart that represents market shares, a callout might be used to provide additional details about a company with a small but significant market share, offering insights into its performance or growth potential.

When using callouts, it is important to strike a balance between providing additional information and maintaining the chart's visual simplicity. Too many callouts can clutter the chart and overwhelm the viewer. It is advisable to use callouts sparingly and only for segments that require further elaboration, such as highlighting an outlier or a particularly important data point.

Callouts can also be styled to align with the overall aesthetic of the chart, using consistent fonts, colors, and borders. This ensures that the callouts do not distract from the chart but rather enhance its readability by drawing attention to key information without overwhelming the viewer.

6.4.3 Alternatives to Direct Labeling: Legends and Interactive Tooltips

While direct labeling is a common and effective approach for pie and donut charts, it is not always the best option, particularly when dealing with charts that contain many segments or require detailed explanations. In these cases, alternatives such as legends and interactive tooltips can help maintain clarity and avoid clutter.

Legends are a traditional method of providing labels for pie and donut charts. They are typically placed to the side or below the chart and consist of a list of categories with corresponding color swatches or patterns that match the segments of the chart. Legends work particularly well when there are too many segments to label directly on the chart without causing visual confusion. However, legends do require the viewer to make an additional mental connection between the legend items and the chart segments, which can slightly slow down the interpretation process. To minimize this cognitive load, ensure that the legend is positioned close to the chart and that the colors or patterns are clearly distinguishable.

Interactive tooltips provide a more modern alternative to both direct labeling and legends, especially in digital or interactive visualizations. Tooltips are pop-up labels that appear when the user hovers over or clicks on a segment of the chart. This allows the chart to remain clean and uncluttered while still providing the user with detailed information on demand. Interactive tooltips are particularly effective in dynamic dashboards or web-based visualizations, where users can explore the data at their own pace.

Example of interactive tooltips in Python:

```python
import plotly.express as px

data = {
    "Category": ["A", "B", "C", "D"],
    "Value": [50, 30, 15, 5]
}

fig = px.pie(data, names='Category', values='Value', hole=0.4)
fig.update_traces(textinfo='percent+label',
    hoverinfo='label+value+percent')
fig.update_layout(title="Interactive Donut with Hover Tooltips")
fig.show()
```

> **♀ Design Tip**
>
> Introduce interactivity using hover effects, tooltips, or clickable segments to keep your audience engaged and allow for deeper data exploration without cluttering the chart.

One of the main advantages of tooltips is that they can contain more detailed information than would be possible with static labels or callouts. In addition to the category name and value, tooltips can provide additional context such as percentages, comparisons to other data points, or even multimedia elements like images or links to further information. This level of interactivity enhances user engagement by allowing the viewer to focus deeper on the data without overwhelming them with too much information upfront.

However, it's important to ensure that tooltips are implemented in a user-friendly way. Tooltips should appear instantly when the user interacts with the chart and should disappear smoothly when the user moves away. They should also be designed in a way that is consistent with the overall visual style of the chart, using appropriate fonts, colors, and layouts.

> **♀ Design Tip**
>
> Be consistent with your color assignments across charts. Assign fixed colors to recurring categories to help users recognize patterns across multiple visualizations.

Labeling and annotations are essential for ensuring that pie and donut charts communicate data clearly and effectively. Whether using direct labels, callouts, legends, or interactive tooltips, designers must prioritize clarity and usability while avoiding clutter. By carefully selecting and placing labels and annotations, designers can create charts that are both informative and visually engaging.

6.5 Donut Charts for Modern Data Storytelling

Donut charts have evolved into a powerful tool for modern data storytelling, combining the simplicity of traditional pie charts with added flexibility and functionality. Their distinct design, with a hollow center, opens up new possibilities for presenting information in a visually appealing and engaging way. Donut charts excel in situations where multiple datasets need to be compared or where the visualization requires a focal point in the center. In this section, we will explore specific use cases for donut charts in data storytelling, how to combine multiple datasets using donut charts, and the aesthetic and functional advantages of the donut hole.

6.5.1 Use Cases for Donut Charts in Storytelling

Donut charts have become increasingly popular in the realm of modern data storytelling because they provide a clean, aesthetically pleasing way to convey key insights. Their structure allows designers to represent proportional data in a simple, digestible format, while also offering flexibility in how additional information is incorporated.

One of the primary use cases for donut charts in storytelling is in the presentation of market or demographic data, where the visual emphasis is placed on the proportional distribution of categories. Donut charts are particularly effective when the goal is to show how a few major categories dominate the whole, while also offering a visual cue for the viewer to explore further. For example, in a dashboard showing market share by company, the central hole of the donut chart can be used to highlight key metrics such as total market size, or the largest company's share.

> **♀ Design Tip**
>
> Use the center of a donut chart to display key metrics such as total value, average, or a category label. This makes use of space effectively and helps contextualize the chart at a glance.

Another use case for donut charts is in progress or performance tracking. Donut charts are well-suited for representing data that shows part-to-whole relationships, such as the completion of a project or the progress toward a target. In these scenarios, the simplicity of the donut chart allows for quick and intuitive understanding of the progress, while the central hole can be used to display the key numerical value or percentage, making the chart both functional and visually engaging.

Donut charts are often used in interactive dashboards and digital reports where data needs to be both accessible and visually appealing. They allow for the presentation of complex data in a format that is less cluttered than traditional pie charts, providing a balance between clarity and aesthetic appeal. The center of the donut chart also provides an opportunity for interactive elements, such as clickable links or additional text, which can enhance the storytelling experience in a digital context.

6.5.2   Combining Multiple Datasets in Donut Charts

One of the key strengths of the donut chart design is its ability to represent multiple datasets within a single chart. By layering multiple rings within the same chart, designers can display different dimensions of data simultaneously while maintaining the overall proportional relationships. This makes donut charts particularly useful in situations where it is necessary to compare datasets or visualize relationships between different categories across time or groups.

For example, a multi-layered donut chart can be used to display sales performance across different product categories over several years. Each ring of the donut could represent a different year, allowing the viewer to see how the proportional distribution of sales changes over time while still keeping all the data within the same visual framework. This type of visualization is highly effective in showing trends and making comparisons at a glance, without the need for multiple charts that might overwhelm the viewer.

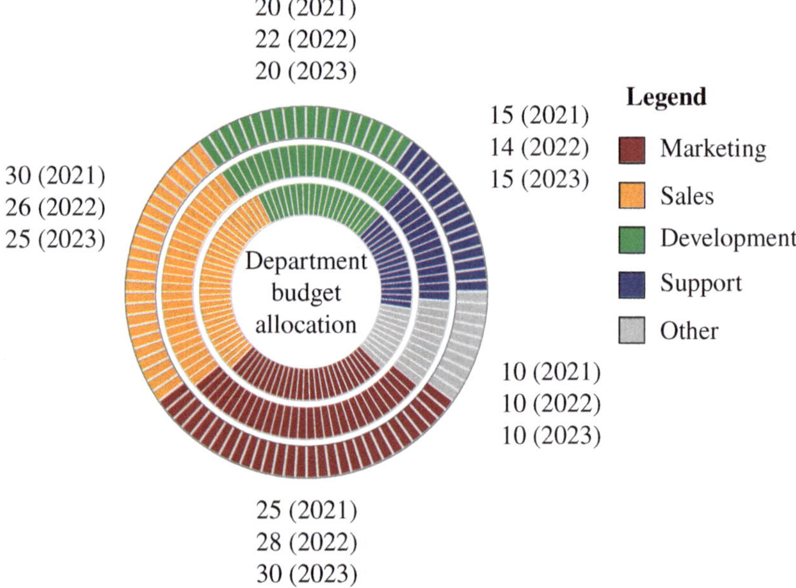

Fig. 6.3: Multi-layered donut chart showing changes in product category sales distribution over three years. Each ring represents a year (2021–2023).

In another scenario, donut charts can be used to compare demographic data across regions or groups. Each ring of the donut chart can represent a different region or group, allowing the viewer to easily see how different populations compare in terms of specific variables like income, age distribution, or employment rates. This format is particularly powerful in multi-dimensional storytelling, where it is necessary to explore how various categories compare without losing the overall sense of proportion within the dataset.

While combining multiple datasets in a single donut chart provides a visually efficient way to display complex information, it is important to avoid overloading the chart with too many rings or too much information. Clarity should always remain the priority. Limiting the number of rings to two or three ensures that the chart remains legible and visually accessible, while still providing the necessary depth of information.

6.5.3 Aesthetic and Functional Advantages of the Donut Hole

The most distinctive feature of the donut chart is the central hole, and this element provides both aesthetic and functional advantages. Aesthetically, the donut hole gives the chart a modern, minimalist look that is often preferred in contemporary design. This design choice reduces visual clutter, allowing the chart to focus more on the data and less on the chart's borders, which can sometimes dominate in traditional pie charts.

Functionally, the donut hole offers valuable space that can be used creatively to enhance the chart's storytelling capacity. One of the most common uses of the donut hole is to display summary statistics or key metrics, such as total values, percentages, or labels that provide context for the chart as a whole. For example, in a chart representing the distribution of customer feedback ratings, the center of the donut chart could display the average rating or the total number of responses, giving the viewer an immediate sense of the overall data without needing to parse through the individual segments.

Another functional advantage of the donut hole is its ability to support interactive design elements in digital environments. In an interactive dashboard, the donut hole can be used to display dynamic content that changes in response to user interactions with the chart. For instance, when a user hovers over different segments, the center of the chart could display additional details related to that segment, such as specific numerical values, comparisons with other segments, or links to deeper data. This interactivity enhances the user experience by allowing for a more engaging and exploratory approach to data analysis.

The donut hole can be used for branding and design purposes, especially in corporate reports or presentations. Designers can use the space to incorporate logos, icons, or other visual elements that reinforce the brand or theme of the presentation. This helps tie the visualization into the broader narrative or identity of the document, adding to the overall coherence of the storytelling experience.

Donut charts offer unique advantages in modern data storytelling, particularly through their ability to present multiple datasets and use the aesthetic and functional potential of the donut hole. Whether used to display key metrics, add interactivity, or provide a clean and minimalist look, the donut chart is a versatile tool that enhances both the clarity and visual appeal of data presentations.

6.6   Animation and Interaction for Dynamic Charts

In the era of digital visualization, static charts are no longer sufficient to engage users fully. Modern data storytelling often benefits from the addition of dynamic elements such as animation and interactivity. Pie and donut charts, with their simple yet powerful visual structure, can be enhanced significantly through the use of animations and interactive features. These additions allow users to explore data more deeply, engage with it in meaningful ways, and gain insights that might not be immediately apparent in static representations. This section will explore how animation can increase user engagement, the benefits of interactive features such as hovering, clicking, and data drilling, and the tools available for creating dynamic pie and donut charts.

6.6.1   Add Animation to Pie and Donut Charts for Engagement

Animation in data visualization serves a dual purpose: it can capture the viewer's attention and also help illustrate changes or comparisons over time. When applied thoughtfully, animation can transform pie and donut charts from static representations into dynamic visual stories that guide users through complex data in a more engaging and intuitive way.

One common use of animation in pie and donut charts is to illustrate transitions between datasets. For example, when visualizing data across time, animations can smoothly transition from one year's data to the next, showing how proportions shift over time. This is particularly effective in storytelling, as it allows the viewer to see patterns and trends emerge gradually rather than having to compare multiple static charts. A simple animation that fades or rotates slices as they grow or shrink can dramatically improve the user's understanding of how categories have changed in relation to each other.

> **♀ Design Tip**
>
> If your chart needs to show changes over time, consider animating transitions between pie or donut chart states. Smooth transitions can help reveal trends and comparisons more effectively.

Animation can also be used to highlight specific segments of the chart. For example, when a viewer hovers over a particular slice, the segment can "pop out" slightly or change color, drawing attention to its importance. This technique can be especially useful when the chart contains numerous segments, and the designer wants to emphasize certain data points without overwhelming the viewer with static labels. In donut charts, the central hole can be used to display animated content, such as dynamically updating values or key insights that change in real-time as the user interacts with different segments.

However, it's important to use animation judiciously. Overuse or overly complex animations can distract from the data and frustrate users, particularly in professional or analytical settings where clarity and precision are paramount. Animations should be smooth, unobtrusive, and directly tied to the data they represent. The goal is to enhance the viewer's understanding, not to create unnecessary visual effects.

6.6.2 Interactive Charts: Hovering, Clicking, and Data Drilling

Incorporating interactivity into pie and donut charts significantly increases their functionality, allowing users to explore the data on their own terms. Interactivity transforms a chart from a static image into a tool for data exploration, where users can hover over, click, or drill down into specific data points for more detailed information.

Hover effects are a simple yet powerful way to enhance pie and donut charts. When the user hovers their cursor over a segment, additional details can be displayed, such as the exact percentage, value, or a more detailed label. This reduces the need for cluttered static labels on the chart itself, keeping the design clean while still providing all necessary information. Hover effects can also change the appearance of the segment–by enlarging it, changing its color, or raising it slightly from the chart–drawing the user's attention to the segment of interest. This interaction helps emphasize particular data points without overwhelming the viewer with too much information upfront.

Clicking is another common interactive feature, allowing users to engage with the data in a more meaningful way. In a pie or donut chart, clicking on a segment could trigger a more detailed breakdown of that category, opening a sub-chart or additional layers of information. For example, in a sales report, clicking on the "North America" segment of a donut chart could reveal a breakdown of sales by country within that region. This hierarchical approach helps keep the chart simple at the top level while still offering access to deeper insights when needed.

Data drilling, a more advanced interactive feature, allows users to explore data across multiple dimensions directly within the same chart. Data drilling involves clicking on a chart segment to dive deeper into the underlying data, often revealing additional layers or visualizations. In donut charts, this could mean transitioning from an overview of market shares to more granular data on individual product categories. Each click provides more detailed data while maintaining a clear visual relationship to the original chart, making it an ideal tool for exploratory analysis in business dashboards and reporting systems.

Interactivity not only enhances user engagement but also empowers users to discover insights on their own, making the data experience more personal and adaptable to the viewer's needs. However, it is essential to ensure that these interactions are intuitive and responsive, with smooth transitions and clear visual cues. This makes it easy for users to navigate through the data without confusion or frustration.

6.6.3　Tools and Libraries for Creating Dynamic Charts

The addition of animation and interactivity to pie and donut charts requires the use of specific tools and libraries designed for dynamic visualizations. These tools offer varying levels of flexibility and complexity, allowing designers and developers to create everything from simple hover effects to fully interactive dashboards.

One of the most popular libraries for creating dynamic charts is `D3.js` (Data-Driven Documents). `D3.js` is a JavaScript library that allows for the creation of highly customizable and interactive data visualizations. It offers full control over the animation and interaction of elements within a chart, making it possible to create smooth transitions, hover effects, clickable segments, and even complex data drilling. However, `D3.js` is a low-level library, meaning that it requires more coding knowledge to implement than some higher-level alternatives.

For users looking for a simpler, more accessible tool, `Chart.js` is a popular option. `Chart.js` is a lightweight JavaScript library that offers a range of pre-built chart types, including pie and donut charts, with built-in animation and interactivity features. It is easier to use than `D3.js`, making it a good choice for users who want to add dynamic elements to their charts without extensive coding. `Chart.js` allows for hover effects, clickable segments, and customizable animations with minimal effort.

Another powerful tool for creating dynamic charts is `Plotly`, which offers both JavaScript and Python libraries for interactive data visualization. `Plotly` provides a range of pre-configured options for adding interactivity to pie and donut charts, such as hover labels, click events, and transitions between datasets. Its user-friendly API makes it a popular choice for creating interactive dashboards and reports, and its ability to integrate with both web-based and Python environments makes it versatile for a wide range of applications.

For users working in Python, `Matplotlib` and `Plotly`'s Python version are excellent options for adding animation and interactivity to pie and donut charts. `Matplotlib` allows for animated transitions between chart states, while `Plotly` provides more sophisticated interactivity, including hover effects and clickable segments. Both libraries integrate seamlessly with Jupyter Notebooks, making them ideal for creating interactive reports and presentations in academic and research settings.

Adding animation and interactivity to pie and donut charts can significantly enhance user engagement and make the data more accessible. By using tools such as `D3.js`, `Chart.js`, and `Plotly`, designers and developers can create dynamic visualizations that bring data to life and empower users to explore it in new and meaningful ways.

6.7 Best Practices

6.7.1 Breakdown of Design Choices and Their Impact

The success of these real-world examples lies in the thoughtful design choices that prioritize clarity, simplicity, and user engagement. Each design element–from color selection to label placement–plays a significant role in ensuring that the viewer can interpret the data accurately and efficiently. Let's break down some of the key design choices and their impact on the overall effectiveness of these charts.

1. **Color Schemes and Contrast** In all of the examples mentioned, color schemes are used strategically to differentiate between segments without overwhelming the viewer. The use of complementary or analogous color palettes helps create visual harmony, while contrast between segments ensures that even smaller categories are easily distinguishable. In the market share example, for instance, the use of a consistent color palette across brands and product lines not only enhances readability but also provides a cohesive visual narrative.

2. **Labeling and Annotations** Clear and concise labeling is critical to the success of pie and donut charts. The placement of labels either inside the chart segments or in a well-organized legend ensures that the viewer can quickly associate data values with their corresponding categories. In cases where the chart is interactive, hover effects provide additional details without cluttering the design. In the investment portfolio example, labels are kept short and positioned neatly outside the chart, with interactive tooltips providing more granular information as needed.

3. **Use of the Donut Hole** The hollow center of donut charts is often utilized to display key metrics or summary information. This not only makes effective use of space but also reinforces the overall message of the visualization. In the financial advisory chart, for example, the total portfolio value is displayed in the center, making it immediately clear to the viewer what the chart represents without having to read through each segment individually. This reinforces the holistic view of the portfolio while still providing detailed insights into individual asset categories.

4. **Interactivity** Many modern visualizations, especially in digital dashboards, incorporate interactive features to enhance user engagement. In the demographic study example, interactive elements such as hover states and drill-down capabilities allow the user to explore different levels of data without overwhelming them with too much information at once. This not only makes the chart more engaging but also empowers the user to tailor their data exploration experience to their specific interests.

5. **Minimalism and Clarity** The best pie and donut charts adhere to the principle of "less is more." Rather than overloading the viewer with too much data or unnecessary design elements, these charts focus on delivering the most critical information in a visually clean and accessible way. The demographic study chart, for instance, uses minimalistic design elements–simple colors, clean lines, and legible fonts–to ensure that the viewer can focus on the data itself, without distraction.

These design choices have a direct impact on how effectively the chart communicates its message. When executed well, they enhance the user's understanding of the data and create a more engaging visualization experience. Conversely, poor design choices–such as using too many colors, overcrowding the chart with labels, or ignoring the importance of interactivity–can lead to confusion, misinterpretation, and disengagement.

6.7.2 Lessons from Successful Visualizations

By examining successful pie and donut charts from various fields, we can extract several valuable lessons that can be applied to future visualizations. These lessons emphasize the importance of simplicity, thoughtful design choices, and user-centered approaches.

1. **Prioritize Simplicity** One of the key takeaways from successful visualizations is the importance of simplicity. Pie and donut charts are most effective when they present a clear, straightforward view of the data. Avoid cluttering the chart with too many segments or overly complex labels. Instead, focus on the key insights that need to be conveyed and design the chart in a way that emphasizes those insights.

2. **Use Interactivity When Appropriate** Interactive features can greatly enhance the user experience, especially in digital environments where users expect to be able to explore the data on their own terms. By incorporating hover effects, clickable segments, or drill-down options, you can allow users to engage with the data more deeply while still maintaining a clean and simple initial presentation.

3. **Use the Donut Hole Effectively** The center space in a donut chart offers a valuable opportunity to highlight key metrics or provide summary information. Rather than leaving this space empty, use it to display important data that complements the chart's overall message. Whether it's a total value, a percentage, or a key insight, this central feature can reinforce the chart's narrative and guide the viewer's attention.

4. **Consider the Context and Audience** Successful visualizations are designed with their specific audience in mind. In academic settings, charts may need to be more detailed and data-rich, while in industry, a high-level overview with interactive features might be more appropriate. Tailoring the design and functionality of your pie and donut charts to the needs and preferences of your audience ensures that the visualization resonates and provides the intended value.

5. **Balance Aesthetics with Functionality** While aesthetic appeal is important in making charts visually engaging, functionality should never be sacrificed for the sake of design. The best visualizations strike a balance between being visually appealing and highly functional. Each design element–color, typography, spacing–should serve a purpose and contribute to the chart's overall effectiveness.

By applying these lessons, designers and data practitioners can create pie and donut charts that not only look stunning but also deliver meaningful, actionable insights. Whether used in industry reports, academic papers, or interactive dashboards, these charts have the potential to enhance data storytelling and improve the user's experience with complex information.

6.8 Tools and Libraries for Pie and Donut Charts

Creating visually stunning and functional pie and donut charts requires the right tools and libraries. With the increasing popularity of data visualization, a wide range of tools are available that offer varying degrees of customization, interactivity, and accessibility. Choosing the best tool for a given project depends on factors such as the complexity of the data, the level of interactivity required, and the intended audience. In this section, we will provide an overview of popular charting tools, guide you through choosing the right tool for your project, and discuss strategies for customizing charts to enhance accessibility and user engagement.

6.8.1 Overview of Popular Charting Tools

There are numerous charting tools and libraries available, each with its own strengths and ideal use cases (Table 6.3). Some of the most popular options include `D3.js`, `Chart.js`, `Plotly`, `Highcharts`, and `Matplotlib`. These tools allow designers and developers to create pie and donut charts ranging from simple static images to highly interactive visualizations. Let's take a closer look at these tools:

1. **D3.js** D3.js (Data-Driven Documents) is one of the most powerful and flexible JavaScript libraries for creating data visualizations. It enables full control over every element of a chart, allowing developers to create highly customized and interactive visualizations. With `D3.js`, you can create animated transitions, interactive tooltips, and advanced data manipulation techniques. While incredibly powerful, `D3.js` requires a solid understanding of JavaScript and is best suited for more complex, custom projects.

2. **Chart.js** Chart.js is a popular open-source JavaScript library that provides a range of pre-built chart types, including pie and donut charts. It is relatively easy to use and allows developers to add animations and interactive features such as hover effects with minimal effort. While it doesn't offer the same level of customization as `D3.js`, it strikes a good balance between ease of use and flexibility, making it a great option for many projects.

3. **Plotly** Plotly is a versatile charting library available in both JavaScript and Python. It allows users to create highly interactive and visually appealing charts with relatively little coding effort. Plotly offers built-in support for pie and donut charts, with extensive options for customization, including hover labels, animations, and interactive legends. Its Python implementation, in particular, makes it a popular choice in academic and research settings, where it integrates well with tools like Jupyter Notebooks.

4. **Highcharts** Highcharts is a commercial JavaScript library known for its ease of use and polished, professional-looking charts. It includes many customization options and offers strong support for pie and donut charts. Highcharts is a great option for developers looking for ready-made solutions with minimal coding effort. The

library is especially popular in enterprise settings where sophisticated visualizations are needed, but customization beyond a certain level may require a commercial license.

5. **Matplotlib** Matplotlib is a widely-used Python library for creating static, animated, and interactive visualizations. It is particularly well-suited for generating pie and donut charts in academic, scientific, or research contexts. While Matplotlib is primarily designed for static charts, it does offer basic interactivity when used in conjunction with tools like Jupyter Notebooks or Plotly's Python bindings. The strength of Matplotlib lies in its integration with Python's data manipulation libraries like Pandas, making it an excellent choice for data-intensive projects.

Table 6.3: Popular Tools and Libraries for Pie and Donut Charts

Tool/Library	Language	Features
D3.js	JavaScript	Highly customizable, full control, steep learning curve
Chart.js	JavaScript	Easy to use, animated charts, limited deep customization
Plotly	Python, JS	Interactive charts, rich tooltips, good for dashboards
Highcharts	JavaScript	Professional quality, responsive, commercial license needed
Matplotlib	Python	Best for static scientific plots, interactivity via extensions

6.8.2   Choosing the Right Tool for Your Project

Selecting the right tool for your project involves considering several factors, including the complexity of the chart, the desired level of interactivity, the target platform, and the skills of the team involved. Here are a few guiding principles to help choose the appropriate tool:

1. **Complexity and Customization Needs** For projects that require highly customized and interactive visualizations, such as dashboards or reports with dynamic data, D3.js is often the best choice due to its extensive customization options. However, it requires a significant amount of JavaScript knowledge. If you need substantial interactivity but want a quicker setup with less coding, Plotly or Highcharts are excellent options, offering pre-built features with customization flexibility.

2. **Ease of Use** If simplicity and ease of use are a priority, Chart.js or Matplotlib are good options. Chart.js is ideal for developers who need quick results with basic interactivity, while Matplotlib is preferred in academic and research settings for creating static or minimally interactive charts with easy integration into Python workflows.

3. **Target Audience and Platform** Consider who will be using or viewing the visualizations and the platform on which they will be presented. For web-based visualizations, D3.js, Chart.js, and Highcharts are ideal, offering seamless

integration with web technologies. If the charts will be part of a Python-based project or research, `Plotly` and `Matplotlib` are better suited.

4. **Interactivity Requirements** If your project requires interactive features like hover effects, clicking to reveal more data, or zooming, tools like `D3.js`, `Plotly`, and `Highcharts` should be considered. For basic interactivity with a user-friendly API, `Chart.js` offers a solid balance between functionality and ease of implementation.

5. **Licensing and Costs** If your project is for commercial purposes, it's essential to consider licensing costs. While open-source tools like `D3.js`, `Chart.js`, and `Plotly` have free versions, commercial use of `Highcharts` requires a paid license. This cost may be justified if the project demands professional support and polished, out-of-the-box visualizations.

Ultimately, the right tool depends on your project's unique needs. Consider the skillset of the team, the complexity of the chart, the required level of interactivity, and the platform on which the chart will be deployed.

6.8.3 Customizing Charts for Accessibility and Engagement

Accessibility is a critical consideration when creating pie and donut charts. Ensuring that your charts are accessible to users with disabilities, such as color vision deficiencies, is essential for providing an inclusive experience. At the same time, customizing your charts to enhance user engagement can make them more interactive, informative, and visually appealing.

1. **Color Accessibility** When choosing colors for your pie and donut charts, consider users who may have difficulty distinguishing between certain colors. To address this, use color palettes that are friendly to individuals with color vision deficiencies. Tools such as `ColorBrewer` can help generate color schemes that are distinguishable by all users. Avoid relying solely on color to differentiate between segments. Use patterns, textures, or labels to ensure that data is conveyed clearly, even for those with color vision challenges.

2. **Labeling and Annotation** To make your pie and donut charts more accessible and engaging, focus on providing clear and concise labels. Ensure that all segments are labeled in a way that is easy to read, either directly on the chart or via a well-organized legend. For interactive charts, tooltips can provide additional information when the user hovers over a segment, reducing clutter while offering deeper insights.

3. **Interactive Features for Engagement** Adding interactive features such as hover effects, clickable segments, and drill-down functionality can significantly enhance user engagement. Interactive pie and donut charts allow users to explore the data in more detail and customize their experience based on their preferences or needs. For instance, adding a hover state that highlights a segment or reveals more detailed information allows users to gain insights without overwhelming them with too much data upfront.

4. **Responsiveness** Ensure that your charts are responsive and perform well on various screen sizes and devices. This is especially important for web-based visualizations, where users may access the charts on different devices such as

smartphones, tablets, and desktops. Tools like Chart.js, D3.js, and Plotly offer options for creating responsive charts that adjust dynamically based on screen size.

5. **Keyboard and Screen Reader Support** For users who rely on keyboard navigation or screen readers, it's important to ensure that your charts are fully accessible. This involves adding alt text for images, making sure that interactive elements are navigable via keyboard, and providing detailed descriptions of the chart's contents that can be read by screen readers.

Customizing pie and donut charts for accessibility and engagement not only ensures that your visualizations are inclusive but also enhances the overall user experience. By choosing accessible color schemes, adding clear labels and annotations, incorporating interactive features, and ensuring responsive design, you can create charts that resonate with a wide audience and provide meaningful insights.

6.9   Summary and Key Takeaways

Throughout this chapter, we have explored the principles, strategies, and tools for designing effective and visually engaging pie and donut charts. These charts, while simple in structure, offer a powerful means of storytelling when used appropriately. In this final section, we will recap the essential design principles, review tools and techniques for successful visual storytelling, and share some final thoughts on the future of pie and donut charts in data visualization.

6.9.1   Recap of Design Principles

The design of pie and donut charts hinges on several key principles that ensure both clarity and engagement. Here is a recap of the most important design considerations:

1. **Simplicity and Clarity** Pie and donut charts should prioritize simplicity. Limit the number of segments to ensure that each part remains distinguishable. Simplicity in design allows users to quickly interpret the data, ensuring clarity in communication.

2. **Appropriate Data Selection** These charts are best suited for illustrating part-to-whole relationships. It is important to ensure that the data categories are mutually exclusive and collectively exhaustive, and that the dataset is small enough to avoid overcrowding.

3. **Color and Contrast** Effective use of color helps differentiate between segments, but it's essential to maintain contrast to enhance readability. For accessibility, use colorblind-friendly palettes and avoid relying solely on color to convey meaning.

4. **Labels and Annotations** Strategic placement of labels and callouts ensures maximum clarity without overwhelming the viewer. Consider alternative labeling techniques, such as legends or interactive tooltips, to reduce clutter while providing detailed information.

5. **Interactivity and Animation** Incorporating dynamic elements like hover effects, clickable segments, and animated transitions can enhance user engagement, making the data more interactive and accessible.

6. **Utilization of the Donut Hole** The donut chart's central space can be creatively used to display key metrics, summaries, or interactive elements. This functional advantage can significantly enhance both the aesthetic appeal and the information delivered.

By adhering to these design principles (Table 6.4), designers can ensure that pie and donut charts remain effective tools for communicating complex data in an intuitive and visually appealing way.

Table 6.4: Essential Design Principles for Pie and Donut Charts

Principle	Description
Simplicity	Keep charts minimal; avoid excessive slices or labels
Clarity	Ensure easy interpretation with spacing and contrast
Color	Use harmonious and accessible palettes
Labeling	Position labels clearly; use legends or tooltips as needed
Interactivity	Allow user exploration with hover, drill-down, and animation
Accessibility	Design for colorblindness, screen readers, and responsive layouts

6.9.2 Tools and Techniques for Effective Visual Storytelling

Successful visual storytelling requires not only thoughtful design but also the right tools and techniques. Here's a recap of the essential tools and techniques for creating compelling pie and donut charts:

1. **Charting Tools** Depending on the project requirements, tools such as `D3.js`, `Chart.js`, `Plotly`, and `Matplotlib` offer various levels of customization and interactivity. Choosing the right tool is crucial for achieving the desired balance between functionality and user engagement. `D3.js` is excellent for highly customized charts, while `Chart.js` and `Plotly` offer simpler, more user-friendly options for most projects.

2. **Interactivity and Animation** Adding interactivity, such as hover states, clickable segments, and drill-down capabilities, enhances user engagement by making the data exploration process dynamic. Animation, when used judiciously, can help illustrate trends and transitions in the data, leading to a more immersive experience.

3. **Customization for Accessibility** Ensuring that your charts are accessible is key to effective storytelling. Use color schemes that accommodate users with color vision deficiencies, provide alternative labels and annotations for screen readers, and ensure that interactive elements are keyboard-friendly.

4. **Storytelling Techniques** Pie and donut charts excel at telling stories when they highlight the most critical parts of the data. Techniques such as using callouts to emphasize important data points, or integrating multiple datasets with layered donut charts, can deepen the storytelling by providing multiple perspectives within a single visualization.

5. **Responsiveness and Flexibility** Today's visualizations must adapt to various devices and screen sizes. Creating responsive charts ensures that they display correctly across platforms, providing users with a seamless experience, whether they are accessing the chart on a desktop, tablet, or smartphone.

The combination of the right tools and storytelling techniques enables the creation of pie and donut charts that go beyond static representations, transforming them into dynamic, engaging, and informative visual stories.

6.9.3 Final Thoughts on the Future of Pie and Donut Charts in Data Visualization

As data visualization continues to evolve, pie and donut charts will remain important tools for conveying proportional relationships, especially in scenarios where simplicity and ease of interpretation are paramount. Their visual appeal, combined with their ability to display part-to-whole relationships effectively, ensures that they will continue to be a staple in business reports, academic research, and interactive dashboards.

Looking ahead, advancements in interactive and dynamic visualization tools will allow pie and donut charts to become even more versatile. We are already seeing these charts being integrated into highly interactive environments where users can explore multiple layers of data with ease. The rise of dashboards, data journalism, and real-time analytics will further drive the demand for interactive donut charts that can present complex datasets in a digestible format.

Another key trend is the increasing emphasis on accessibility and inclusivity in design. Future iterations of pie and donut charts will need to focus on making data visualizations more accessible to a wider audience, including those with disabilities. This could involve using new techniques in responsive design, color accessibility, and screen reader compatibility, ensuring that everyone can benefit from data insights.

Finally, the future of pie and donut charts will likely see continued experimentation with hybrid visualizations, where pie or donut charts are combined with other types of visualizations to provide a richer, multi-dimensional perspective. By pairing pie charts with bar charts, line graphs, or even map-based visualizations, designers will be able to create more comprehensive narratives that guide the viewer through complex datasets. Pie and donut charts have evolved from simple data visualizations into sophisticated tools for storytelling and interaction. With the right design principles, tools, and techniques, these charts will continue to play a critical role in the future of data visualization, helping to make complex data more accessible and understandable to a global audience.

Suggested Readings

1. Brewer, C.A.: Colorbrewer 2.0: Color advice for cartography (2024).   URL https://colorbrewer2.org

2. Cairo, A.: How Charts Lie: Getting Smarter about Visual Information. W. W. Norton & Company (2019)
3. Chart.js Contributors: Chart.js: Simple yet flexible JavaScript charting (2024). URL https://www.chartjs.org
4. Few, S.: Show Me the Numbers: Designing Tables and Graphs to Enlighten, 2nd edn. Analytics Press (2012)
5. Knaflic, C.N.: Storytelling with Data: A Data Visualization Guide for Business Professionals. Wiley (2015)
6. Plotly Technologies Inc.: Plotly Python Graphing Library (2024). URL https://plotly.com/python/
7. Tufte, E.R.: The Visual Display of Quantitative Information. Graphics Press, Cheshire, Connecticut (1983)

7 Creating Enchanting Spider Plots

Abstract

This chapter explores the design and implementation of spider plots, also known as radar charts, which are powerful tools for visualizing multi-dimensional data across multiple categories. Spider plots are widely used in performance metrics, capability assessments, and multi-criteria decision-making due to their ability to present complex data in a compact and intuitive format. The chapter begins by introducing the purpose and applications of spider plots, highlighting their strengths and limitations. It then discusses essential design principles, including selecting meaningful metrics, maintaining readability, and balancing data representation with visual appeal. Techniques for enhancing visual appeal are covered, such as adding color, smoothing lines, and incorporating interactive features like hover tooltips and responsive design. Advanced design techniques, such as layering multiple datasets and adding annotations, are discussed to elevate the effectiveness of spider plots for comparative analysis.

Aims

After reading this chapter, you should be able to:

- ➢ Understand the purpose of spider plots and identify scenarios where they are an effective visualization tool.

- ➢ Recognize the advantages and limitations of spider plots, especially in representing multi-dimensional data.

- ➢ Select appropriate categories and metrics to construct readable and informative spider plots.

- ➢ Apply design principles to balance data clarity and visual appeal, ensuring that spider plots are engaging yet easily interpretable.

- ➢ Troubleshoot common design issues in spider plots, such as overcrowding and scaling challenges, to maintain visual clarity.

R. Damaševičius, *Human-Centred Scientific Data Visualisation*,
Undergraduate Topics in Computer Science,
https://doi.org/10.1007/978-3-032-01606-5_7

7.1   Introduction to Spider Plots

Spider plots , also known as radar charts, serve as an effective method for visualizing and comparing multiple variables across several categories. These plots take their name from their characteristic web-like structure, in which each axis radiates from a central point. Each axis represents a distinct variable, while data points plotted along these axes are connected to form a polygonal shape. This configuration provides an intuitive and visually engaging way to compare multidimensional data at a glance, allowing users to observe the overall shape of the data distribution and assess relative strengths or weaknesses across categories. Spider plots are commonly used in fields that require the visualization of performance metrics, multi-criteria decision-making, and capability assessments, as they are particularly adept at displaying the trade-offs between various metrics.

7.1.1   Anatomy of A Spider Plot

Spider plots , also known as radar charts, are a compelling method for visualizing multivariate data across multiple categories in a circular layout. They are widely used in performance evaluations, capability assessments, multi-criteria decision-making, and visual comparisons of multiple datasets. Their appeal lies in their ability to compactly and intuitively represent multiple variables simultaneously, using a visually engaging, web-like structure.

Each axis in a spider plot represents a unique variable radiating from a central point. The value of each variable is plotted along its corresponding axis, and these points are then connected to form a closed polygon. The resulting shape allows for an immediate visual interpretation of the data pattern, making spider plots particularly useful for identifying strengths, weaknesses, and performance gaps across dimensions.

Fig. 7.1 provides a visual breakdown of the core elements of a spider plot. From the central origin point to the radial axes, category labels, gridlines, and polygonal data areas, each component plays a crucial role in constructing an effective and interpretable chart.

To support this visual overview, Table 7.1 summarizes the anatomy of a basic spider plot. It defines each component's function and highlights its role in enhancing the clarity and usability of the visualization.

7.1.2   Purpose of Spider Plots

The primary value of spider plots lies in their ability to display multiple variables simultaneously, making them a versatile tool for data comparison across diverse contexts. In the realm of performance metrics, spider plots allow users to evaluate multiple dimensions of performance in a single, cohesive graphic. For example, a spider plot can represent various key performance indicators (KPIs) for an organization or team, such as productivity, quality, efficiency, and customer satisfaction. By

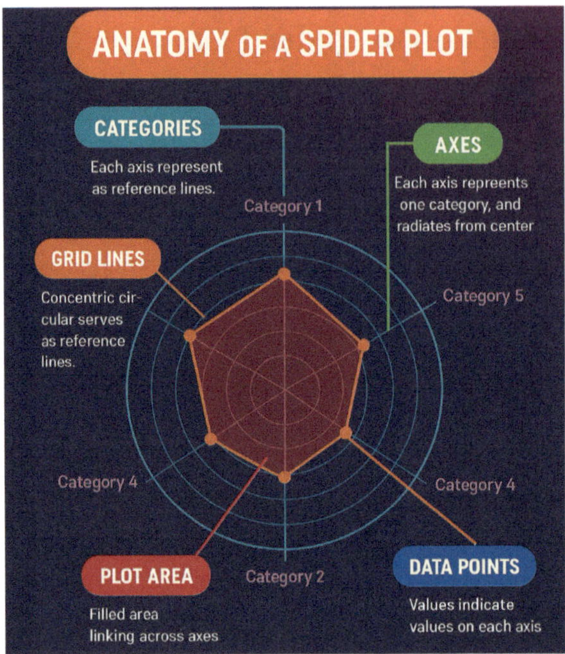

Fig. 7.1: Anatomy of a Spider Plot: A visual breakdown of a basic radar chart, highlighting key components such as axes, grid lines, categories, data points, and the plot area. This infographic serves as a foundational reference for understanding spider plot structure and design.

plotting these metrics on the same chart, users can easily compare performance across dimensions, identifying areas of strength or highlighting aspects that require improvement. This comparative aspect is especially useful in strategic decision-making, where understanding relative strengths and weaknesses is crucial.

Spider plots also play an important role in multi-criteria decision-making. Decision-makers often evaluate multiple options across several criteria, and spider plots provide a straightforward means of visualizing these comparisons. For instance, in selecting a vendor, a decision-maker may compare options based on criteria such as cost, reliability, speed, and customer support. Representing each vendor as a polygon within a spider plot allows stakeholders to visually assess which option best meets their criteria in a balanced way, thus facilitating a more informed decision-making process. Spider plots are frequently employed in capability assessments to visualize the relative competencies of individuals, teams, or organizations. In this context, each axis may represent a different capability, such as technical expertise, creativity, or communication skills. By visualizing competencies across these dimensions, stakeholders can identify developmental areas or highlight the strengths of a team or

Table 7.1: Anatomy of a Spider Plot

Component	Description
Center Point	The origin from which all axes radiate. It represents the baseline or zero level for all variables.
Axes	Radial lines extending from the center, each representing a different category or variable.
Categories	Labels at the ends of each axis indicating what each axis measures (e.g., Speed, Accuracy, Cost).
Grid Lines	Concentric circles or polygons serving as reference lines for value estimation along the axes.
Data Points	The values plotted on each axis based on the metric value, typically connected by lines.
Plot Area	The filled polygonal area connecting all data points, visually representing the shape and magnitude of a profile.
Legend	A reference that explains the color or line style for each dataset (used in comparative spider plots).
Reference Scale	The radial scale that defines the value range along each axis (e.g., 0 to 5 or 0% to 100%).

individual. This application is particularly useful in performance reviews, training needs assessments, and team-building initiatives.

7.1.3 Advantages and Limitations

Spider plots offer several unique advantages, making them a valuable tool in scenarios requiring multidimensional comparison. One of their most significant strengths is the ability to present high-dimensional data in a compact, easily interpretable form. Unlike traditional bar or line charts, which may require multiple panels or separate charts to represent various dimensions, spider plots consolidate numerous variables into a single graphic. This characteristic makes them particularly useful in contexts where quick, high-level data comprehension is essential. The distinct polygonal shapes formed by different datasets within a spider plot provide an immediate visual impression of relative strengths and weaknesses, enabling users to draw comparisons without extensive cognitive processing. This holistic view is beneficial in presenting complex data to non-specialist audiences or stakeholders who may not be familiar with the nuances of data analysis, as the shape alone can convey valuable insights.

> **⚠ Warning**
>
> Do not use spider plots for precise numerical comparisons. They are best suited for visual overviews and relative trends, not exact values.

However, spider plots also have notable limitations, particularly when applied to high dimensional data. As the number of categories or variables increases, the readability of spider plots can decrease significantly. When too many axes are plotted, the resulting shape may become convoluted, and the overlapping lines can obscure individual data points, making it challenging to distinguish between categories. This effect is especially problematic when multiple datasets are plotted on the same spider plot, as overlapping polygons can create visual clutter. Spider plots can suffer from scale distortion, as data near the plot's center appears minimized compared to data points near the edges. Without careful scaling, users may overemphasize data points located farther from the center, leading to skewed interpretations.

> **⚠ Warning**
>
> Spider plots are not ideal for high-dimensional datasets. Consider alternative visualizations like parallel coordinates or heatmaps if you exceed 10 dimensions.

Spider plots may not be ideal for precise quantitative comparisons.They provide a quick visual overview of relative values, but they lack the accuracy and detail of other chart types, such as bar charts or scatter plots, which allow for precise value reading and facilitate statistical analysis. As a result, spider plots are best used as exploratory tools or summary visuals rather than as primary sources for rigorous data interpretation. Recognizing these limitations is essential for data analysts and UX designers, as it ensures that spider plots are applied in appropriate contexts where their strengths outweigh their limitations.

Spider plots offer a visually engaging and compact means of comparing multiple variables across categories, making them well-suited for applications in performance evaluation, decision-making, and capability assessment. While their unique structure provides immediate visual insights, spider plots must be used judiciously to avoid readability issues associated with high-dimensional data. By understanding both the advantages and constraints of spider plots, analysts and decision-makers can use these tools to effectively communicate complex, multidimensional data.

7.2   Design Principles for Effective Spider Plots

Spider plots , with their unique radial structure, require careful design considerations to remain effective and visually engaging. When designing spider plots, it is essential to maintain a balance between data richness and readability. Cluttered or overly complex spider plots can lead to cognitive overload, reducing the plot's usefulness for users. This section outlines key design principles to maximize the effectiveness of spider plots by optimizing the selection of categories and metrics, maintaining visual balance, and ensuring clarity.

7.2.1   Selecting Categories and Metrics

Selecting appropriate categories and metrics is crucial for designing spider plots that are both informative and easy to interpret. The purpose of a spider plot is to reveal the relative strengths and weaknesses of multiple variables within a single visual frame. However, when too many categories are included, the resulting plot can become overwhelming, reducing its interpretability. To avoid this, it is essential to apply a selective approach when choosing metrics, prioritizing those that are most meaningful to the plot's purpose.

Guideline 1: Define the Primary Objective of the Plot. Before selecting categories, clarify the primary objective of the spider plot. Determine what insights you want the user to gain from the visualization, as this will shape the choice of metrics. For instance, if the objective is to compare employee competencies, prioritize categories directly related to job performance, such as technical skill, communication ability, and time management. Focusing on the objective helps avoid the inclusion of extraneous metrics that do not add significant value to the analysis.

Guideline 2: Limit the Number of Categories. A general recommendation is to limit the number of axes to between five and ten categories. With fewer than five categories, a spider plot may lack the visual complexity needed for comparison; however, more than ten categories can result in excessive overlap and visual clutter. When there is a need to present a large number of metrics, consider grouping related metrics into broader categories or creating multiple spider plots for different dimensions. This approach can simplify the visualization while maintaining depth in the analysis.

Guideline 3: Prioritize High-Impact Metrics. When deciding which metrics to include, prioritize those that will have the greatest impact on the analysis and decision-making process. High-impact metrics are those that vary significantly across the data or are particularly relevant to the comparison at hand. For instance, if evaluating product performance, focus on primary indicators such as durability, cost-efficiency, and user satisfaction, rather than secondary metrics that may not be directly relevant to user goals. Prioritizing high-impact metrics not only improves clarity but also directs user focus to the most critical insights.

Guideline 4: Standardize Metrics for Consistent Comparison. To ensure an accurate and meaningful comparison, metrics should be standardized across categories. Each category should share a consistent scale, as disparities in units or scales can distort the interpretation. For instance, if one axis represents a metric measured in percentages while another is in dollar amounts, the plot may not accurately reflect relative strengths and weaknesses. Standardization can be achieved by converting metrics to a common scale or through normalization techniques, such as rescaling values to a range from 0 to 100.

Guideline 5: Ensure Distinct and Interpretable Labels. Categories should be labeled in a way that is immediately interpretable, especially for non-expert users. Avoid jargon or technical terms that may not be universally understood. Labels should clearly describe each metric without excessive wording, as long labels can clutter the plot. If a category label is particularly complex, consider using tooltips or a legend to provide further clarification.

Guideline 6: Avoid Redundancy in Metrics. Including multiple metrics that convey the same information can create redundancy, reducing the interpretability of the spider plot. For instance, if both "customer satisfaction" and "net promoter score" are included as axes, they may overlap conceptually and visually, leading to a less distinct plot shape. Instead, select one representative metric for each conceptual dimension to maintain clarity and avoid unnecessary repetition.

Guideline 7: Test for Readability in Prototype Phase. Before finalizing a spider plot, create a prototype and evaluate its readability with a sample audience. Testing with users can reveal whether the chosen metrics and categories are meaningful and if any adjustments are needed. This step is especially valuable in UX contexts where the spider plot will be used by a diverse audience, as user feedback can guide refinements that improve the plot's overall clarity and usability.

By adhering to these guidelines, designers can create spider plots that effectively balance complexity with clarity, enabling users to extract insights without encountering cognitive overload. When meaningful, well-selected metrics are combined with a limited number of distinct categories, spider plots become powerful tools for comparative analysis, empowering users to quickly understand multidimensional data in a cohesive visual format. Thoughtful selection and prioritization of metrics ensure that spider plots remain accessible, interpretable, and aligned with the goals of the data analysis or decision-making process.

7.2.2 Balancing Data and Visual Appeal

Creating a spider plot that is both visually engaging and easy to interpret requires careful attention to balance. A well-designed spider plot effectively combines aesthetic appeal with functional clarity, allowing users to quickly grasp insights without unnecessary visual distractions. However, the challenge lies in making the plot visually appealing while preserving its ability to convey accurate and meaningful data. This section outlines strategies for achieving this balance, including the use of smooth lines, appropriate scaling, and optimized label spacing.

Use Smooth Lines for a Cohesive Look. One of the simplest ways to improve the visual appeal of a spider plot is by using smooth lines instead of harsh, angular edges. This is particularly relevant when plotting continuous data, as smooth curves create a flowing, cohesive appearance that enhances the plot's overall aesthetics. Many visualization tools, such as Python's `Matplotlib` and R's `ggplot2`, allow for line smoothing options that can be applied directly to the spider plot's radial lines. Line thickness and transparency can be adjusted to avoid overwhelming the plot, particularly in multi-series plots. Lighter, semi-transparent lines create a layered effect that allows multiple datasets to coexist without cluttering the visual space.

Select Appropriate Scales to Ensure Interpretability. Scaling is critical in spider plots, as it directly impacts users' perception of each category's relative strength. When the scales of individual axes are not properly aligned, the resulting shape may distort the data, leading to misinterpretation. Standardizing the axes by using a consistent scale across all categories is essential, as it ensures that each axis contributes equally to the overall shape. For cases where different metrics inherently use distinct units or ranges, consider normalizing the data, such as by converting each value to a $0 - 100$ scale. This approach allows each category to be compared on a common basis without compromising data accuracy. Applying proportional scaling ensures that all data points are visible within the plot's bounds, preventing overlap or underutilization of the available space.

Optimize Label Placement for Clarity. Labels play a vital role in guiding users' interpretation of each axis, yet poorly placed labels can detract from the plot's visual appeal. For optimal readability, labels should be well-spaced and strategically positioned to minimize overlap and clutter. Radial labels that fan out from the center often provide a cleaner look, as they prevent overlap near the plot's outer edges. In cases with long labels, consider abbreviations or concise phrasing to maintain clarity. Interactive plots can benefit from hover-over tooltips to display full category names without occupying static space. Designers should also experiment with font sizes and styles that complement the overall visual style, ensuring readability without dominating the plot.

Use a Thoughtful Color Palette to Enhance Visual Appeal. Color selection is essential for both visual engagement and data distinction. A well-chosen color palette can make a spider plot appear more cohesive, especially when using multiple layers or comparing multiple datasets. Monochromatic or analogous color schemes (colors next to each other on the color wheel) are effective for single-series plots, as they create a subtle, harmonious look. In multi-series plots, use contrasting but complementary colors to differentiate between datasets while avoiding excessively vibrant or clashing tones. When transparency is applied to overlapping polygons, colors should be carefully selected to ensure that overlapping areas remain distinguishable and do not visually blend into an ambiguous shade.

Employ Background Gridlines Sparingly. Background gridlines can enhance the readability of spider plots by helping users estimate values along each axis. However, excessive or dark gridlines can distract from the data itself, leading to visual clutter. For a minimalist design, use light, unobtrusive gridlines with wider spacing to provide visual reference without overwhelming the plot. Alternatively, subtle concentric circles, rather than full radial gridlines, can offer a sleek and less obtrusive guide for value estimation.

Use White Space for Visual Balance. White space, or the empty space around the plot, is crucial in creating a balanced and visually appealing design. Crowded spider plots with minimal white space can appear overwhelming, making it challenging for users to focus on individual data points. Adequate white space, both within and around the plot, allows users to view each axis distinctly and perceive the overall plot shape more easily. Expanding margins or increasing plot size can help in distributing elements effectively, creating a clean and inviting visualization.

Minimize Visual Noise by Simplifying Lines and Markers. Too many line styles, markers, or data points can distract from the main message of the plot. For single-series plots, opt for a single line style and avoid additional markers unless they serve a specific purpose, such as highlighting critical thresholds. In multi-series plots, subtle differences in line styles, such as dashed or dotted lines, can effectively distinguish datasets without cluttering the visual. Avoid using overly complex patterns, as these can detract from both readability and aesthetic coherence.

By applying these strategies, designers can achieve a balance between data clarity and visual appeal in spider plots. A well-balanced spider plot enhances user engagement, directing attention to the data's structure and key insights while maintaining a cohesive and aesthetically pleasing visual experience. Ultimately, thoughtful design choices help transform spider plots from simple data representations into effective, user-centered visualizations.

7.3   Choosing the Right Tools for Spider Plots

Selecting an appropriate tool for creating spider plots is essential for achieving both aesthetic quality and functional effectiveness. The right tool depends largely on the intended purpose of the plot, the level of interactivity required, and the audience's needs. Popular tools for creating spider plots include Python's `matplotlib`, R's `ggplot2`, and Plotly. Each offers unique strengths that make it suitable for different scenarios, from static publication-ready visuals to dynamic, interactive charts.

7.3.1   Tool Comparison

Python's `matplotlib`: Matplotlib is one of the most versatile and widely-used libraries in Python for creating static visualizations, including spider plots. Known for its extensive customization options, `matplotlib` allows users to control every aspect of a plot's design, from colors and line styles to axis scaling and label formatting. This level of detail makes it particularly suitable for users who need high-quality static spider plots for reports, publications, or presentations. However, `matplotlib` does not support interactivity natively, making it less ideal for web applications or dynamic user interfaces.

When to Use: Choose `matplotlib` for creating static spider plots that require fine-tuned customization, such as academic publications or detailed analytical reports. It is an excellent choice when precise control over visual elements is essential.

R's `ggplot2`: In R, `ggplot2` is the primary visualization library, known for its "grammar of graphics" approach, which emphasizes creating layered, declarative plots. Although `ggplot2` does not natively support spider plots, users can achieve this effect by using packages like `fmsb`. With `ggplot2`, users can create polished, aesthetically pleasing spider plots that are suitable for academic and business contexts. `ggplot2` is particularly valuable for users who work within R's statistical environment and need to produce high-quality visuals with minimal code. However, `ggplot2` spider plots are static and may require additional packages for full customization.

When to Use: Choose `ggplot2` when creating spider plots as part of a larger data analysis workflow in R. Its ability to produce publication-quality visuals with concise code makes it ideal for users who prioritize aesthetic consistency and are working in a statistical environment.

Plotly: Plotly is a versatile, cross-platform library that supports interactive, web-based visualizations, making it ideal for spider plots intended for dashboards or digital presentations. With Plotly, users can create highly interactive spider plots with features such as tooltips, zooming, and responsive design for various screen sizes. Plotly is accessible both as a Python and R package, which broadens its applicability for data scientists working in either language. Plotly's interactivity enhances the user experience, making it a strong choice for applications that involve exploratory data analysis or audience engagement. However, Plotly requires more computing resources and may be overkill for static visuals.

When to Use: Use Plotly for interactive, web-based spider plots that require user engagement or dynamic exploration. It is well-suited for dashboards, web applications, and presentations where interactivity adds value.

7.3.2 Summary of Tool Comparison

Each tool has its strengths that make it suited to different types of spider plots:

- ➢ **Matplotlib**: Best for static, highly customizable spider plots in Python, ideal for detailed reports and academic visuals.

- ➢ **ggplot2**: Suitable for creating high-quality, static spider plots in R, particularly within statistical workflows.

- ➢ **Plotly**: Excellent for interactive spider plots on web platforms, suitable for dashboards and data exploration.

Choosing the right tool depends on the balance between customization needs, interactivity, and the intended audience (Table 7.2). By selecting the most appropriate platform, users can create spider plots that not only convey the necessary data insights but also engage and serve the intended audience effectively.

Table 7.2: Comparison of Tools for Creating Spider Plots

Tool	Strengths	Best Use Case
Matplotlib (Python)	High customizability, static plots	Ideal for publication-quality spider plots requiring fine-grained control.
ggplot2 + fmsb (R)	Aesthetic visual quality, statistical capabilities	Suitable for R-based data analysis, generating static plots with a polished look.
Plotly (Python/R)	Interactivity, responsive design	Ideal for web-based, interactive spider plots with hover effects and tooltips.

7.3.3 Creating Spider Plots in Python with Matplotlib

Python's matplotlib library is a robust tool for creating static spider plots. Although matplotlib does not have a dedicated function for spider plots, its polar coordinate system can be utilized to achieve a similar effect. This section provides a step-by-step guide for creating a basic spider plot in matplotlib, followed by customization techniques to enhance the visual appeal and clarity of the plot.

Creating a basic spider plot in matplotlib requires a few initial steps to set up the polar coordinate system and plot data points along each axis. The following

Python code demonstrates how to construct a spider plot from scratch, using a sample dataset to illustrate the process.

```python
import numpy as np
import matplotlib.pyplot as plt

# Sample data and labels
labels = ['Metric A', 'Metric B', 'Metric C', 'Metric D', 'Metric E']
values = [4, 3, 2, 5, 4]
num_vars = len(labels)

# Compute angle of each axis
angles = np.linspace(0, 2 * np.pi, num_vars, endpoint=False).tolist()

# The plot is circular, so we append the start to the end
values += values[:1]
angles += angles[:1]

# Initialize the plot in polar coordinates
fig, ax = plt.subplots(figsize=(6, 6), subplot_kw=dict(polar=True))

# Draw the plot
ax.plot(angles, values, linewidth=1, linestyle='solid')
ax.fill(angles, values, color='skyblue', alpha=0.4)

# Add labels for each axis
ax.set_xticks(angles[:-1])
ax.set_xticklabels(labels)

# Display the plot
plt.title('Basic Spider Plot')
plt.show()
```

In this example, we begin by defining a list of labels and corresponding values for each axis. Next, we calculate the angles for each axis using `numpy.linspace`, ensuring the angles are evenly spaced around the circle. We then initialize the plot with a polar coordinate system, which allows the plot to radiate from a central point. Finally, the values are plotted with a solid line, and the area inside the plot is filled with a semi-transparent color to enhance visual clarity. This simple spider plot provides a foundation upon which more advanced customizations can be built.

Customizing the spider plot enhances its readability and aesthetic appeal, particularly when comparing multiple datasets. `Matplotlib` offers several options for adjusting colors, line styles, and transparency to make each dataset distinct. In multi-dataset plots, these elements are essential for helping users quickly differentiate between overlapping polygons.

Adding Colors and Transparency. Adding colors to spider plots helps convey data distinctions and creates visual interest. `Matplotlib` supports various color formats, including named colors (e.g., 'red'), hex codes (e.g., '#FF5733'), and color maps for gradient effects. Transparency can be adjusted using the `alpha` parameter, which ranges from 0 (fully transparent) to 1 (fully opaque). Transparency is particularly useful when displaying multiple datasets, as it allows overlapping areas to remain visible.

```python
import numpy as np
import matplotlib.pyplot as plt

# Sample data for two datasets
labels = ['Metric A', 'Metric B', 'Metric C', 'Metric D', 'Metric E']
values1 = [4, 3, 2, 5, 4]
values2 = [2, 4, 5, 3, 4]
num_vars = len(labels)

# Compute angle of each axis
angles = np.linspace(0, 2 * np.pi, num_vars, endpoint=False).tolist()

# The plot is circular, so we append the start to the end
values1 += values1[:1]
values2 += values2[:1]
angles += angles[:1]

# Initialize the plot in polar coordinates
fig, ax = plt.subplots(figsize=(6, 6), subplot_kw=dict(polar=True))

# Plot each dataset with unique color and transparency
ax.plot(angles, values1, linewidth=1, linestyle='solid', color='blue')
ax.fill(angles, values1, color='blue', alpha=0.2)

ax.plot(angles, values2, linewidth=1, linestyle='solid', color='orange')
ax.fill(angles, values2, color='orange', alpha=0.3)

# Add labels for each axis
ax.set_xticks(angles[:-1])
ax.set_xticklabels(labels)

# Display the plot
plt.title('Customized Spider Plot with Color and Transparency')
plt.show()
```

In this example, two datasets are plotted on the same spider plot using different colors. The `alpha` parameter is adjusted to add transparency, enabling both datasets to be visible even where they overlap. The combination of color and transparency effectively differentiates the datasets, facilitating comparison without overcrowding the visual.

Adjusting Line Styles and Marker Types. Line styles and markers are valuable for enhancing clarity, especially in multi-dataset spider plots. `Matplotlib` supports various line styles (e.g., solid, dashed) and markers (e.g., circles, triangles), which can be applied to distinguish different data series. For example, a solid line might represent one dataset while a dashed line represents another, with distinct markers indicating data points.

```
1  # Additional customization with line styles and markers
2  ax.plot(angles, values1, linewidth=1, linestyle='dashed', color='blue',
   ↪  marker='o')
3  ax.fill(angles, values1, color='blue', alpha=0.2)
4
5  ax.plot(angles, values2, linewidth=1, linestyle='solid', color='orange',
   ↪  marker='s')
6  ax.fill(angles, values2, color='orange', alpha=0.3)
```

This additional customization allows the line style and marker type to further differentiate datasets, creating a more readable spider plot with enhanced interpretability.

Optimizing Axis Labels and Tick Marks. Well-placed axis labels and tick marks are essential for usability, as they guide users in interpreting each dimension of the plot. `Matplotlib` allows users to customize label positions, font sizes, and spacing to ensure clarity. Labels can be rotated or abbreviated if space is limited, and tick marks can be adjusted to provide clear scaling for each axis (Table 7.3).

By applying these color, style, and label customizations, designers can transform a basic spider plot into a visually engaging and highly functional visualization. Customization enhances user experience by making data distinctions clearer and facilitating comparisons, while maintaining an organized and aesthetically pleasing presentation.

Table 7.3: Scaling and Normalization Techniques for Spider Plots

Technique	Description and Use Case
Min-Max Normalization	Scales each variable to a 0-1 range by subtracting the minimum value and dividing by the range (max - min). Useful for data with different units to ensure comparability.
Z-Score Normalization	Standardizes data by subtracting the mean and dividing by the standard deviation, resulting in a mean of 0 and standard deviation of 1. Best for variables with different distributions.
Fixed Range Scaling	Adjusts all data to a set range, such as $0 - 100$, by dividing each value by the maximum possible score. Effective for data with known upper limits.
Logarithmic Scaling	Applies a logarithmic transformation to compress large value ranges. Useful for skewed data or data with large outliers .
Percentage Scaling	Converts values into percentages of a maximum score or sum, making each variable directly comparable as a proportion of 100%. Suitable for balanced and consistent categories.

7.3.4   Enhancing Spider Plots with Interactivity Using `Plotly`

The `Plotly` library in Python is an exceptional tool for creating interactive data visualizations, including spider plots (or radar charts). Unlike static visualizations,

`Plotly` enables dynamic interactions such as hover effects, tooltips, and responsive scaling, which enhance user engagement and facilitate deeper data exploration. This section provides a step-by-step guide for creating interactive spider plots using `Plotly`, along with techniques for incorporating interactivity to improve the user experience in web-based applications.

Creating an interactive spider plot in `Plotly` is straightforward and requires only a few steps to set up the data and define the plot's appearance. The following Python code example demonstrates how to build a basic spider plot with `Plotly`, using sample data to illustrate the process.

```python
import plotly.express as px
import pandas as pd

# Sample data for the spider plot
categories = ['Metric A', 'Metric B', 'Metric C', 'Metric D', 'Metric E']
values = [4, 3, 2, 5, 4]

# Create a DataFrame with repeated first value to close the plot
df = pd.DataFrame(dict(r=values + [values[0]], theta=categories +
    [categories[0]]))

# Create the spider plot using Plotly \index{Plotly} Express
fig = px.line_polar(df, r='r', theta='theta', line_close=True,
    title='Interactive Spider Plot')
fig.update_traces(fill='toself', fillcolor='skyblue', opacity=0.6)

# Display the plot
fig.show()
```

In this example, we first define the categories (labels) and values for each axis. The data is then structured into a DataFrame, with an additional value added to each list to close the circular shape of the spider plot. Using `px.line_polar` in `Plotly` `Express`, we plot the data in a polar coordinate system and apply basic styling with `fill` to color the area inside the polygon. The interactive plot allows users to hover over each point to view its value, providing a more engaging and informative experience.

A strength of `Plotly` is its built-in interactivity, which can be used to add features like hover effects and tooltips. These interactive elements are especially useful in web-based applications or dashboards, where they enable users to explore data in greater depth without additional visual clutter. The following example demonstrates how to create a multi-series interactive spider plot with tooltips for a more detailed data comparison.

```python
import plotly.graph_objects as go

# Define categories and values for two datasets
categories = ['Metric A', 'Metric B', 'Metric C', 'Metric D', 'Metric E']
```

```python
values1 = [4, 3, 2, 5, 4]
values2 = [2, 4, 5, 3, 4]

# Add the first value to close the plot
values1 += [values1[0]]
values2 += [values2[0]]
categories += [categories[0]]

# Create the spider plot using Plotly Graph Objects
fig = go.Fig.()

# Add the first dataset with hover information
fig.add_trace(go.Scatterpolar(
    r=values1,
    theta=categories,
    fill='toself',
    name='Dataset 1',
    hoverinfo='text',
    text=['Metric A: 4', 'Metric B: 3', 'Metric C: 2', 'Metric D: 5',
    ↪    'Metric E: 4', 'Metric A: 4'],
    line=dict(color='blue')))

# Add the second dataset with hover information
fig.add_trace(go.Scatterpolar(
    r=values2,
    theta=categories,
    fill='toself',
    name='Dataset 2',
    hoverinfo='text',
    text=['Metric A: 2', 'Metric B: 4', 'Metric C: 5', 'Metric D: 3',
    ↪    'Metric E: 4', 'Metric A: 2'],
    line=dict(color='orange')))

# Update plot layout for better presentation
fig.update_layout(
    polar=dict(radialaxis=dict(visible=True, range=[0, 5])),
    title='Interactive Multi-Series Spider Plot',
    showlegend=True)

# Display the interactive plot
fig.show()
```

In this example, we use `Plotly Graph Objects` to create a multi-series spider plot with two datasets. Each dataset is added as a separate trace using `go.Scatterpolar`, allowing for distinct coloring and styling. The `hoverinfo` parameter is set to display text upon hovering, and we provide custom tooltips with descriptive labels for each data point. This interactive feature (Table 7.4) enables users to view precise metric values by hovering over each point, enhancing the usability of the spider plot for comparative analysis.

Customizing Tooltips and Hover Effects. To maximize the effectiveness of tooltips, `Plotly` allows customization of the hover text format. Using the `text`

Table 7.4: Interactive Features for Spider Plots in `Plotly`

Feature	Description and Use Case
Hover Tooltips	Display detailed information about each data point when hovered over, useful for complex plots with multiple datasets.
Dynamic Legends	Allow users to toggle datasets on or off by clicking legend entries, helping manage visual clutter on multi-layered plots.
Zoom and Pan	Enable users to zoom into specific areas and pan across the plot, improving navigation in dense or high-dimensional spider plots.
Customizable Colors and Transparency	Adjust colors and transparency dynamically to enhance readability and compare overlapping data series.
Responsive Layout	Ensures that the plot adapts to various screen sizes, enhancing usability on mobile devices or embedded in web applications.

attribute, users can specify detailed labels for each data point, making it easy to convey context or additional information about each metric. The hover effect, combined with interactive legends, helps users navigate complex datasets without overwhelming the visual space.

Responsive Design for Web Applications. Spider plots created in `Plotly` are inherently responsive, making them ideal for web-based applications where they may need to adapt to various screen sizes. For optimal usability on mobile devices, interactive plots can adjust their layout based on the available screen real estate, ensuring that elements remain visible and accessible across devices.

Adjusting Aesthetic Elements. `Plotly` provides numerous customization options to adjust the visual appearance of spider plots, such as colors, transparency levels, and line thickness. By fine-tuning these elements, users can create spider plots that are both visually appealing and functionally clear, with a focus on delivering an optimal user experience in interactive settings.

`Plotly` enhances spider plots with a range of interactive features that facilitate data exploration and engagement. The ability to add hover effects, tooltips, and responsive layouts makes `Plotly` a powerful choice for applications where interactivity is essential. These features not only improve the user experience but also provide a richer, more informative visualization that supports deeper insights.

7.3.5 Customizing Spider Plots in R with `ggplot2` and `fmsb`

In R, the `fmsb` package provides a convenient way to create spider plots (or radar charts) with `ggplot2`-based styling. While `ggplot2` itself does not natively support spider plots, `fmsb` extends its capabilities to offer an interface for generating visually appealing and customizable spider plots. This section introduces the `fmsb` package and provides a step-by-step guide to creating and customizing spider plots in R.

The `fmsb` package is designed to facilitate the creation of spider plots in R. It provides functions for plotting radar charts with easy customization options for colors, line styles, axis labels, and more. This package is particularly useful for R users looking to create publication-quality spider plots within the R environment. With fmsb, users can quickly define data, specify ranges, and apply color schemes to highlight multiple datasets effectively.

To get started with fmsb, users should install the package if it is not already available in their R environment. The following command installs the package:

```r
# Install the fmsb package if not already installed
install.packages("fmsb")
```

Once installed, fmsb provides functions for constructing and customizing spider plots. Users can specify the maximum and minimum values for each axis to ensure consistency across metrics, making it easier to compare multiple datasets. The package also supports color customization for both the polygon area and the borders, enhancing readability and visual appeal.

The following guide demonstrates how to create a basic spider plot using fmsb, along with customization techniques to improve clarity and aesthetics.

```r
# Load the fmsb package
library(fmsb)

# Define the data for the spider plot
data <- data.frame(
  Metric_A = c(5, 4, 3),
  Metric_B = c(2, 3, 4),
  Metric_C = c(4, 5, 3),
  Metric_D = c(4, 2, 5),
  Metric_E = c(3, 3, 4)
)

# Define the max and min values for each axis
data <- rbind(rep(5, 5), rep(1, 5), data)

# Create the basic spider plot
radarchart(data,
           axistype=1,
           title="Basic Spider Plot with fmsb",
           pcol=c("blue", "red", "green"),
           pfcol=c(rgb(0.1,0.4,0.8,0.3),
           rgb(0.8,0.2,0.2,0.3),
           rgb(0.2,0.6,0.3,0.3)),
           plwd=2,
           cglcol="grey", cglty=1, axislabcol="black",
           cglwd=0.8,
           vlcex=0.8)
```

In this example, we first define a data frame containing the metrics for each category (in this case, Metric_A through Metric_E). Each row in the data frame corresponds to a different dataset, and two additional rows are added at the beginning to specify the maximum and minimum values for each axis. The `radarchart` function from the `fmsb` package generates the spider plot, with options to customize the color and transparency of the polygons.

Customizing Colors and Transparency. Custom colors and transparency levels can be specified using the `pcol` and `pfcol` parameters. In the above example, each dataset has a unique color, with transparency adjusted through the `rgb()` function. This approach allows overlapping datasets to remain visible, making it easy to distinguish between them without excessive visual clutter.

Adjusting Axis and Gridline Styles. The `fmsb` package offers options to customize axis and gridline styles, enhancing readability and organization. The `axistype` parameter specifies how the axes are displayed (e.g., radial or circular), while the `cglcol` and `cglty` parameters control the color and line type of the gridlines. These adjustments can improve clarity, especially in complex spider plots with multiple datasets.

Adding Labels and Titles. Axis labels and plot titles can be modified to provide additional context. The `vlcex` parameter controls the size of the axis labels, ensuring that they are readable without overwhelming the plot. Titles can be added using the `title` parameter, allowing users to give context to the visualization.

```r
# Enhanced spider plot with custom labels and gridline styles
radarchart(data,
           axistype=1,
           title="Customized Spider Plot with fmsb",
           pcol=c("darkblue", "darkred", "darkgreen"),
           pfcol=c(rgb(0.1,0.4,0.8,0.2),
           rgb(0.8,0.2,0.2,0.2),
           rgb(0.2,0.6,0.3,0.2)),
           plwd=2,
           cglcol="lightgrey", cglty=1, axislabcol="black",
           cglwd=0.8,
           vlcex=0.8)
```

In this enhanced example, we apply darker colors for each polygon outline and reduce the transparency slightly to create a more professional look. Gridlines are adjusted to a lighter grey, and labels are resized to optimize readability. Such customizations provide greater control over the plot's appearance, making it suitable for presentations or publication.

Using Multiple Datasets. The `fmsb` package supports multiple datasets within a single spider plot, allowing users to compare different groups or entities across categories. By specifying unique colors for each dataset and adjusting transparency, the visualization remains clear and engaging even with overlapping polygons.

By following these steps, R users can create spider plots that are both visually appealing and informative. The `fmsb` package offers a range of customization options to tailor each plot to specific analytical needs, making it a valuable tool for creating effective spider plots within the R environment.

7.4   Advanced Design Techniques for Spider Plots

To elevate the readability and visual appeal of spider plots, advanced design techniques can be used to enrich the viewer's experience. Layering multiple datasets and incorporating strategic annotations can transform basic plots into powerful comparative tools, enhancing their ability to communicate insights effectively. This section explores how to apply layering and shading for comparative analysis and offers best practices for adding annotations and callouts to highlight key data points.

7.4.1   Layering and Comparative Visuals

Layering multiple datasets within a single spider plot is an effective technique for comparative analysis. This approach allows viewers to examine differences and overlaps across various groups or entities within a unified visual. Using distinct colors, line styles, and shading for each dataset enhances the plot's readability, while transparency ensures that overlapping areas remain visible, facilitating direct comparisons.

Example of Layering with Shading. In Python's `matplotlib`, the `fill` option allows users to add shading to polygons representing different datasets, with adjustable `alpha` values for transparency. The following example demonstrates how to layer two datasets within the same spider plot, each with unique shading and line colors to aid visual differentiation.

```python
import numpy as np
import matplotlib.pyplot as plt

# Define categories and values for two datasets
labels = ['Metric A', 'Metric B', 'Metric C', 'Metric D', 'Metric E']
values1 = [4, 3, 2, 5, 4]
values2 = [2, 4, 5, 3, 4]

# Add the first value to close the plot
values1 += [values1[0]]
values2 += [values2[0]]
angles = np.linspace(0, 2 * np.pi, len(values1), endpoint=False).tolist()

# Initialize the plot in polar coordinates
fig, ax = plt.subplots(figsize=(6, 6), subplot_kw=dict(polar=True))

# Plot each dataset with unique color and transparency
ax.plot(angles, values1, linewidth=2, linestyle='solid', color='blue',
    label='Dataset 1')
ax.fill(angles, values1, color='blue', alpha=0.2)

```

```python
21  ax.plot(angles, values2, linewidth=2, linestyle='dashed', color='orange',
    ↪  label='Dataset 2')
22  ax.fill(angles, values2, color='orange', alpha=0.3)
23
24  # Add labels for each axis
25  ax.set_xticks(angles[:-1])
26  ax.set_xticklabels(labels)
27
28  # Add a legend and title
29  plt.legend(loc='upper right', bbox_to_anchor=(1.1, 1.1))
30  plt.title('Layered Spider Plot with Shading')
31  plt.show()
```

In this example, two datasets are plotted using different colors and line styles, with one dataset represented by a solid line and the other by a dashed line. Transparency (alpha) is applied to each dataset's shaded area, allowing the viewer to easily identify overlaps and unique areas within each dataset. This layered approach highlights comparative differences across metrics and enables users to assess relative performance or distribution effectively.

Using Contrasting Colors for Better Comparisons. When layering datasets, using contrasting but complementary colors enhances the distinction between each layer. For instance, choosing colors from opposite ends of a color wheel (such as blue and orange) provides a clear visual separation that aids comparative analysis. It's important to select colors that are accessible to viewers with color vision deficiencies, as this ensures the plot remains informative for a broader audience.

Adjusting Line Styles and Thickness for Clarity. Varying line styles (e.g., solid vs. dashed) and thickness can further differentiate each dataset, particularly in cases where color alone may not be sufficient. Thicker lines are effective for emphasizing primary datasets, while thinner lines can represent secondary datasets without overpowering the visual.

7.4.2 Adding Annotation and Callouts

Annotations and callouts are essential tools for directing attention to specific areas of a spider plot, particularly in complex visualizations. Strategic annotations enhance the plot's narrative by highlighting key data points, unusual trends, or critical thresholds, making it easier for viewers to interpret the data.

Best Practices for Adding Annotations. Effective annotations are concise, relevant, and positioned to avoid overlapping with key data areas. Each annotation should provide context without distracting from the plot's overall layout. In spider plots, annotations can be placed directly near the data point they describe or, alternatively, connected via lines (callouts) to reduce visual clutter.

Example of Annotating Key Data Points. The following example demonstrates how to add annotations to highlight specific metrics within a spider plot, using matplotlib to add text and arrows for emphasis.

```python
import numpy as np
import matplotlib.pyplot as plt

# Define categories and values
labels = ['Metric A', 'Metric B', 'Metric C', 'Metric D', 'Metric E']
values = [4, 3, 2, 5, 4]
values += [values[0]]
angles = np.linspace(0, 2 * np.pi, len(values), endpoint=False).tolist()

# Initialize the plot in polar coordinates
fig, ax = plt.subplots(figsize=(6, 6), subplot_kw=dict(polar=True))

# Plot the data with shading
ax.plot(angles, values, linewidth=2, linestyle='solid', color='purple')
ax.fill(angles, values, color='purple', alpha=0.2)

# Add labels for each axis
ax.set_xticks(angles[:-1])
ax.set_xticklabels(labels)

# Add annotations for key data points
ax.annotate('Peak', xy=(angles[3], values[3]), xytext=(angles[3]+0.1,
    values[3]+0.5),
            arrowprops=dict(facecolor='black', arrowstyle="->"),
            fontsize=10, color='black')
ax.annotate('Low Point', xy=(angles[2], values[2]), xytext=(angles[2]-0.2,
    values[2]+0.5),
            arrowprops=dict(facecolor='black', arrowstyle="->"),
            fontsize=10, color='black')

# Add a title
plt.title('Annotated Spider Plot')
plt.show()
```

In this example, annotations (Table 7.5) are used to highlight the highest and lowest data points within the plot. Arrows, created with `arrowprops`, point directly to the relevant data points, while labels such as "Peak" and "Low Point" provide context. This approach directs the viewer's attention to significant metrics, enabling a more focused interpretation of the plot.

Adding Callouts for Key Areas. Callouts are especially useful in spider plots with dense or overlapping data, where direct annotations might clutter the visualization. Callouts can be connected to their respective data points using lines, keeping the labels outside the primary plot area. This method is particularly helpful in presentations, as it maintains clarity without detracting from the overall visual.

Using Color and Font for Emphasis. Choosing the right font size, style, and color for annotations is critical. Text should be readable without overshadowing other

Table 7.5: Best Practices for Adding Annotations and Callouts in Spider Plots

Practice	Description and Use Case
Concise Text	Keep annotations short and to the point. Long text can clutter the plot; limit annotations to key insights only.
Use Arrows and Callouts	Use arrows to direct viewers' attention to specific data points, especially on complex or multi-layered plots. Callouts keep labels out of the main plot area.
Consistent Font and Color	Ensure font size and color for annotations match the plot's style for readability. Use bold or italic fonts sparingly for emphasis.
Position Labels Outside Plot	For crowded plots, position labels outside the radial plot area, connected by lines to the relevant data points. This minimizes overlap and maintains clarity.
Highlight Significant Differences	Focus annotations on metrics with notable differences or outliers , as this provides viewers with a quick understanding of critical insights.

elements, and colors should complement the plot's palette. Avoid using excessively bright or large text, as it can detract from the data rather than support it.

7.5 Accessibility Considerations

Ensuring that spider plots are accessible to all users, including those with visual impairments or color vision deficiencies, is essential for effective data visualization. Accessibility not only makes the visualization more inclusive but also enhances readability for all users by applying clear, well-designed elements. This section addresses key accessibility considerations, focusing on color selection for color-blind users, readability adjustments, and optimal label placement to avoid clutter.

7.5.1 Color and Readability

Color plays a crucial role in making spider plots both engaging and accessible. However, reliance on color alone can pose challenges for individuals with color vision deficiencies, such as red-green color blindness, the most common form of color blindness. Ensuring that spider plots remain distinguishable and clear for all users requires thoughtful color selection and attention to readability elements like line width.

Choosing Color-Blind Friendly Palettes. When designing spider plots, use color palettes that are accessible to color-blind users. Color-blind friendly palettes, such as those provided by `viridis`, `plasma`, and `cividis` color maps in Python, are designed to be easily distinguishable by all users, regardless of color perception. These palettes offer a range of colors with high contrast, avoiding problematic hues such as red and green, which are often indistinguishable to color-blind individuals. Using monochromatic schemes (shades of a single color) can provide distinctiveness when combined with varying levels of transparency, enhancing accessibility without overwhelming the viewer.

```python
import numpy as np
import matplotlib.pyplot as plt

# Define data
labels = ['Metric A', 'Metric B', 'Metric C', 'Metric D', 'Metric E']
values = [4, 3, 2, 5, 4]
values += [values[0]]
angles = np.linspace(0, 2 * np.pi, len(values), endpoint=False).tolist()

# Initialize the plot in polar coordinates
fig, ax = plt.subplots(figsize=(6, 6), subplot_kw=dict(polar=True))

# Plot with a color-blind\index{Color-Blind Friendly Palette} friendly
#   ↪ color
ax.plot(angles, values, linewidth=2, linestyle='solid',
    ↪ color=plt.cm.viridis(0.7))
ax.fill(angles, values, color=plt.cm.viridis(0.7), alpha=0.4)

# Add labels and title
ax.set_xticks(angles[:-1])
ax.set_xticklabels(labels)
plt.title('Color-Blind Friendly Spider Plot')
plt.show()
```

In this example, a color-blind friendly color is chosen from the `viridis` color map, enhancing accessibility without sacrificing visual appeal. The use of a consistent, easily distinguishable color scheme ensures that all viewers can accurately interpret the data.

Table 7.6: Color-Blind Friendly Palettes for Spider Plots

Palette	Description
Viridis	A perceptually uniform palette transitioning from dark blue to yellow, ideal for color-blind accessibility.
Plasma	A vibrant, high-contrast palette transitioning from purple to yellow, useful for visibility on complex plots.
Cividis	A low-contrast, color-blind friendly palette moving from blue to yellow, reducing visual strain.
Inferno	A high-contrast palette from dark purple to bright yellow, suitable for highlighting extreme values.
Magma	A palette transitioning from dark purple to orange-yellow, providing a warm, high-contrast gradient.

Adjusting Line Widths for Readability. Line width is an important aspect of accessibility, as thicker lines improve readability, especially on small screens or when viewed by individuals with reduced visual acuity. In spider plots, consider using slightly thicker lines (e.g., 1.5 – 2 pixels) for primary data, as this helps each dataset stand out more clearly. For multi-layered spider plots, differentiate lines with varied thicknesses, thicker lines for primary datasets and thinner lines for secondary datasets, to maintain hierarchy without creating visual clutter.

Using Patterns and Textures for Distinction. For users who may not benefit fully from color distinctions, adding patterns or textures to filled areas within a spider plot can improve interpretability. Simple patterns like stripes, dots, or cross-hatching provide an alternative way of distinguishing between datasets, especially in monochromatic designs. This is particularly valuable in grayscale printing or where colors may be misinterpreted.

7.5.2 Label Placement and Sizing

Clear, well-placed labels are essential for making spider plots easy to read and interpret. Poorly positioned or overlapping labels can create confusion, detracting from the plot's effectiveness. To enhance label clarity, it is important to optimize label placement and size, especially in spider plots with multiple datasets or a large number of axes.

> **♀ Design Tip**
>
> Limit the number of axes in your spider plot to between 5 and 10. Too many dimensions lead to visual clutter and reduced interpretability.

Optimizing Label Placement to Prevent Overlap. Labels should be placed in positions that avoid overlap with data lines or other labels. In polar coordinate systems, placing labels slightly outside the plot's main area reduces the chance of overlap, especially when data points are close to the axis edges. Rotating labels to align with the radial axes can also improve readability, as it follows the natural direction of each axis. For crowded plots, use abbreviated labels or consider placing a legend outside the plot area to describe each metric.

Example of Optimized Label Placement. The following code example demonstrates how to adjust label placement in `matplotlib`, positioning each label outside the plot area to enhance readability.

```python
import numpy as np
import matplotlib.pyplot as plt

# Define data and labels
labels = ['Metric A', 'Metric B', 'Metric C', 'Metric D', 'Metric E']
values = [4, 3, 2, 5, 4]
values += [values[0]]
angles = np.linspace(0, 2 * np.pi, len(values), endpoint=False).tolist()
```

```python
 9
10  # Initialize the plot in polar coordinates
11  fig, ax = plt.subplots(figsize=(6, 6), subplot_kw=dict(polar=True))
12
13  # Plot the data
14  ax.plot(angles, values, linewidth=2, linestyle='solid', color='purple')
15  ax.fill(angles, values, color='purple', alpha=0.2)
16
17  # Add labels slightly outside the plot area for clarity
18  ax.set_xticks(angles[:-1])
19  ax.set_xticklabels(labels, fontsize=10, ha='center', va='center')
20  for label, angle in zip(ax.get_xticklabels(), angles):
21      label.set_horizontalalignment('right' if angle < np.pi else 'left')
22
23  # Add title
24  plt.title('Spider Plot with Optimized Label Placement')
25  plt.show()
```

This example positions labels outside the primary plot area, ensuring that they remain readable without overlapping with the data lines. Adjusting label alignment based on their position along the plot's circumference further improves legibility.

Sizing Labels for Readability Across Devices. Label font size is another factor that significantly impacts readability. Labels should be large enough to be easily readable on different devices, from desktops to mobile screens. A good starting point is a font size of 10 – 12 points, but this may vary depending on the overall plot size and presentation format. Avoid bold or highly stylized fonts, as these can reduce readability in complex visualizations.

Minimizing Label Overload with Tooltips or Legends. For spider plots with numerous axes or overlapping datasets, labels can quickly overcrowd the plot. To maintain clarity, consider using a legend or interactive tooltips in digital contexts, where additional information can be displayed on hover. This approach reduces visual clutter while still providing access to detailed information.

Accessibility in spider plots is achieved through thoughtful color selection, line width adjustments, and optimized label placement. By implementing color-blind friendly palettes, readable line widths, and strategic label positioning, designers can create spider plots that are clear, inclusive, and suitable for a wide range of users. These accessibility considerations not only improve the user experience but also ensure that spider plots remain effective tools for data communication across diverse audiences.

7.6   Troubleshooting Spider Plot Design

Despite the advantages of spider plots for visualizing multidimensional data, certain design challenges can hinder their effectiveness. Overcrowding and scaling issues are common problems in spider plot design, especially when dealing with high-dimensional data. This section provides practical tips for managing overcrowding and discusses techniques for scaling and normalizing data to enhance the clarity and interpretability of spider plots.

7.6.1 Overcrowding

Overcrowding occurs when too many variables or datasets are included in a single spider plot, causing the plot to appear cluttered and difficult to read. Excessive overlap of lines, labels, and shaded areas can obscure key insights, making it challenging for viewers to distinguish between categories. To address overcrowding, several techniques can be employed to simplify the plot and improve readability.

Limit the Number of Categories. One of the simplest ways to prevent overcrowding is to limit the number of categories (axes) in the spider plot. A general recommendation is to include between 5 and 10 categories for optimal readability. When there is a need to represent more variables, consider grouping related metrics into broader categories or creating multiple spider plots to display different dimensions of the data. This approach maintains clarity while still presenting comprehensive insights.

Use Multiple Spider Plots for Complex Comparisons. When comparing multiple datasets, it may be more effective to use separate spider plots rather than layering all datasets in a single plot. Separate plots can display each dataset individually, reducing visual clutter and allowing for more focused comparisons. This method works particularly well in cases where each dataset has distinct characteristics that need to be examined without interference from overlapping visuals.

Vary Line Styles and Colors for Distinction. For spider plots that require multiple datasets within a single plot, using distinct line styles (e.g., solid, dashed, or dotted lines) and color schemes helps differentiate each dataset. Applying transparency to shaded areas also minimizes visual overlap, allowing viewers to see underlying structures without excessive visual interference.

Simplify Labels and Legends. Clear, concise labeling is essential for preventing overcrowding. Use short, descriptive labels, and avoid adding extraneous information that might increase visual density. For complex plots, consider moving legends outside the primary plot area, where they remain accessible without crowding the central plot. Legends with distinct color and line style representations for each dataset can improve clarity, especially in high-dimensional spider plots.

```python
import numpy as np
import matplotlib.pyplot as plt

# Define categories and two datasets with distinct line styles
labels = ['Metric A', 'Metric B', 'Metric C', 'Metric D', 'Metric E']
values1 = [4, 3, 2, 5, 4]
values2 = [2, 4, 5, 3, 4]
values1 += [values1[0]]
values2 += [values2[0]]
angles = np.linspace(0, 2 * np.pi, len(values1), endpoint=False).tolist()

# Initialize the plot in polar coordinates
```

```python
13  fig, ax = plt.subplots(figsize=(6, 6), subplot_kw=dict(polar=True))
14
15  # Plot with distinct line styles and colors to reduce clutter
16  ax.plot(angles, values1, linewidth=1.5, linestyle='solid', color='blue',
    ↪ label='Dataset 1')
17  ax.fill(angles, values1, color='blue', alpha=0.2)
18  ax.plot(angles, values2, linewidth=1.5, linestyle='dashed', color='orange',
    ↪ label='Dataset 2')
19  ax.fill(angles, values2, color='orange', alpha=0.2)
20
21  # Add labels and legend outside the main plot area
22  ax.set_xticks(angles[:-1])
23  ax.set_xticklabels(labels)
24  plt.legend(loc='upper right', bbox_to_anchor=(1.3, 1.1))
25  plt.title('Spider Plot with Reduced Overcrowding')
26  plt.show()
```

In this example, two datasets are represented with distinct line styles and colors. The legend is positioned outside the main plot area to reduce clutter, enhancing the plot's readability.

7.6.2 Scaling and Normalization

Scaling and normalization are essential techniques for spider plots, as they help ensure that data is displayed proportionally and that small differences between categories remain visible. Without proper scaling, data points near the plot's center may appear minimized, while those near the edges may be overemphasized. This distortion can mislead viewers, obscuring true patterns in the data.

Normalize Data to a Common Scale. When variables are measured on different scales, normalization is crucial for achieving a fair comparison. Normalizing data to a common scale, such as 0 to 1 or 0 to 100, ensures that each axis contributes equally to the overall shape of the plot. This can be achieved by dividing each data value by the maximum value within its category, resulting in standardized values that fit uniformly within the plot.

> **♀ Design Tip**
>
> Normalize your data to a common scale (e.g., 0–1 or 0–100) before plotting to ensure all axes contribute equally to the polygon shape.

```python
1  # Sample data with different scales
2  raw_values = {
3      'Metric A': [4, 3],
4      'Metric B': [60, 80],   # Higher scale than other metrics
5      'Metric C': [3, 5],
6      'Metric D': [10, 15],   # Higher scale than other metrics
7      'Metric E': [4, 4]
8  }
```

```python
9
10  # Normalize data to a 0-1 scale
11  normalized_values = {key: [val / max(raw_values[key]) for val in
    ↪   raw_values[key]] for key in raw_values}
12  values1 = [normalized_values[key][0] for key in raw_values]
13  values2 = [normalized_values[key][1] for key in raw_values]
14
15  # Repeat the first value to close the plot
16  values1 += [values1[0]]
17  values2 += [values2[0]]
18  angles = np.linspace(0, 2 * np.pi, len(values1), endpoint=False).tolist()
19
20  # Initialize the plot in polar coordinates
21  fig, ax = plt.subplots(figsize=(6, 6), subplot_kw=dict(polar=True))
22
23  # Plot the normalized data
24  ax.plot(angles, values1, linewidth=1.5, linestyle='solid', color='blue',
    ↪   label='Dataset 1')
25  ax.fill(angles, values1, color='blue', alpha=0.2)
26  ax.plot(angles, values2, linewidth=1.5, linestyle='dashed', color='orange',
    ↪   label='Dataset 2')
27  ax.fill(angles, values2, color='orange', alpha=0.2)
28
29  # Add labels and title
30  ax.set_xticks(angles[:-1])
31  ax.set_xticklabels(list(raw_values.keys()))
32  plt.legend(loc='upper right', bbox_to_anchor=(1.3, 1.1))
33  plt.title('Normalized Spider Plot for Consistent Scaling')
34  plt.show()
```

In this example, each metric is normalized to a $0-1$ scale, ensuring that variations between datasets are accurately represented without distortion due to differing scales.

Adjusting Axis Ranges for Optimal Display. To prevent overemphasis or minimization of certain data points, consider setting axis ranges that suit the dataset. For example, use radial axis limits that capture the full range of the data without extending beyond necessary bounds. This can be especially important in multi-layered spider plots, where scaling inconsistencies may obscure differences between datasets.

> **⚠ Warning**
>
> Inconsistent scaling across axes will distort your spider plot and lead to misleading interpretations. Always normalize or standardize your data.

Use Concentric Circles for Enhanced Proportionality. Concentric circles within the spider plot can provide a visual reference for interpreting relative values along each axis. This technique helps viewers gauge distances from the center, adding a layer of proportionality that clarifies how each data point relates to others within the plot.

> **💡 Design Tip**
>
> Use concentric gridlines or radial labels sparingly and ensure they support, not distract from, the core data visualization.

By applying these scaling and normalization techniques, designers can improve the accuracy and readability of spider plots, ensuring that each axis and dataset is represented in a way that conveys meaningful, undistorted insights. Effective scaling enhances the interpretability of spider plots, making them more reliable tools for data comparison.

7.7 Common Mistakes in Spider Plot Design

While spider plots are valuable tools for visualizing multidimensional data, certain common mistakes can reduce their effectiveness and clarity. Understanding these pitfalls allows designers to avoid them, creating spider plots that are not only visually appealing but also accessible and informative. This section highlights frequent mistakes encountered in spider plot design and offers guidance on how to address them.

7.7.1 Overcrowding with Too Many Categories

Including too many categories in a single spider plot is a common mistake that results in a cluttered, hard-to-read visualization. When too many axes are present, the plot may become visually overwhelming, and individual data points can be difficult to interpret. To avoid overcrowding, limit the number of categories to between 5 and 10. If additional metrics are necessary, consider creating separate spider plots for different data dimensions or grouping related metrics into broader categories.

> **⚠ Warning**
>
> Avoid including too many categories in a single spider plot. More than 10 axes can result in severe visual clutter and hinder readability.

7.7.2 Inconsistent Scaling of Axes

Inconsistent scaling across axes can lead to a distorted view of the data, where certain categories appear disproportionately prominent or minimized. This misalignment occurs when each axis has different minimum and maximum values, causing the shape of the spider plot to misrepresent relative values. To prevent this, normalize all data to a common scale, such as $0 - 1$ or $0 - 100$, ensuring proportionality across categories. Standardized scaling helps convey accurate comparisons, enabling users to interpret the data correctly.

7.7.3 Using Inaccessible Color Schemes

Relying solely on color distinctions without considering accessibility can render spider plots difficult to interpret for color-blind users or viewers with low contrast sensitivity. Choosing colors that are indistinguishable to certain viewers, such as red and green, limits accessibility. Instead, use color-blind friendly palettes (e.g., `viridis` or `plasma`) and apply varying line styles or patterns to make distinctions clear for all users. Using contrasting colors and applying transparency to overlapping datasets can also enhance readability.

7.7.4 Poor Label Placement and Sizing

Improper label placement is another common issue that can detract from a spider plot's readability. Labels positioned too close to the data lines or overlapping with other labels make it difficult for viewers to interpret each axis accurately. Position labels slightly outside the plot area and consider using abbreviated terms or a legend to maintain clarity. Ensure that font size is sufficient for readability across different screen sizes, especially for web-based applications or presentations.

7.7.5 Overuse of Layers and Transparency

While layering multiple datasets within a single plot can provide valuable comparative insights, overuse of transparency and layers can lead to a chaotic, cluttered appearance. When multiple datasets overlap excessively, it becomes challenging to distinguish between them. Limit the number of datasets per plot or vary line styles and colors for differentiation. In cases with many datasets, consider using interactive features (e.g., toggling datasets on and off) to reduce visual noise.

7.7.6 Neglecting User Accessibility Needs

A common oversight in spider plot design is neglecting to consider the diverse needs of users, particularly those with visual impairments. Beyond color accessibility, factors such as line width, marker size, and font choice play significant roles in readability. Thicker lines and appropriately sized markers improve visibility for low-vision users, while clear, high-contrast text aids comprehension. Testing the plot for readability on different devices and screen sizes is essential to ensure accessibility for all users.

7.7.7 Failing to Highlight Key Insights

Spider plots can be difficult to interpret without specific guidance, especially for users unfamiliar with the data. Failing to highlight key insights, such as significant data points or trends, may lead to a loss of important information. Use annotations, callouts, or contrasting colors to draw attention to notable data points, enabling viewers to quickly grasp the main insights. Concise annotations provide context, helping users interpret complex plots more easily.

> **💡 Design Tip**
>
> Avoid overlapping annotations and data lines by placing labels slightly outside the plot and using clear callouts when needed.

7.7.8 Checklist of Common Mistakes to Avoid

Spider plots, while visually striking and powerful for multivariate analysis, are prone to specific design pitfalls that can undermine their effectiveness and interpretability. As spider plots increase in popularity across scientific, business, and public data communication contexts, it becomes essential to follow best practices and avoid common errors. Table 7.7 presents a comprehensive checklist of the most frequent mistakes encountered in spider plot design, accompanied by practical solutions for each. One of the most critical issues is *overcrowding*, where designers attempt to display too many categories in a single plot. This leads to visual clutter and cognitive overload. A recommended solution is to limit the number of axes to between five and ten or to group related metrics into separate subplots.

Another prevalent issue is *inconsistent axis scaling*, which can mislead viewers by distorting data patterns. Ensuring a uniform normalization range (e.g., 0–1 or 0–100) is essential for accurate comparison. The table also addresses the importance of accessible design. For instance, using *color-blind friendly palettes* and high-contrast styles ensures inclusivity for users with visual impairments.

Table 7.7: Checklist of Common Mistakes in Spider Plot Design

Mistake	Solution
Overcrowding with too many categories	Limit categories to 5 – 10; group related metrics or use multiple plots.
Inconsistent scaling of axes	Normalize data to a consistent scale (e.g., 0 – 1 or 0 – 100).
Using inaccessible color schemes	Apply color-blind friendly palettes and contrasting line styles.
Poor label placement and sizing	Position labels outside the plot area; use abbreviations or legends if needed.
Overuse of layers and transparency	Limit the number of datasets per plot or use interactive toggles for comparison.
Neglecting user accessibility needs	Ensure high-contrast colors, adequate line width, and readable text across devices.
Failing to highlight key insights	Use annotations, callouts, or color contrasts to emphasize important data points.

Poor label placement and overly dense layering of datasets can impair readability. Proper placement of axis labels outside the plot area and the use of interactive features for toggling datasets can significantly enhance clarity. Attention should also be paid to *highlighting key insights* through annotations and visual emphasis, as spider plots can otherwise appear as abstract shapes devoid of narrative.

Ultimately, this checklist serves as a diagnostic and preventative tool for anyone creating spider plots. By systematically addressing these design concerns, authors and analysts can create clear, accessible, and insightful visualizations that fulfill their communicative purpose.

Avoiding these common mistakes can greatly improve the effectiveness of spider plots, ensuring that they are accessible, easy to read, and capable of conveying meaningful insights. By following best practices and checking for these frequent issues, designers can create spider plots that serve as powerful tools for data visualization and comparative analysis.

7.8 Summary and Best Practices

Spider plots are valuable tools for visualizing multidimensional data, offering an intuitive way to compare multiple variables across categories. However, their effectiveness hinges on thoughtful design and the application of best practices that enhance clarity, accessibility, and interpretability. This section summarizes the key practices covered in this guide and provides a checklist to help readers create spider plots that are both functional and visually captivating.

1. Define Clear Objectives for the Spider Plot. Before designing a spider plot, determine its primary purpose. Understanding the main insights you wish to convey helps in selecting appropriate metrics and structuring the plot to focus on these objectives.

2. Limit the Number of Categories. Overcrowding is a common issue in spider plots. To prevent visual clutter, aim to include no more than 5 – 10 categories. For more complex datasets, consider grouping metrics or using multiple plots to maintain readability.

3. Select Accessible, High-Contrast Color Palettes. Choose color schemes that are accessible to users with color vision deficiencies. Use color-blind friendly palettes such as `viridis` or `plasma`, and ensure that overlapping datasets are distinct by using contrasting colors and varying transparency levels.

4. Normalize Data for Consistent Scaling. When metrics vary in scale, normalize the data to a common range, such as 0 to 1 or 0 to 100, to ensure proportional representation. Consistent scaling allows viewers to accurately interpret the plot and compare values across categories.

5. Employ Layering and Shading for Comparative Visuals. To differentiate datasets within a single plot, use layering techniques with distinct line styles, colors, and levels of transparency. Shading can add depth and clarity, making overlaps visible without obscuring the underlying data.

6. Position Labels to Avoid Overlap. Labels are crucial for understanding each category, but poorly positioned labels can cause confusion. Place labels slightly outside the plot area, and consider using shorter labels or a legend to maintain clarity.

7. Add Annotations for Key Insights. Highlight significant data points or unusual patterns with concise annotations or callouts. Annotations draw attention to important aspects of the data and can enhance the narrative conveyed by the plot.

8. Adjust Line Widths and Marker Sizes for Readability. Use thicker lines for primary datasets and thinner lines for secondary data, ensuring that each dataset is easily distinguishable. For interactive plots, adjust marker sizes for visibility across different screen sizes.

9. Test for Accessibility and Cross-Device Compatibility. Ensure that spider plot is readable on various devices, from desktops to mobile screens. For web-based applications, use responsive design to maintain functionality and clarity across screen sizes.

10. Simplify and Refine for Aesthetic Appeal. Avoid unnecessary complexity or decorative elements that may detract from the plot's clarity. Strive for a clean, organized layout that directs focus toward the data itself, creating a visually appealing presentation.

7.8.1   Checklist for Creating Enchanting Spider Plots

The following checklist summarizes these best practices, serving as a guide to designing spider plots that are both effective and engaging:

- ➤ **Objective Clarity**: Ensure that the plot's purpose and primary insights are clearly defined.

- ➤ **Category Limitation**: Limit the number of categories to prevent overcrowding.

- ➤ **Accessible Colors**: Use color-blind friendly palettes and ensure sufficient contrast.

- ➤ **Data Normalization**: Normalize data to a common scale for consistent comparisons.

- ➤ **Layering for Comparison**: Use different colors, line styles, and transparency for multiple datasets.

- ➤ **Optimized Label Placement**: Position labels to avoid overlap and enhance readability.

- ➤ **Effective Annotations**: Add annotations or callouts to highlight significant points.

- ➤ **Readability Adjustments**: Set line widths and marker sizes that are easy to view on all devices.

- ➤ **Cross-Device Testing**: Test the plot's readability and functionality on multiple devices.

> **Visual Simplicity**: Maintain a clean, focused layout without unnecessary decoration.

Table 7.8: Best Practices for Designing Effective Spider Plots

Best Practice	Description
Objective Clarity	Clearly define the plot's purpose and the primary insights it aims to convey.
Category Limitation	Limit categories to 5 – 10 to avoid overcrowding; use grouping for complex data.
Accessible Colors	Use color-blind friendly palettes with high contrast to ensure readability.
Data Normalization	Normalize data to a consistent scale (e.g., 0 – 1 or 0 – 100) to facilitate comparison.
Layering for Comparison	Differentiate datasets with distinct colors, line styles, and transparency levels.
Optimized Label Placement	Position labels outside the plot area to avoid overlap and maintain clarity.
Effective Annotations	Use concise annotations to highlight key data points or insights.
Readability Adjustments	Adjust line width and marker size for visibility across devices.
Cross-Device Testing	Test plot readability on multiple devices, using responsive design if needed.
Visual Simplicity	Keep layout clean and free from unnecessary decorative elements.

By following these best practices and using the checklist as a design guide, readers can create spider plots that communicate complex data effectively, offering a balance of functionality and visual appeal. With attention to accessibility, clarity, and aesthetic coherence, spider plots become powerful tools for data storytelling, helping audiences derive insights from multidimensional information.

Suggested Readings

1. Al-Ghuwairi, A.R., Al-Fraihat, D., Sharrab, Y., Alrashidi, H., Almujally, N., Kittaneh, A., Ali, A.: Visualizing software refactoring using radar charts. Scientific Reports **13**(1) (2023). DOI 10.1038/s41598-023-44281-6
2. Casals, M., Daunis-i Estadella, P.: Violinboxplot and enhanced radar plot as components of effective graphical dashboards: An educational example of sports analytics. International Journal of Sports Science and Coaching **18**(2), 572–583 (2023). DOI 10.1177/17479541221099638
3. Chang, Y.C., Chang, C.J., Chen, K.T., Lei, C.L.: Radar chart: Scanning for satisfactory qoe in qos dimensions. IEEE Network **26**(4), 25–31 (2012). DOI 10.1109/MNET.2012.6246749
4. Chen, Y., Chen, X.K., Li, Z.Y., Lin, Y.: Method of radar chart comprehensive evaluation with uniqueness feature. Beijing Ligong Daxue Xuebao/Transaction of Beijing Institute of Technology **30**(12), 1409–1412 (2010)

8 Creating Vibrant Ridgeline Plots

Abstract

This chapter explores the design of ridgeline plots, a visualization technique for displaying the distribution of datasets in a compact, layered format. The chapter begins by introducing the purpose of ridgeline plots, highlighting their ability to reveal patterns and trends while conserving visual space. Key design principles are discussed, including choosing appropriate scales to prevent overlap, using color gradients and transparency to differentiate layers, and arranging categories to enhance interpretability. Practical guidance is provided on customizing ridgeline plots to suit various data contexts. Step-by-step examples demonstrate how to create vibrant ridgeline plots, apply custom color palettes, and integrate annotations to highlight key insights. The chapter concludes with advanced techniques, including layering ridgeline plots with additional data visualizations, smoothing distributions for cleaner aesthetics, and incorporating interactive elements for exploratory analysis.

Aims

After reading this chapter, you should be able to:

- ➤ Understand the purpose and applications of ridgeline plots, and identify scenarios where they are the most effective visualization tool.

- ➤ Select and organize variables effectively to create clear and informative ridgeline plots.

- ➤ Use aesthetic design techniques to make ridgeline plots visually engaging while maintaining data clarity.

- ➤ Incorporate accessibility considerations by choosing color-blind friendly palettes, adjusting line widths, and optimizing spacing.

- ➤ Annotate key data points and add informative legends to enhance interpretability and highlight significant trends within the ridgeline plot.

- ➤ Troubleshoot common design challenges in ridgeline plots.

8.1   Introduction to Ridgeline Plots

Ridgeline plots, also known as joy plots, have become an increasingly popular visualization tool for displaying the distribution of continuous variables across multiple categories, time periods, or conditions. Named for their resemblance to mountainous ridges, ridgeline plots layer multiple density curves or histograms, allowing each distribution to be visualized in close proximity to the others. This design creates a "stacked" appearance that is both visually appealing and highly functional, as it facilitates quick comparison between multiple distributions while preserving the shape and spread of each dataset. Originally developed in the context of music data by the artist Joy Division, the ridgeline plot has evolved to serve a broad range of applications across disciplines, from environmental studies and finance to social sciences and medical research.

Ridgeline plots provide a unique ability to reveal patterns and trends across sequential or grouped data, making them a valuable tool for comparative analysis. By aligning multiple distributions along a single axis, ridgeline plots allow viewers to observe both individual and collective changes in data over time, across categories, or between different experimental conditions. Unlike traditional histograms or box plots, which often require separate panels or side-by-side comparison, ridgeline plots enable overlapping visualizations, thus maximizing space and creating a cohesive visual that reveals nuanced details. This overlap is not merely an aesthetic choice; it serves a practical purpose by enhancing the plot's capacity to show relationships between distributions, such as shifts, peaks, and variability. Ridgeline plots effectively convey the shape of each distribution, including multimodality, skewness, and the presence of outliers , which are often important for in-depth data analysis.

8.1.1   The Anatomy of A Ridgeline Plot

A ridgeline plot–also known as a joy plot–is composed of several carefully layered graphical elements that work together to display the distribution of a continuous variable across multiple categories in a compact, interpretable format. Understanding the anatomy of a ridgeline plot is essential for effective design, as each component plays a specific role in conveying the underlying data structure and insights.

Fig. 8.1 presents a comprehensive visual breakdown of a typical ridgeline plot, highlighting the major graphical components. Each density curve represents a smoothed estimate of the distribution of values for a specific category. These curves are vertically stacked and horizontally aligned along a shared baseline axis, allowing for comparison between distributions. Vertical spacing ensures that curves do not overlap excessively, while transparency (alpha) helps reveal overlaps and density patterns without creating visual clutter. Filled gradients or distinct color palettes distinguish categories and emphasize regions of higher or lower density.

Category labels are typically positioned adjacent to each curve, and optional legends may be used to reinforce category identification. The outline of each curve enhances definition and readability, especially when layers are closely packed. Together, these elements create a powerful tool for analyzing temporal changes, category-based variation, or experimental conditions.

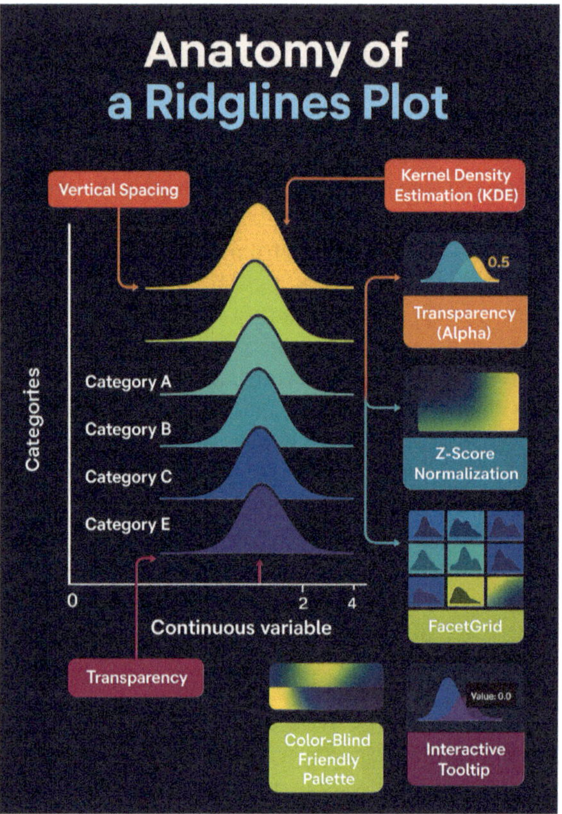

Fig. 8.1: Anatomy of a Ridgeline Plot: A visual breakdown highlighting key components such as density curves, vertical spacing, category labels, fill gradients, and transparency settings used to construct a standard ridgeline plot.

Table 8.1 provides a detailed explanation of each anatomical component depicted in the figure. It describes the functional role of each visual element and offers guidance on how and why it is used in practice. This structured understanding of the plot's anatomy supports the creation of clear, accessible, and analytically valuable ridgeline visualizations.

8.1.2 Purpose and Applications

The primary purpose of ridgeline plots is to provide a compact and coherent visualization of multiple distributions, especially when examining a continuous

Table 8.1: Anatomy of a Ridgeline Plot

Component	Description
Density Curve	A smoothed representation of the distribution of a continuous variable using kernel density estimation (KDE). Each curve corresponds to a different category.
Baseline Axis	The horizontal axis along which each density curve is plotted; typically represents the value range of the continuous variable.
Category Labels	Text annotations aligned with each curve to indicate the category it represents (e.g., years, groups, conditions).
Vertical Spacing	The distance between each stacked curve, controlled by the `scale` parameter to avoid excessive overlap.
Fill Gradient	A color fill applied to the area under each curve, often representing density intensity or category distinctions.
Transparency (Alpha)	Controls the opacity of filled areas to ensure that overlapping curves remain distinguishable.
Curve Outline	A border line surrounding the density curve, often black or dark-colored, to define boundaries and enhance contrast.
Legend (Optional)	Explains the meaning of colors or gradients used; may be omitted if category labels are embedded directly.

variable across various categories or time points. By displaying these distributions in a layered format, ridgeline plots offer a visually efficient means of understanding changes, patterns, and trends across dimensions without overwhelming the viewer with separate plots for each category. In contrast to other plot types, such as line charts or stacked histograms, ridgeline plots retain the full shape of each distribution, making them particularly useful for datasets where the shape of the data carries meaningful information. For instance, in ecological studies, ridgeline plots can be used to illustrate the density of species sightings over time, showing seasonal trends or changes in population density. Similarly, in finance, they can represent the distribution of stock returns across different time periods, revealing volatility, shifts in average returns, or unusual outliers . In social sciences, ridgeline plots are valuable for survey data, where the distributions of responses across multiple demographic groups can be analyzed to understand variations in opinion, behavior, or sentiment.

One of the unique aspects of ridgeline plots is their ability to present overlapping distributions while preserving readability and clarity, which is a significant advantage when examining datasets with numerous categories or time points. For example, in a study of temperature changes over multiple years, a ridgeline plot can display the daily temperature distributions for each year, allowing researchers to detect seasonal patterns, shifts in climate, and anomalies, all within a single visual frame. The overlapping structure of ridgeline plots facilitates direct comparison, enabling the viewer to assess how one distribution differs from another in terms of spread, central tendency, or skewness. By using density or kernel density estimation (KDE) curves, ridgeline plots smooth out variations in data, making underlying trends more discernible and aiding the identification of broader patterns without sacrificing important distributional details. This makes them particularly useful in exploratory data analysis, where understanding data distribution and variability is key to generating hypotheses or making preliminary interpretations.

8.1.3 When to Use Ridgeline Plots

Ridgeline plots are particularly effective when visualizing time series data, category comparisons, or trends across multiple distributions. They are ideally suited for datasets with a continuous variable measured across several discrete categories, such as years, groups, or conditions. In time series analysis, ridgeline plots offer an innovative way to visualize temporal trends by stacking density curves for each time period along a common axis. For example, a ridgeline plot depicting daily air pollution levels over multiple years could reveal both the annual variation and any overarching trends in pollution levels, such as increased emissions during certain seasons or declining trends due to regulatory changes. Similarly, in environmental science, ridgeline plots can illustrate variations in rainfall or temperature distributions across different seasons or decades, allowing researchers to analyze climate trends over time.

In category comparisons, ridgeline plots excel at providing a clear view of how distributions differ across groups without fragmenting the data into multiple subplots. This application is particularly useful in fields like social sciences, where researchers may want to compare the distribution of survey responses across demographic groups, such as age, income level, or educational attainment. Ridgeline plots allow all groups to be viewed within a single frame, facilitating the identification of distinct peaks, outliers , or skewness specific to each demographic category. The simultaneous visualization of multiple distributions makes it easy to detect patterns such as bimodal distributions, shifts in central tendencies, or variability across groups, which are often of interest in behavioral studies or public opinion research.

Ridgeline plots are also effective in contexts that require visualizing overlapping distributions with minimal visual clutter. For example, in finance, they can be used to show the distribution of returns for different financial assets or portfolios, highlighting differences in volatility, risk, and average returns. In biology and medical research, ridgeline plots may be used to compare distributions of physiological measurements across treatment groups, helping to identify patterns in response to interventions or treatments. The ability to layer distributions with varying levels of transparency and color intensity further enhances the plot's flexibility, allowing researchers to create informative visuals that communicate complex data in an accessible format.

Overall, ridgeline plots serve as an effective tool for comparing distributions across categories or time points, offering unique advantages for visualizing trends, density, and distributional shapes. Their ability to present overlapping data without sacrificing readability, combined with their flexibility in representing complex data, makes them a preferred choice for researchers and analysts in a wide range of fields. When used appropriately, ridgeline plots reveal relationships, trends, and insights that may be difficult to detect with other visualization methods, making them a valuable addition to any data analyst's toolkit.

8.2 Design Principles for Effective Ridgeline Plots

Ridgeline plots offer a unique and visually compelling means of displaying distributions across multiple categories, allowing viewers to perceive trends and relationships at a glance. However, the effectiveness of ridgeline plots hinges on thoughtful design principles that ensure clarity, readability, and interpretability. When poorly designed, ridgeline plots can become cluttered and confusing, undermining their potential as a comparative analysis tool. This section outlines foundational principles for designing effective ridgeline plots, focusing on the careful selection of variables and categories, as well as strategies for balancing data density with readability. These principles are essential for creating ridgeline plots that are not only visually engaging but also provide meaningful insights to the viewer.

8.2.1 Selecting Variables and Categories

The selection of variables and categories is a critical first step in designing a ridgeline plot that is both informative and accessible. Ridgeline plots are particularly suited for visualizing continuous variables across a series of categories, such as time points, demographic groups, or experimental conditions. However, the utility of these plots diminishes when too many categories or irrelevant variables are included. Therefore, selecting variables that convey meaningful trends and focusing on key categories is essential for maintaining plot clarity.

When choosing variables, it is important to identify those that reflect significant trends or patterns within the data. Continuous variables with notable variability or those that represent essential dimensions of the dataset are ideal for ridgeline plots. For example, in environmental studies, temperature or rainfall distributions across months or years are meaningful variables, as they reveal seasonal patterns and climate trends. In financial analysis, stock return distributions across different periods or asset classes can provide insights into market volatility and investment risk. Variables with subtle trends or those that do not vary significantly across categories may not add substantial value in a ridgeline plot; thus, they should be omitted to prevent clutter and preserve focus.

Similarly, it is advisable to limit the number of categories displayed in a ridgeline plot. Including too many categories can lead to excessive overlap and visual complexity, making it difficult for viewers to interpret each distribution individually. A general recommendation is to include between five and ten categories, as this range typically allows for a balance between detailed insight and readability. For instance, in time series data, a ridgeline plot could include distributions for each year over a five- or ten-year period, rather than attempting to display monthly distributions over several decades. If additional categories are essential to the analysis, it may be beneficial to create separate plots for different subsets of data or to group related categories, such as combining seasons into a single distribution per year rather than displaying each month individually. This approach maintains the plot's interpretability while providing the necessary granularity to reveal trends.

> **⚠ Warning**
>
> Displaying too many categories in a single ridgeline plot can overwhelm the reader and obscure important distributional patterns. For readability, limit the number of layers to approximately 5–10 or group related categories into broader aggregates.

In cases where multiple variables are available for visualization, it is crucial to select those that highlight the most relevant aspects of the data. Focusing on key variables ensures that the ridgeline plot conveys clear, targeted insights rather than overwhelming the viewer with superfluous information. For instance, in a study on consumer behavior, visualizing distributions of purchase amounts across customer age groups may reveal valuable insights about spending patterns, while displaying less pertinent metrics, such as individual product ratings across those same groups, could dilute the plot's focus. Selecting variables and categories that align closely with the plot's purpose not only improves clarity but also strengthens the overall impact of the visualization.

8.2.2 Balancing Data Density and Readability

A major challenge in designing ridgeline plots lies in balancing data density with readability, as these plots inherently involve layering multiple distributions. When not carefully managed, overlapping density curves or histograms can lead to a cluttered visualization that obscures rather than reveals trends. Effective ridgeline plots maintain a balance between providing detailed information and preserving visual clarity, allowing viewers to identify patterns without difficulty. Several strategies, including spacing adjustments, transparency settings, and line weight modifications, can enhance the readability of ridgeline plots without compromising the richness of the data (Table 8.2).

Adjusting the vertical spacing between layers is one of the most effective methods for preventing overcrowding in ridgeline plots. By increasing the distance between individual distributions, designers can reduce overlap, making it easier to distinguish each layer. However, excessive spacing can cause the plot to lose its cohesive "ridgeline" appearance, so finding the right balance is key. Moderate spacing typically achieves the best results, providing enough separation to distinguish each distribution while preserving the visual continuity of the plot. In some cases, dynamic spacing, where distributions with larger variances or peaks are given additional room, can further enhance readability, as it prevents high-density areas from obscuring neighboring curves.

Transparency is another essential tool for managing overlapping layers in ridgeline plots. By setting appropriate transparency levels (often represented as an "alpha" value in visualization software), designers can layer multiple distributions without creating visual congestion. Transparency allows overlapping areas to remain visible, providing depth to the plot and enabling viewers to see multiple distributions simultaneously. For example, a transparency setting of 0.3 to 0.5 often works well for ridgeline plots, as it provides a blend of color intensity and visibility that enhances interpretability. Transparency is especially beneficial when working with categorical data where certain groups may dominate the visualization; by making dominant layers slightly transparent, secondary distributions become more visible, facilitating a more balanced view of the data.

> **⚠ Warning**
>
> Setting the transparency (`alpha`) too low or too high can hinder interpretability. Too much transparency may make curves invisible, while too little may create visual clutter in overlapping areas. Aim for an `alpha` value between 0.4 and 0.7.

In addition to spacing and transparency, adjusting line weights can further enhance the clarity of ridgeline plots. Line weights should be chosen carefully to distinguish boundaries without overpowering the visual elements of the plot. Thicker lines may be appropriate for highlighting primary distributions or critical trends, while thinner lines can be used for secondary distributions or less important data layers. Consistent line weights across categories can provide a uniform appearance, but selectively varying line thicknesses based on importance or category hierarchy can direct viewer attention to the most relevant elements of the plot. Line style, such as solid or dashed lines, can also aid in differentiation when multiple groups or conditions are represented within a single plot.

Choosing a suitable color palette is another crucial factor in balancing data density and readability in ridgeline plots. Colors should be chosen to provide contrast between overlapping layers while remaining visually harmonious. Gradient color schemes, which transition from light to dark shades within a single hue, are particularly effective for ridgeline plots, as they allow each layer to be distinguishable without introducing excessive color variation. When displaying time series data, gradient schemes that increase in intensity over time can also suggest progression, helping viewers intuitively interpret chronological data. For category-based ridgeline plots, using distinct colors for each category, while adjusting transparency as needed, allows clear differentiation without overwhelming the viewer with color. Color-blind friendly palettes are recommended to ensure accessibility for all viewers.

> **⚠ Warning**
>
> Using non-accessible color palettes (e.g., red-green) can exclude color-blind users and reduce visual clarity. Always opt for color-blind-friendly palettes such as `viridis`, `plasma`, or `cividis` to ensure inclusivity.

By applying these strategies, careful selection of spacing, transparency, line weights, and color schemes, designers can create ridgeline plots that balance data density with readability. When effectively designed, ridgeline plots reveal meaningful insights across multiple distributions without overwhelming the viewer, offering a powerful visualization tool for comparative and exploratory analysis.

Table 8.2: Design Choices for Ridgeline Plots

Design Aspect	Best Practice
Variable Selection	Choose variables that highlight meaningful trends. Avoid over-crowding by limiting the number of categories.
Layer Transparency	Use moderate transparency (0.4 to 0.7) to allow overlapping layers to remain visible.
Vertical Spacing	Adjust vertical spacing to reduce visual clutter and improve readability, especially in dense plots.
Color Scheme	Apply accessible color palettes like `viridis` or `cividis` for color-blind-friendly visualizations.
Annotation	Use concise labels and callouts to highlight significant data points, avoiding overlap with data layers.

8.3 Creating Ridgeline Plots in Python with `seaborn` and `plotly`

Ridgeline plots can be created in Python using various libraries, with `seaborn` being particularly popular for creating static ridgeline plots. The `seaborn` library, with its built-in functions for density estimation and grid-based plots, enables users to generate visually appealing ridgeline plots with a few lines of code. This section provides a step-by-step guide to creating static ridgeline plots in Python using `seaborn` and demonstrates how to enhance the plot with color gradients, transparency, and density adjustments for improved readability.

8.3.1 Using `seaborn` for Static Ridgeline Plots

To create a basic ridgeline plot in `seaborn`, we can use the `kdeplot` function in combination with `FacetGrid` The `kdeplot` function allows us to generate kernel density estimates, which can then be layered to create the ridgeline effect. The following example demonstrates how to set up a basic ridgeline plot in Python.

```python
import seaborn as sns
import matplotlib.pyplot as plt
import pandas as pd

```

```python
5   # Sample data
6   data = pd.DataFrame({
7       'value': [1, 2, 3, 4, 2, 3, 5, 7, 6, 3, 4, 3, 2, 5, 6, 7, 5, 3],
8       'category': ['A', 'A', 'A', 'A', 'B', 'B', 'B', 'B', 'B', 'C', 'C',
        ↪    'C', 'D', 'D', 'D', 'D', 'D', 'D']
9   })
10
11  # Initialize a FacetGrid for ridgeline plot
12  g = sns.FacetGrid(data, row="category", hue="category", aspect=4,
    ↪   height=1.2)
13  g.map(sns.kdeplot, "value", fill=True)
14  g.set_titles("{row_name}")
15  g.set(yticks=[], ylabel="")
16  g.despine(left=True)
17
18  # Display the plot
19  plt.show()
```

In this example, we use a sample dataset with a continuous variable `value` and a categorical variable `category` We initialize a `FacetGrid` object, setting `row="category"` to create a separate density plot for each category. The `hue="category"` argument applies a different color to each category. The `aspect` and `height` parameters control the dimensions of each plot, providing a cohesive layout for the ridgeline plot. Finally, `kdeplot` is used to plot density estimates, with the `fill=True` parameter creating filled density curves for a classic ridgeline appearance.

8.3.2 Adding Color and Customization in seaborn

Customizing color, transparency, and density settings in `seaborn` can enhance the readability and visual appeal of ridgeline plots, especially when multiple categories are displayed. The following example shows how to apply gradient color schemes, set transparency levels, and adjust density settings to create a more polished and visually informative ridgeline plot.

```python
1   import seaborn as sns
2   import matplotlib.pyplot as plt
3   import pandas as pd
4
5   # Sample data
6   data = pd.DataFrame({
7       'value': [1, 2, 3, 4, 2, 3, 5, 7, 6, 3, 4, 3, 2, 5, 6, 7, 5, 3],
8       'category': ['A', 'A', 'A', 'A', 'B', 'B', 'B', 'B', 'B', 'C', 'C',
        ↪    'C', 'D', 'D', 'D', 'D', 'D', 'D']
9   })
10
11  # Custom color palette\index{Color Palette} for gradient effect
12  palette = sns.cubehelix_palette(len(data['category'].unique()), start=.5,
    ↪   rot=-.75, light=0.8, dark-0.3)
```

```
13
14   # Initialize a FacetGrid with customized colors and transparency
15   g = sns.FacetGrid(data, row="category", hue="category", aspect=4,
     ↪  height=1.2, palette=palette)
16   g.map(sns.kdeplot, "value", fill=True, alpha=0.6)  # Adjust alpha for
     ↪  transparency
17   g.set_titles("{row_name}")
18   g.set(yticks=[], ylabel="")
19   g.despine(left=True)
20
21   # Display the customized ridgeline plot
22   plt.show()
```

In this example, we create a gradient color palette using `sns.cubehelix_palette`, which provides smooth transitions in color intensity across categories. This palette is passed to the `FacetGrid` object through the `palette` parameter, allowing each distribution to be visually distinct while maintaining a harmonious gradient effect. The `alpha=0.6` argument within the `kdeplot` function applies transparency to the density curves, enabling the viewer to see overlapping areas without overwhelming the plot. This transparency level (typically between 0.3 and 0.7) strikes a balance between color vibrancy and layer separation, improving interpretability.

```
1   # Adjust bandwidth\index{Bandwidth} for finer control over density
    ↪  estimation
2   g = sns.FacetGrid(data, row="category", hue="category", aspect=4,
    ↪  height=1.2, palette=palette)
3   g.map(sns.kdeplot, "value", fill=True, alpha=0.6, bw_adjust=0.8)  #
    ↪  Decrease bw_adjust for more detail
4   g.set_titles("{row_name}")
5   g.set(yticks=[], ylabel="")
6   g.despine(left=True)
7
8   plt.show()
```

In this example, we set `bw_adjust=0.8` within the `kdeplot` function, decreasing the bandwidth to add more detail to each density curve. Lower bandwidth values (e.g., 0.5–0.8) reveal finer distinctions within the data, which is useful when analyzing categories with subtle distributional differences. However, overly small bandwidth values can lead to jagged curves, so it is important to experiment to find a balance that best represents the data.

By customizing color, transparency, and density settings, `seaborn` allows users to design ridgeline plots that are visually appealing and easy to interpret. These customization options not only enhance readability but also enable the creation of ridgeline plots that align with specific analytical or aesthetic goals, providing a powerful visualization tool for comparative data analysis.

8.3.3   Interactive Ridgeline Plots with `plotly`

Creating interactive ridgeline plots allows users to engage with the data more dynamically, offering features like hover effects, interactive legends, and zooming options. `plotly`, a powerful library for interactive visualizations in Python, is particularly well-suited for building interactive ridgeline plots. In addition to providing detailed views of overlapping distributions, `plotly` enables real-time exploration, making it ideal for web-based applications or dashboards where user engagement is essential. This section provides a walkthrough for creating interactive ridgeline plots in `plotly`, with code examples illustrating the use of hover effects, legends, and other interactive components.

```python
import plotly.express as px
import pandas as pd

# Sample data
data = pd.DataFrame({
    'value': [1, 2, 3, 4, 2, 3, 5, 7, 6, 3, 4, 3, 2, 5, 6, 7, 5, 3],
    'category': ['A', 'A', 'A', 'A', 'B', 'B', 'B', 'B', 'B', 'C', 'C',
      'C', 'D', 'D', 'D', 'D', 'D', 'D']
})

# Create the ridgeline plot using Plotly \index{Plotly} Express
fig = px.violin(data, x="value", y="category", color="category",
  orientation="h",
                points="all", hover_data=["value"], box=True, width=800,
                  height=400)

# Customize hover template for detailed interaction
fig.update_traces(hovertemplate="Category: %{y}<br>Value: %{x}")

# Update layout for better readability
fig.update_layout(title="Interactive Ridgeline Plot", yaxis_title="",
  xaxis_title="Value",
                showlegend=True, template="plotly_white")

# Display the interactive plot
fig.show()
```

In this example, we use `plotly.express.violin` to create an interactive ridgeline plot where each category's distribution is visualized horizontally. The `color="category"` parameter applies distinct colors to each category, allowing for easy differentiation. The `points="all"` argument displays individual data points alongside the distribution, while `box=True` adds a box plot within each category, providing additional insights into central tendency and spread. The `hovertemplate` parameter is used to format the hover information, allowing users to view category labels and values interactively.

```
1  fig.update_traces(hovertemplate="<b>Category</b>: %{y}<br><b>Value</b>:
2  %{x}<extra></extra>")
3  fig.show()
```

In this example, we add a custom hover template with bold labels for "Category" and "Value." The `<extra></extra>` tag removes default information from the hover box, leaving only the specified content, which simplifies the appearance and makes it easier to read.

```
1   # Enable interactive legend
2   fig.update_layout(
3       title="Interactive Ridgeline Plot with Dynamic Legend",
4       legend_title="Categories",
5       yaxis_title="",
6       xaxis_title="Value",
7       showlegend=True,
8       template="plotly_white"
9   )
10
11  fig.show()
```

The interactive legend, enabled by default in `plotly`, allows viewers to click on category names to toggle visibility. This feature is particularly helpful for ridgeline plots with numerous categories, as it reduces clutter and enables viewers to focus on relevant data.

```
1   # Adjust color and transparency for clarity
2   fig.update_traces(marker=dict(opacity=0.6), line=dict(color="black",
    ↪  width=1))
3
4   fig.show()
```

By setting `marker.opacity=0.6`, we add transparency to the individual points within each category, enhancing visibility of overlapping layers. The `line=dict(color="black", width=1)` parameter applies a thin outline to each distribution, which improves the definition of each layer and helps viewers distinguish individual categories.

```
1   # Enable zoom and responsive layout
2   fig.update_layout(
3       title="Responsive and Interactive Ridgeline Plot",
4       autosize=True,
```

```
5      width=800,
6      height=500,
7      template="plotly_white"
8  )
9
10   fig.show()
```

Here, the `autosize=True` option ensures that the plot adapts to different screen dimensions, providing a responsive design suitable for dashboards or mobile applications. Users can also zoom in and out to inspect specific regions of the plot, enhancing the interactivity and accessibility of the visualization.

In summary, `plotly` offers a range of interactive features for ridgeline plots, from hover effects and dynamic legends to color customization and responsive layouts. These elements collectively improve the user experience, making ridgeline plots more engaging and informative. With `plotly`, designers can create interactive visualizations that cater to various analytical needs, whether for exploratory analysis, presentations, or interactive dashboards.

8.4 Customizing Ridgeline Plots in R with `ggplot2` and `ggridges`

R offers robust tools for creating and customizing ridgeline plots, with the `ggridges` package providing extensive functionality specifically for ridgeline visualizations. Built on the `ggplot2` framework, `ggridges` enables users to produce visually appealing ridgeline plots that integrate seamlessly with other `ggplot2` customizations. This section introduces the `ggridges` package, covering its key functions and options for customizing ridgeline plots to create effective, informative visuals.

8.4.1 Introduction to the `ggridges` Package

The `ggridges` package, developed by Claus Wilke, is designed to generate ridgeline plots within the `ggplot2` ecosystem. It provides a `geom_density_ridge` function that creates layered density plots, with each layer corresponding to a specific category or group. This layered approach allows distributions to be stacked along a vertical axis, resulting in the characteristic "ridgeline" appearance. By integrating `ggridges` with `ggplot2`, users can apply a wide range of customization options, from color gradients and transparency to axis modifications and thematic adjustments.

To get started with `ggridges`, users need to install and load the package if it is not already available in their R environment:

```
1   # Install ggridges package if not already installed
2   install.packages("ggridges")
3
4   # Load the required libraries
```

```r
5  library(ggplot2)
6  library(ggridges)
```

Once loaded, the `ggridges` package can be used to create basic ridgeline plots with minimal code. The following example demonstrates a simple ridgeline plot using `ggridges`, where each category's density distribution is layered to form a coherent visualization.

```r
1  # Sample data
2  data <- data.frame(
3      value = c(rnorm(100, mean=3), rnorm(100, mean=5), rnorm(100, mean=7),
4      rnorm(100, mean=9)),
5      category = rep(c("A", "B", "C", "D"), each = 100)
6  )
7
8  # Basic ridgeline plot
9  ggplot(data, aes(x = value, y = category, fill = category)) +
10     geom_density_ridge(alpha = 0.6) +
11     labs(title = "Basic Ridgeline Plot", x = "Value",
12     y = "Category") + theme_minimal()
```

In this example, we create a basic ridgeline plot with a continuous variable `value` and a categorical variable `category` The `geom_density_ridge` function is used to generate the density curves for each category, with `alpha = 0.6` applying transparency to the fills for better visual separation. Also, `theme_minimal()` is applied to create a clean, minimalistic appearance.

8.4.1.1 Key Customization Options in ggridges

```r
1  ggplot(data, aes(x = value, y = category, fill =..x..)) +
2      geom_density_ridge_gradient(scale = 1.5, rel_min_height = 0.01) +
3      scale_fill_viridis_c(option = "C") +
4      labs(title = "Ridgeline Plot with Gradient Fill", x = "Value",
5      y = "Category") +
6      theme_minimal()
```

In this example, `geom_density_ridge_gradient` is used to create a gradient fill based on the `value` variable within each category, providing visual depth and helping to highlight distributional shapes. The `scale_fill_viridis_c` function applies a colorblind-friendly gradient from the `viridis` palette, which is particularly useful for accessibility. Adjusting `scale` and `rel_min_height` parameters fine-tunes the appearance, with `scale` controlling the height of each ridge and `rel_min_height` defining the minimum relative height to retain smaller density features.

> **⚠ Warning**
>
> Avoid plotting distributions with vastly different scales without normalization
> . This can result in misleading comparisons and distort visual perception.
> Apply z-score or min-max normalization to ensure consistent interpretation
> across categories.

```r
ggplot(data, aes(x = value, y = category, fill = category)) +
    geom_density_ridge(scale = 2, alpha = 0.5) +
    labs(title = "Ridgeline Plot with Increased Spacing", x = "Value",
    y = "Category") +
    theme_minimal()
```

In this code, `scale = 2` increases the spacing between distributions, reducing
the degree of overlap. This setting is useful when distributions have similar shapes
or when the dataset includes many categories, as it prevents excessive clutter and
maintains visual clarity.

```r
ggplot(data, aes(x = value, y = category, fill = category)) +
    geom_density_ridge(color = "black", alpha = 0.4, size = 0.8) +
    labs(title = "Ridgeline Plot with Outlines", x = "Value", y = "Category") +
    theme_minimal()
```

In this example, `color = "black"` adds an outline to each ridge, while `alpha =
0.4` introduces transparency. The `size = 0.8` parameter controls the line thickness,
providing a clear boundary around each density curve and making each layer
more distinguishable. These customizations are particularly helpful when multiple
distributions overlap, as they maintain the plot's cohesion without sacrificing detail.

```r
ggplot(data, aes(x = value, y = category, fill = category)) +
    geom_density_ridge(alpha = 0.5) +
    scale_fill_brewer(palette = "Set3") +
    labs(title = "Custom-Themed Ridgeline Plot", x = "Measurement",
    y = "Group") +
    theme_classic() +
    theme(
        plot.title = element_text(size = 16, face = "bold"),
        axis.title = element_text(size = 14)
    )
```

In this example, we use `scale_fill_brewer` to apply a qualitative color palette
from `RColorBrewer`, enhancing visual differentiation between categories. The
`theme_classic()` function provides a clean aesthetic, while custom settings within
`theme()` allow for control over title and axis text styles, improving readability and
visual impact.

By using these customization options in `ggridges`, users can create ridgeline plots that are visually distinctive, accessible, and tailored to specific analytical goals. The integration of `ggridges` with `ggplot2` offers a flexible framework for designing ridgeline plots that convey complex distributional data with clarity and precision.

8.4.2   Step-by-Step Guide to Creating Ridgeline Plots in R

Creating a ridgeline plot in R using the `ggridges` package is a straightforward process, but designing an effective plot often involves several steps, including data preparation, color customization, and axis adjustments. This section provides a comprehensive guide to setting up a ridgeline plot in R, using sample data to demonstrate the process from start to finish. By following these steps, users can create an informative and visually appealing ridgeline plot with customized colors, spacing, and axis settings.

```r
# Load required libraries
library(ggplot2)
library(ggridges)

# Create sample data
set.seed(123)
data <- data.frame(
    value = c(rnorm(100, mean=3, sd=1), rnorm(100, mean=5, sd=1.2),
    rnorm(100, mean=7, sd=1.5), rnorm(100, mean=9, sd=1.1)),
    category = rep(c("Category A", "Category B", "Category C", "Category D"),
    each = 100)
)

# Preview the data
head(data)
```

In Step 1, we load the `ggplot2` and `ggridges` packages, which are necessary for creating ridgeline plots in R. We then generate a sample dataset consisting of a continuous variable, `value`, and a categorical variable, `category` Each category has 100 data points with different mean values, which will create visually distinct ridgelines.

```r
# Basic ridgeline plot
ggplot(data, aes(x = value, y = category)) +
    geom_density_ridge(fill = "skyblue", color = "black", alpha = 0.6) +
    labs(title = "Basic Ridgeline Plot", x = "Value", y = "Category") +
    theme_minimal()
```

In Step 2, we create a basic ridgeline plot by calling `geom_density_ridge` The `fill` parameter is set to `"skyblue"` to provide a light blue color for the ridgelines, while `color = "black"` adds a black outline around each curve, enhancing contrast. `alpha = 0.6` sets the transparency level, which helps reduce overlap and makes the plot easier to read. Finally, `labs` and `theme_minimal()` are used to add titles and create a clean layout.

```r
# Customize ridgeline plot with gradient fill and spacing
ggplot(data, aes(x = value, y = category, fill =..x..)) +
    geom_density_ridge_gradient(scale = 1.5, rel_min_height = 0.01) +
    scale_fill_viridis_c(option = "C") +
    labs(title = "Ridgeline Plot with Gradient Fill", x = "Value",
    y = "Category") +
    theme_minimal()
```

In Step 3, we enhance the plot by applying a gradient fill using `geom_density_ridge_gradient`, which colors each ridgeline according to its `value` variable, resulting in a visually appealing gradient effect. `scale_fill_viridis_c` applies the `viridis` color palette, which is both visually attractive and accessible to color-blind users. Also, `scale = 1.5` increases the vertical separation between ridgelines, and `rel_min_height = 0.01` ensures that minor density features are retained.

```r
# Final ridgeline plot with axis adjustments and customized labels
ggplot(data, aes(x = value, y = category, fill = category)) +
    geom_density_ridge(alpha = 0.7, color = "black") +
    scale_fill_brewer(palette = "Set3") +
    scale_x_continuous(breaks = seq(0, 10, by = 2), limits = c(0, 10)) +
    labs(title = "Final Ridgeline Plot with Custom Axis",
    x = "Measurement Value", y = "Category") +
    theme_minimal() +
    theme(
        plot.title = element_text(size = 16, face = "bold"),
        axis.title.x = element_text(size = 14, margin = margin(t = 10)),
        axis.title.y = element_text(size = 14, margin = margin(r = 10))
    )
```

In Step 4, we make final adjustments to improve the plot's readability and aesthetic appeal. By setting `scale_fill_brewer(palette = "Set3")`, we apply a color scheme from RColorBrewer, providing distinct colors for each category. `scale_x_continuous` specifies the axis ticks and limits, controlling the visible range and positioning of the x-axis labels. Finally, `theme` customizations adjust the font size and position of titles and axis labels, ensuring that the plot is easy to interpret and visually balanced.

8.4.3 Advanced Customization with `ggplot2`

In addition to the basic functionalities provided by `ggridges`, `ggplot2` offers a wide range of customization options that allow for more advanced fine-tuning of ridgeline plots. These customizations enhance readability, create a polished appearance, and enable the visualization to align with specific analytical or aesthetic goals. This section covers techniques for advanced customization, including changing line styles, adjusting spacing, and incorporating custom color palettes to create ridgeline plots that are both visually engaging and informative.

The appearance of line borders around each ridgeline can be customized to emphasize certain categories or add clarity to overlapping distributions. `ggridges` supports several line style options, such as solid, dashed, and dotted lines, which can be specified using the `linetype` argument. Adjusting line thickness with the `size` parameter further enhances the distinction between ridgelines, particularly in plots with high data density.

```r
ggplot(data, aes(x = value, y = category, fill = category)) +
    geom_density_ridge(color = "black", linetype = "dashed", size = 0.8,
    alpha = 0.6) +
    labs(title = "Ridgeline Plot with Custom Line Style", x = "Value",
    y = "Category") +
    theme_minimal()
```

In this example, `linetype = "dashed"` creates a dashed border around each ridgeline, while `size = 0.8` adjusts the line thickness. These adjustments provide a visually distinct appearance, making each category easier to distinguish. Dashed or dotted lines are particularly useful for secondary categories, as they subtly differentiate less prominent data without overwhelming the primary focus.

The spacing and vertical positioning of ridgelines play a crucial role in managing visual clarity, especially when displaying many categories or overlapping distributions. The `scale` parameter in `geom_density_ridge` controls the vertical separation between ridgelines, where higher values increase the distance between each ridge.

```r
ggplot(data, aes(x = value, y = category, fill = category)) +
    geom_density_ridge(scale = 2, alpha = 0.5, color = "black") +
    labs(title = "Ridgeline Plot with Increased Spacing",
    ↪  x = "Measurement Value",
    y = "Category") +
    theme_minimal()
```

In this example, setting `scale = 2` increases the vertical separation between distributions, reducing overlap and enhancing readability. Adjusting spacing is particularly helpful when working with categories that have similar shapes or densities, as it allows each distribution to stand out individually without visual clutter.

The use of color is essential for distinguishing categories in a ridgeline plot, and `ggplot2` provides flexible options for applying custom color schemes. By using `scale_fill_manual`, users can specify a custom color palette tailored to the data, ensuring visual coherence and adherence to accessibility standards. Color gradients can be applied through `scale_fill_gradient` or `scale_fill_viridis_c`, providing a continuous color scale that reflects data intensity or other variables.

```r
# Define custom colors for each category
custom_colors <- c("Category A" = "#4CAF50", "Category B" = "#FF5722",
```

```
3                    "Category C" = "#2196F3", "Category D" = "#FFC107")
4
5  ggplot(data, aes(x = value, y = category, fill = category)) +
6      geom_density_ridge(alpha = 0.6, color = "black") +
7      scale_fill_manual(values = custom_colors) +
8      labs(title = "Ridgeline Plot with Custom Colors", x = "Value",
9      y = "Category") +
10     theme_minimal()
```

In this example, we define a custom color palette using hexadecimal color codes, then apply it to each category using `scale_fill_manual(values = custom_colors)` This approach enables precise control over the colors used, allowing the plot to match specific branding or thematic guidelines. Custom color palettes can also improve accessibility by choosing high-contrast colors that are distinguishable for color-blind users.

Gradient fills can enhance the interpretability of ridgeline plots by indicating changes in density or intensity within each distribution. The `ggridges` package supports gradient fills through the `geom_density_ridge_gradient` function, which varies the fill color based on the **x** or **y** values.

```
1  ggplot(data, aes(x = value, y = category, fill =..x..)) +
2      geom_density_ridge_gradient(scale = 1.5, rel_min_height = 0.01) +
3      scale_fill_viridis_c(option = "D") +
4      labs(title = "Ridgeline Plot with Density-Based Gradient", x = "Value",
5      y = "Category") +
6      theme_minimal()
```

In this example, `geom_density_ridge_gradient` is used to create a gradient fill based on the **x** (value) variable within each category. The `scale_fill_viridis_c` function applies the `viridis` color palette, which is color-blind friendly and provides smooth transitions. This approach highlights density variations within each category, making it easier to interpret the distribution shapes and emphasizing peaks or clusters in the data.

Clear labeling and consistent theming enhance the readability of ridgeline plots, especially when used in reports or presentations. In `ggplot2`, the `theme` function allows customization of text size, font style, and element positioning, creating a cohesive aesthetic that aligns with professional standards.

```
1  ggplot(data, aes(x = value, y = category, fill = category)) +
2      geom_density_ridge(alpha = 0.6, color = "black") +
3      scale_fill_brewer(palette = "Set2") +
4      labs(title = "Ridgeline Plot with Custom Labels", x = "Measurement Value",
5      y = "Category") +
6      theme_minimal() +
7      theme(
8          plot.title = element_text(size = 18, face = "bold", hjust = 0.5),
9          axis.title.x = element_text(size = 14, margin = margin(t = 10)),
```

```
10        axis.title.y = element_text(size = 14, margin = margin(r = 10)),
11        axis.text.y = element_text(size = 12),
12        legend.position = "none"
13     )
```

In this example, `element_text` within `theme()` customizes the font size and styling for titles and axis labels, ensuring that text is legible and well-positioned. By setting `hjust = 0.5`, the plot title is centered, providing a balanced appearance. Additional adjustments, such as increasing the `margin` around axis titles, create visual separation, while `legend.position = "none"` hides the legend to reduce clutter when not needed.

> **⚠ Warning**
>
> Crowded or overlapping labels can confuse readers and obscure key insights. Avoid placing text directly over ridgelines, and use external legends, callouts, or tooltips to maintain a clean layout and enhance readability.

Through these advanced customization techniques, `ggplot2` and `ggridges` empower users to create ridgeline plots that are highly tailored to specific data visualization needs. Whether adjusting line styles, applying gradient fills, or fine-tuning labels and themes, these options enable the creation of polished, professional ridgeline plots that effectively communicate complex data distributions.

8.5 Enhancing Ridgeline Plots with Aesthetic Design Techniques

To create ridgeline plots that are not only informative but also visually appealing, designers can employ a range of aesthetic techniques. Effective use of color gradients, opacity, patterns, and annotation can transform a basic ridgeline plot into a sophisticated visualization that communicates trends and insights with clarity and style. This section explores key aesthetic design techniques, focusing on color gradients and opacity adjustments, the use of patterns and textures, and best practices for annotating key data points and adding legends.

8.5.1 Using Color Gradients and Opacity

Color gradients are a powerful tool for indicating trends or intensity levels within ridgeline plots. By transitioning colors smoothly across a continuous scale, gradients can convey subtle differences in data density, making it easier for viewers to identify peaks, clusters, or trends across categories. In ridgeline plots, gradients can be applied to individual ridges based on their values, enabling comparisons between distributions without relying solely on shape or size.

For example, applying a gradient from light to dark within each category can indicate regions of higher density, such as peak values, where darker shades represent more frequent occurrences. This technique is especially effective when visualizing time series data, as gradients can imply progression and allow users to intuitively follow changes over time. In R, `ggridges` supports gradient fills through the `geom_density_ridge_gradient` function, which applies a continuous color scale along the x-axis values or another variable.

```r
ggplot(data, aes(x = value, y = category, fill =..x..)) +
    geom_density_ridge_gradient(scale = 1.5, rel_min_height = 0.01) +
    scale_fill_viridis_c(option = "plasma") +
    labs(title = "Ridgeline Plot with Color Gradient", x = "Value",
    y = "Category") +
    theme_minimal()
```

In this example, we use `scale_fill_viridis_c` with the `plasma` color option to apply a color gradient to the ridgelines. The gradient transition enhances the plot's interpretability by visually distinguishing areas of high and low density. `viridis` color palettes are also colorblind-friendly, making this approach accessible to a broader audience.

Adjusting opacity is another critical technique for maintaining visual separation across overlapping layers. Opacity levels can be controlled with the `alpha` parameter, allowing multiple layers to be visible without excessive blending. This is particularly useful when dealing with categories that have similar shapes or overlapping distributions, as transparency reduces clutter while preserving layer details.

```r
ggplot(data, aes(x = value, y = category, fill = category)) +
    geom_density_ridge(alpha = 0.5, color = "black") +
    labs(title = "Ridgeline Plot with Adjusted Opacity", x = "Measurement",
    y = "Category") +
    theme_minimal()
```

In this example, `alpha = 0.5` sets the transparency level, providing a balanced view of overlapping ridgelines. This adjustment is essential for maintaining readability in plots with high data density, as it allows viewers to see each layer without obscuring the underlying data.

8.5.2 Incorporating Patterns and Textures

In cases where color may not be effective or suitable, such as monochromatic designs or grayscale printing, patterns and textures can serve as an alternative method for distinguishing layers within a ridgeline plot. Patterns, such as stripes, dots, or cross-hatching, help differentiate categories by adding visual variation without relying on color. This technique is particularly useful when creating visualizations for printed materials, where color distinctions may be lost or limited.

While `ggridges` does not natively support patterned fills, patterns can be applied by overlaying additional layers or using specialized graphics tools. For example, a lightly dotted or striped overlay can be added to certain ridges in a design software post-export, or users can layer textures in `ggplot2` by drawing additional elements over specific areas. This approach is effective in settings where accessibility is a concern or when producing graphics that need to remain readable in black-and-white format.

To approximate the use of textures, consider applying varied line styles or dash patterns to the outlines of each ridge. In ggplot2, the `linetype` argument supports different line patterns (e.g., `"solid"`, `"dashed"`, or `"dotted"`) that can indicate specific categories or groups without relying on color.

```r
ggplot(data, aes(x = value, y = category, fill = category)) +
    geom_density_ridge(linetype = "dashed", size = 0.8, alpha = 0.6) +
    labs(title = "Ridgeline Plot with Line Styles", x = "Value",
    y = "Category") +
    theme_minimal()
```

In this example, `linetype = "dashed"` applies a dashed border to each ridge, effectively adding a textural element that aids in distinguishing categories. This technique allows for more flexible use of color while enhancing the plot's aesthetic quality.

8.5.3 Annotating Key Points and Adding Legends

Annotations and legends are essential tools for guiding viewers through ridgeline plots, especially in complex visualizations with multiple categories. Well-placed annotations can draw attention to specific data points, trends, or anomalies, helping viewers interpret critical insights. Legends, meanwhile, provide context by identifying color codes or line patterns associated with each category, enabling users to understand the plot's structure at a glance.

When annotating key points, it is important to ensure that labels are concise and strategically positioned to avoid overlapping with data lines or other elements. ggplot2's `annotate` function allows for flexible placement of text and arrows, making it easy to highlight peaks, troughs, or specific distributions within the ridgeline plot.

```r
ggplot(data, aes(x = value, y = category, fill = category)) +
    geom_density_ridge(alpha = 0.6, color = "black") +
    annotate("text", x = 7, y = "Category C", label = "High Peak", size = 4,
    hjust = 1, color = "red") +
    labs(title = "Ridgeline Plot with Annotations", x = "Value",
    y = "Category") +
    theme_minimal()
```

In this example, the `annotate` function places a text label next to a high peak in `Category C`, drawing attention to an area of interest. By setting `color = "red"`, the annotation is visually distinct, guiding viewers to the highlighted data point without disrupting the overall layout.

Adding a legend further supports interpretability by clarifying the color scheme, patterns, or line styles used in the plot. In ggplot2, legends are automatically generated based on the fill or color attributes, but they can be customized for clarity and aesthetics. For example, setting legend.position = "top" or "bottom" can improve readability by minimizing interference with the main plot area.

```r
ggplot(data, aes(x = value, y = category, fill = category)) +
    geom_density_ridge(alpha = 0.6, color = "black") +
    labs(title = "Ridgeline Plot with Custom Legend", x = "Value",
    y = "Category") +
    theme_minimal() +
    theme(legend.position = "top", legend.title = element_blank())
```

In this example, setting legend.position = "top" moves the legend above the plot, keeping the main area uncluttered. Removing the legend title with element_blank() streamlines the layout, creating a clean, organized appearance that supports the plot's readability.

By applying these aesthetic techniques, color gradients, opacity adjustments, patterns, annotations, and legend customization, designers can create ridgeline plots that are both visually appealing and highly informative. These enhancements not only improve the plot's aesthetic appeal but also aid viewers in interpreting complex data, making ridgeline plots more accessible and engaging for a wide audience.

8.6 Accessibility Considerations

Creating accessible ridgeline plots is essential to ensure that visualizations are understandable and usable by a diverse audience, including individuals with color vision deficiencies and other visual impairments. Accessibility in data visualization not only enhances readability but also promotes inclusivity by making insights available to all viewers. This section outlines key considerations for improving accessibility in ridgeline plots, focusing on the selection of color palettes for color-blind users and the adjustment of line widths and spacing to accommodate varying levels of visual acuity.

8.6.1 Color Selection for Accessibility

Color is a critical component of ridgeline plots, as it helps differentiate layers, categories, and data trends. However, reliance on color alone can exclude viewers with color vision deficiencies, such as red-green color blindness (deuteranopia and protanopia) or blue-yellow color blindness (tritanopia). To make ridgeline plots accessible, it is essential to select color palettes that provide sufficient contrast and are distinguishable by color-blind users.

Color-blind friendly palettes, such as those offered by viridis, cividis, and plasma in the viridis package, are particularly effective for accessible data visualization. These palettes are designed with perceptually uniform color transitions, which improve readability and help viewers discern layers even when colors overlap. For example, the viridis palette progresses from dark purple to yellow, offering high contrast without relying on problematic hues like red and green.

```
library(viridis)
ggplot(data, aes(x = value, y = category, fill =..x..)) +
    geom_density_ridge_gradient(scale = 1.5, rel_min_height = 0.01) +
    scale_fill_viridis_c(option = "viridis") +
    labs(title = "Accessible Ridgeline Plot with Color-Blind Friendly Palette",
    x = "Value", y = "Category") +
    theme_minimal()
```

In this example, scale_fill_viridis_c(option = "viridis") applies the viridis color palette, ensuring that the plot remains accessible to color-blind viewers. The continuous gradient also enhances interpretability by emphasizing areas of high density within each ridgeline.

It is also advisable to maintain a sufficient contrast ratio between layers, especially when multiple ridgelines overlap. High contrast ratios help distinguish each layer, particularly for viewers with low vision or in environments with limited lighting. Avoiding low-contrast color combinations, such as light shades of green and yellow or red and blue, enhances readability across a range of viewing conditions. Testing visualizations with color-blind simulation tools, such as Color Oracle or Adobe's color-blind proofing feature, can provide additional assurance that the color choices are accessible to all users.

8.6.2 Adjusting Line Widths and Spacing

Adjusting line widths and spacing is another effective strategy for improving accessibility in ridgeline plots. Thicker lines and increased spacing can make each layer more distinct, helping viewers with visual impairments to interpret individual distributions without difficulty. When ridgelines are closely packed, their overlap may create visual clutter, making it challenging to identify specific layers. Adjustments to line thickness and vertical spacing ensure that ridgelines remain clear and distinguishable.

```
ggplot(data, aes(x = value, y = category, fill = category)) +
    geom_density_ridge(size = 1.2, alpha = 0.6, color = "black") +
    labs(title = "Ridgeline Plot with Increased Line Width",
    x = "Measurement", y = "Category") +
    theme_minimal()
```

In this example, size = 1.2 increases the line thickness, making the edges of each ridgeline more prominent. This approach is particularly effective in plots with overlapping distributions, as the thicker lines define the boundaries more clearly, enhancing accessibility for viewers with low vision.

```
ggplot(data, aes(x = value, y = category, fill = category)) +
    geom_density_ridge(scale = 1.8, alpha = 0.6, color = "black") +
    labs(title = "Ridgeline Plot with Adjusted Spacing", x = "Value",
    y = "Category") +
    theme_minimal()
```

In this example, `scale = 1.8` increases the spacing between ridgelines, ensuring that each layer remains visually distinct. This adjustment is particularly useful in datasets with high category counts or when categories have similar distributions, as it prevents overlapping curves from blending together. By providing sufficient space between layers, the plot becomes easier to interpret, even for viewers with visual impairments.

```
ggplot(data, aes(x = value, y = category, fill = category)) +
    geom_density_ridge(size = 1.2, scale = 1.8, alpha = 0.6, color = "black") +
    labs(title = "Accessible Ridgeline Plot with Line Width and
    ↪    Spacing Adjustments",
    x = "Value", y = "Category") +
    theme_minimal()
```

In this final example, both `size = 1.2` and `scale = 1.8` are applied to enhance accessibility. This approach creates a well-balanced ridgeline plot that accommodates diverse visual needs, making the data accessible and interpretable for a broader audience.

By following these accessibility considerations (Table 8.3), selecting color-blind friendly palettes, ensuring sufficient contrast, and adjusting line widths and spacing, designers can create ridgeline plots that are inclusive and accessible. These techniques not only improve readability for viewers with visual impairments but also enhance the overall clarity of the visualization, making it more effective for all users.

8.7 Case Studies and Practical Applications

Ridgeline plots are versatile tools for visualizing complex datasets, particularly those involving changes over time, comparisons across categories, or patterns across multiple groups. By displaying distributions in an overlapping yet visually distinct manner, ridgeline plots provide insights into temporal trends, category comparisons, and multi-group experimental results. This section presents case studies and practical applications using benchmark datasets, demonstrating how ridgeline plots can be used to visualize seasonal data, analyze social trends over time, and compare experimental results across groups.

Table 8.3: Accessibility Enhancements for Ridgeline Plots

Enhancement	Description
Color Palette	Use color-blind-friendly palettes like viridis or plasma for accessible color differentiation.
Contrast	Ensure high contrast between layers for better distinction, particularly for overlapping distributions.
Line Thickness	Increase line thickness to define boundaries clearly, aiding users with low vision.
Spacing	Increase vertical spacing to avoid overlap and clutter, especially in high-density plots.
Simulation Testing	Use color-blind simulators to confirm that color choices are accessible to a broad audience.

8.7.1 Real-World Examples

The following examples highlight specific use cases where ridgeline plots are particularly effective. Each case study demonstrates how ridgeline plots can simplify the visualization of intricate data patterns, revealing underlying trends and providing clarity for decision-makers.

Ridgeline plots are highly effective for visualizing seasonal patterns in climate data, where they allow for the comparison of temperature distributions across multiple years or months. The NOAA (National Oceanic and Atmospheric Administration) dataset, which includes historical monthly temperature data, is a prime example of how ridgeline plots can showcase seasonal variation and long-term climate trends.

```r
# Load required libraries
library(ggplot2)
library(ggridges)
library(dplyr)

# Example NOAA dataset with monthly temperature data
data <- data.frame(
    temperature = c(rnorm(30, mean=5), rnorm(30, mean=10), rnorm(30, mean=15),
    rnorm(30, mean=20)),
    month = rep(c("January", "April", "July", "October"), each = 30),
    year = rep(2010:2019, times = 12)
)

# Ridgeline plot showing seasonal temperature variation
ggplot(data, aes(x = temperature, y = month, fill =..x..)) +
    geom_density_ridge_gradient(scale = 2, rel_min_height = 0.01) +
    scale_fill_viridis_c(option = "plasma") +
    labs(title = "Seasonal Temperature Variation (NOAA Data)",
```

```
19      x = "Temperature (°C)", y = "Month") +
20      theme_minimal()
```

In this example, a ridgeline plot is used to display the monthly temperature distributions across multiple years. The `scale_fill_viridis_c` function applies a gradient fill, highlighting areas of higher temperature. This type of visualization reveals seasonal patterns, such as higher temperature distributions in the summer months and lower ones in winter, and allows viewers to quickly identify any unusual shifts in temperature across different years.

Social science researchers often analyze public opinion trends across time to understand shifts in attitudes, behaviors, or preferences. For example, survey datasets like the General Social Survey (GSS) track responses to questions on social and political issues over decades. Ridgeline plots can help visualize these changes by showing the distribution of responses over time, highlighting trends in public opinion.

```
1    # Sample data representing survey responses across years
2    data <- data.frame(
3        response = c(rnorm(100, mean=3), rnorm(100, mean=4), rnorm(100, mean=2),
4        rnorm(100, mean=5)),
5        year = rep(c("1980", "1990", "2000", "2010"), each = 100)
6    )
7
8    # Ridgeline plot of public opinion changes over time
9    ggplot(data, aes(x = response, y = year, fill = year)) +
10       geom_density_ridge(alpha = 0.7, color = "black") +
11       scale_fill_brewer(palette = "Set2") +
12       labs(title = "Changes in Public Opinion Over Time (GSS Data)",
13       x = "Response Scale", y = "Year") +
14       theme_minimal()
```

This ridgeline plot represents the distribution of survey responses across four decades. The plot allows viewers to track shifts in opinion over time, providing a high-level view of how responses have evolved. For example, the trend may show increasing support for a particular issue or changes in public attitudes over decades. This application of ridgeline plots is particularly valuable in policy research, where understanding social trends is essential for effective decision-making.

In experimental research, especially in the life sciences, it is common to compare the distribution of outcomes across different treatment groups. Ridgeline plots are well-suited for visualizing experimental results across multiple groups, as they allow for a quick comparison of distributions while retaining the individual characteristics of each group's data. The `iris` dataset, a well-known benchmark dataset in machine learning, provides a straightforward example for comparing sepal length distributions across species.

```
1    # Load iris dataset
2    data <- iris
```

```r
# Ridgeline plot of sepal length across species
ggplot(data, aes(x = Sepal.Length, y = Species, fill = Species)) +
    geom_density_ridge(alpha = 0.6, color = "black") +
    scale_fill_brewer(palette = "Dark2") +
    labs(title = "Comparison of Sepal Length Across Iris Species",
    x = "Sepal Length (cm)", y = "Species") +
    theme_minimal()
```

In this example, the ridgeline plot displays the sepal length distributions for each species of iris. The plot reveals differences in distribution shape, central tendency, and spread among the species, making it easy to identify which species have longer or shorter sepal lengths. This approach can be extended to other experimental datasets to compare group-level results, providing insights into variations in responses across conditions or treatments.

Ridgeline plots are also useful in economic data analysis, where they can visualize changes in distributions over time for indicators such as inflation rates, GDP growth, or income levels. The U.S. Census Bureau's Current Population Survey (CPS), for instance, includes data on household incomes across years. Ridgeline plots can show the distribution of income levels over time, revealing trends such as income growth or inequality.

```r
# Sample data representing household income levels across years
data <- data.frame(
    income = c(rnorm(100, mean=50000),
    rnorm(100, mean=55000), rnorm(100, mean=60000),
    rnorm(100, mean=65000)),
    year = rep(c("2000", "2005", "2010", "2015"), each = 100)
)

# Ridgeline plot of income levels over time
ggplot(data, aes(x = income, y = year, fill = year)) +
    geom_density_ridge(alpha = 0.6, color = "black") +
    scale_fill_brewer(palette = "Spectral") +
    labs(title = "Household Income Distribution Over Time (CPS Data)",
    x = "Income ($)", y = "Year") +
    theme_minimal()
```

In this ridgeline plot, we observe income distributions for four different years, with each ridge representing the household income levels for a specific year. This plot can reveal trends such as increasing income levels or changes in income distribution patterns over time. Economic analysts and policymakers can use such visualizations to monitor economic conditions, assess policy impacts, or identify trends related to income inequality.

8.7.2 Comparison with Other Plot Types

Ridgeline plots are one of several options for visualizing distributions, particularly when there are multiple categories or time points to compare. Other common plot types, such as histograms, violin plots, and heatmaps, each have strengths and specific applications; however, ridgeline plots offer unique advantages in situations that require comparing multiple distributions in a compact and visually appealing format. This section discusses the advantages of ridgeline plots over other methods, providing guidance on when they are preferable for representing distributions.

Histograms are a fundamental tool for visualizing the distribution of a single variable by grouping data points into bins, which are then represented as bars. While histograms are effective for displaying a single distribution or comparing two distributions side-by-side, they can become cluttered and difficult to interpret when applied to multiple categories or time points. In cases where distributions across numerous categories or time periods need to be visualized, ridgeline plots offer a cleaner, more organized alternative.

Ridgeline plots allow for the layering of multiple distributions along a vertical axis, which preserves the shape of each distribution without requiring individual panels or excessive space. The overlapping nature of ridgeline plots also enables viewers to quickly identify patterns, such as shifts in central tendency or changes in distribution spread, across categories. For instance, while comparing monthly temperature distributions across several years, a histogram would require multiple panels or color-coded bars within a single plot, which can be visually overwhelming. Ridgeline plots, by contrast, provide a clear visual flow, facilitating comparisons by reducing visual clutter.

Violin plots are another option for displaying distributions across categories. They combine aspects of box plots and kernel density plots, presenting the density of each category's data around a central axis. Violin plots are particularly useful when comparing a limited number of categories, as they show both the distribution's shape and central tendency in a single plot. However, they have limitations when dealing with numerous categories or time series data, as they require ample horizontal space and can become cluttered if too many violins are displayed.

Ridgeline plots are preferable to violin plots when dealing with a larger number of categories, as they layer the distributions vertically, conserving horizontal space. This layout also allows ridgeline plots to emphasize trends across categories, which is particularly useful in time series analysis or when examining data with a natural ordering (e.g., years or age groups). Ridgeline plots offer more flexibility in handling overlapping distributions through transparency and color gradients, allowing users to see the relative density of each distribution without the need for symmetry around a central axis. This makes ridgeline plots especially advantageous in cases where distributions have asymmetric shapes or exhibit gradual changes over categories.

Heatmaps are commonly used to visualize the intensity or density of data across two variables, typically using color gradients to represent values. While heatmaps are highly effective for representing large datasets, particularly when values are on a grid or spatially distributed, they are limited in their ability to convey the shape and spread of distributions. Heatmaps present aggregate data as cells in a matrix, which can obscure finer details, especially in datasets where distribution shape is meaningful, such as in multimodal data or data with skewed distributions.

Ridgeline plots, on the other hand, preserve the distributional shape for each category, allowing for a detailed view of each data group. This is particularly useful when analyzing the underlying characteristics of each distribution rather than simply its intensity. For example, in an environmental study analyzing monthly temperature variations over several years, a heatmap would show high-level temperature ranges by color intensity, but a ridgeline plot would reveal seasonal peaks, skewness, or spread in each month's temperature distribution. Thus, ridgeline plots are advantageous when the goal is to compare the shape of distributions across categories, as they provide more depth of information than a heatmap's aggregated cell-based format.

Ridgeline plots are most suitable when the primary goal is to compare multiple distributions across categories or time periods, especially in datasets where shape and spread convey important insights. They are ideal for:

> **Time Series Analysis**: Ridgeline plots excel at visualizing changes over time, making them ideal for seasonal data, economic trends, or survey data tracked over multiple years.

> **Category Comparison**: For datasets with numerous categories, such as demographic groups or experimental conditions, ridgeline plots provide a clear comparison of distributions without requiring excessive space.

> **Visualizing Distribution Shape**: When the shape of the distribution is as important as central tendency, ridgeline plots offer detailed insights, as they preserve density features such as skewness, bimodality, and outliers .

> **Compact Data Representation**: Ridgeline plots efficiently display many distributions in a compact format, which is helpful when screen space or plot area is limited.

While other plot types like histograms, violin plots, and heatmaps have their own applications, ridgeline plots are preferable when comparing multiple distributions with overlapping layers, especially if there is an order to the categories. Their unique design allows for effective visual comparisons and trend analysis across multiple groups, making them a powerful tool for exploring complex datasets.

8.8   Troubleshooting Common Design Issues

While ridgeline plots are effective tools for visualizing distributions across multiple categories or time periods, they can present certain design challenges, particularly in cases where there are many overlapping layers or wide variability in data scales. Overcoming these challenges is essential for creating clear, interpretable ridgeline plots. This section addresses common issues in ridgeline plot design, including managing overlapping layers and visual clutter, handling scaling and normalization across categories, and addressing labeling challenges in crowded plots.

8.8.1 Overlapping Layers and Clutter

One of the most frequent issues in ridgeline plots is excessive overlap between layers, which can result in visual clutter and make it difficult for viewers to distinguish individual distributions. This problem often arises when there are numerous categories or when the distributions are closely aligned. To address this issue, designers can use a combination of transparency adjustments, vertical spacing, and density controls.

```r
ggplot(data, aes(x = value, y = category, fill = category)) +
    geom_density_ridge(alpha = 0.5, color = "black") +
    labs(title = "Ridgeline Plot with Adjusted Transparency", x = "Value",
    y = "Category") +
    theme_minimal()
```

In this example, `alpha = 0.5` applies a moderate level of transparency, allowing each layer to be visible even in areas of overlap. Adjusting transparency is especially effective when comparing distributions with minor variations, as it reduces visual clutter while preserving data integrity.

```r
ggplot(data, aes(x = value, y = category, fill = category)) +
    geom_density_ridge(scale = 1.8, alpha = 0.5, color = "black") +
    labs(title = "Ridgeline Plot with Increased Vertical Spacing",
    x = "Value", y = "Category") +
    theme_minimal()
```

In this example, `scale = 1.8` increases the spacing between layers, preventing distributions from overlapping excessively. This approach is particularly helpful for large datasets with many categories, as it reduces visual clutter and improves readability.

```r
ggplot(data, aes(x = value, y = category, fill = category)) +
    geom_density_ridge(alpha = 0.6, bw = 0.8, color = "black") +
    labs(title = "Ridgeline Plot with Adjusted Bandwidth", x = "Value",
    y = "Category") +
    theme_minimal()
```

In this example, `bw = 0.8` decreases the bandwidth, providing a more refined density estimation that reduces curve overlap. Experimenting with bandwidth settings allows designers to balance smoothness and detail, enhancing the clarity of ridgeline plots with densely packed layers.

8.8.2 Scaling and Normalization

When categories in a ridgeline plot have different ranges or units, scaling and normalization become essential to ensure that distributions are comparable. Without proper scaling, certain categories may dominate the plot, distorting the viewer's perception of the data. Techniques such as min-max scaling and z-score normalization can standardize distributions across categories, enabling more meaningful comparisons.

```r
# Normalize data using min-max scaling
data <- data %>%
    group_by(category) %>%
    mutate(scaled_value = (value - min(value)) / (max(value) - min(value)))

# Ridgeline plot with normalized data
ggplot(data, aes(x = scaled_value, y = category, fill = category)) +
    geom_density_ridge(alpha = 0.6, color = "black") +
    labs(title = "Ridgeline Plot with Min-Max Scaled Data",
    x = "Normalized Value", y = "Category") +
    theme_minimal()
```

In this example, min-max scaling is applied within each category to normalize values. The normalized ridgeline plot allows for direct comparison between categories, as each distribution now shares a common scale.

```r
# Normalize data using z-score normalization \index{Normalization}
data <- data %>%
    group_by(category) %>%
    mutate(z_scaled_value = (value - mean(value)) / sd(value))

# Ridgeline plot with z-score normalized data
ggplot(data, aes(x = z_scaled_value, y = category, fill = category)) +
    geom_density_ridge(alpha = 0.6, color = "black") +
    labs(title = "Ridgeline Plot with Z-Score Normalized Data",
    x = "Z-Score", y = "Category") +
    theme_minimal()
```

Here, z-score normalization centers each distribution around zero and scales it by the standard deviation. This technique is particularly effective for identifying deviations from the mean and comparing distributions with different variances.

8.8.3 Labeling Challenges

In ridgeline plots with numerous categories or overlapping layers, label placement can be challenging. Crowded labels may overlap with data lines, reducing readability and detracting from the plot's visual appeal. Techniques such as callouts, interactive tooltips, or separate legends can help mitigate these challenges and improve interpretability.

```r
ggplot(data, aes(x = value, y = category, fill = category)) +
    geom_density_ridge(alpha = 0.6, color = "black") +
    annotate("text", x = 6.5, y = "Category B", label = "Key Point",
    hjust = -0.1, color = "red") +
    labs(title = "Ridgeline Plot with Callout Annotation", x = "Value",
    y = "Category") +
    theme_minimal()
```

In this example, the `annotate` function is used to place a label outside the main plot area, connected by a line to a specific data point. This approach keeps the plot clean while drawing attention to important information.

```r
library(plotly)
fig <- ggplot(data, aes(x = value, y = category, fill = category)) +
    geom_density_ridge(alpha = 0.6, color = "black") +
    labs(title = "Interactive Ridgeline Plot", x = "Value", y = "Category")

# Convert ggplot to interactive plotly plot
ggplotly(fig, tooltip = c("x", "y"))
```

In this example, `ggplotly` converts the `ggplot` object into an interactive plot where viewers can hover over ridges to see detailed information. This feature is particularly useful in dense plots, as it provides details on demand without cluttering the visualization.

```r
ggplot(data, aes(x = value, y = category, fill = category)) +
    geom_density_ridge(alpha = 0.6, color = "black") +
    labs(title = "Ridgeline Plot with External Legend", x = "Value",
    y = "Category") +
    theme_minimal() +
    theme(legend.position = "right")
```

In this example, the `theme(legend.position = "right")` argument positions the legend outside the main plot area on the right side. This layout keeps the plot area clean and unobstructed, particularly when working with numerous categories. External legends are useful when the plot contains a high number of categories, as they provide a clear reference without causing label overlap or clutter within the main visualization.

These troubleshooting techniques, adjusting transparency, spacing, and density for overlapping layers; scaling and normalizing data for comparability; and using callouts, tooltips, and legends for labeling, address common challenges in ridgeline plot design (Table 8.4). By applying these methods, designers can create ridgeline plots that remain visually clean, accessible, and informative, even in cases with complex, high-density data. These techniques ensure that the plot remains easy to interpret, providing an effective tool for data visualization across various fields and applications.

Table 8.4: Solutions to Common Ridgeline Plot Design Challenges

Challenge	Solution
Overlapping Layers	Adjust transparency, increase vertical spacing, and fine-tune density settings to reduce clutter.
Scaling and Normalization	Use min-max scaling or z-score normalization to ensure comparability across categories.
Label Placement	Use callouts, interactive tooltips, or external legends to improve readability in crowded plots.
Excessive Density	Adjust bandwidth to manage curve density, balancing detail with clarity.

8.9 Summary and Best Practices

Designing effective ridgeline plots requires a thoughtful approach that balances data clarity, accessibility, and visual appeal. Ridgeline plots are a versatile tool for representing multiple distributions, especially when comparisons across categories or time points are necessary. By following best practices, designers can ensure that ridgeline plots are not only aesthetically pleasing but also clear and accessible to a wide audience. This section summarizes the key takeaways for creating effective ridgeline plots and provides a checklist to guide readers through the design process.

8.9.1 Key Takeaways

Creating vibrant and accessible ridgeline plots involves carefully considering aspects such as data clarity, readability, and design aesthetics. The following key takeaways summarize best practices for designing ridgeline plots that communicate insights effectively:

1. **Select Meaningful Variables and Categories:** Focus on variables that offer valuable insights, and limit the number of categories to maintain readability. When comparing distributions across many categories, group related categories or use multiple plots to avoid overcrowding.

2. **Balance Data Density and Readability:** Adjust transparency, spacing, and line weights to manage overlapping layers, which is essential for preventing visual clutter. Proper spacing and layer separation improve interpretability, especially in plots with numerous categories.

3. **Use Accessible Color Palettes:** Choose color palettes that are distinguishable by color-blind users, such as those provided by the `viridis` package, and ensure adequate contrast between layers. Accessible color choices expand the usability of ridgeline plots for diverse audiences.

4. **Apply Gradient Fills and Opacity Strategically:** Use color gradients to indicate density or intensity within distributions, and apply transparency to maintain clarity when layers overlap. These techniques enhance visual appeal without sacrificing data clarity.

5. **Ensure Proper Scaling and Normalization:** When categories have different ranges or units, apply normalization techniques such as min-max scaling or z-score normalization to standardize the data. Proper scaling ensures that distributions are comparable, avoiding misinterpretation.

6. **Address Labeling Challenges Thoughtfully:** Use callouts, interactive tooltips, or external legends to manage label placement in crowded plots. Clear labeling is essential for viewer interpretation, especially when there are numerous categories or data points.

7. **Customize Line Styles and Spacing:** Adjust line thickness, line style, and vertical spacing to create a well-organized plot. Thicker lines, distinct line styles, and adequate spacing improve readability, particularly for viewers with visual impairments.

8. **Test for Accessibility:** Use color-blind simulation tools to check that colors and contrasts are accessible, and test readability on different screen sizes if the plot will be used in digital formats. Accessible design practices increase the reach and impact of the visualization.

Following these best practices ensures that ridgeline plots remain functional, visually appealing, and accessible to a wide audience, allowing viewers to explore and understand complex data patterns with ease.

8.9.2 Checklist for Creating Effective Ridgeline Plots

To aid in the creation of clear and impactful ridgeline plots, the following checklist outlines essential steps for the design process, from data selection to final layout adjustments:

➤ **Data Selection and Preparation**

  ➤ Identify relevant variables and categories that provide meaningful insights.

  ➤ Limit the number of categories to maintain clarity; group categories if necessary.

  ➤ Normalize or scale data if categories have different ranges or units.

➤ **Plot Setup and Layer Management**

  ➤ Use transparency (`alpha`) to manage overlapping layers and reduce visual clutter.

Table 8.5: Checklist for Creating Effective Ridgeline Plots

Step	Description
Data Selection	Choose relevant variables and limit categories to maintain clarity. Normalize data if categories have different ranges.
Layer Management	Adjust transparency, spacing, and line thickness to avoid clutter and ensure clarity.
Color and Aesthetics	Use accessible color palettes, apply gradient fills for density, and maintain high contrast.
Labeling	Use callouts, external legends, or tooltips to manage label placement in crowded plots.
Accessibility Testing	Check for color contrast, readability, and usability across different screen sizes.

- ➤ Adjust vertical spacing between ridgelines using the `scale` parameter to improve readability.
- ➤ Set line thickness (`size`) and line style (`linetype`) for clear boundary definition.

➤ **Color and Aesthetic Design**

- ➤ Apply a color-blind friendly palette (e.g., `viridis`) to enhance accessibility.
- ➤ Use gradient fills to represent density or intensity within distributions.
- ➤ Ensure sufficient contrast between layers for visual clarity.

➤ **Labeling and Annotation**

- ➤ Place labels strategically to avoid overlap with ridgelines.
- ➤ Use callouts or external legends to label crowded plots.
- ➤ For interactive environments, add tooltips for on-demand information.

➤ **Final Layout and Accessibility Testing**

- ➤ Check color contrast and readability for color-blind viewers using simulation tools.
- ➤ Test the plot on various screen sizes to ensure legibility in digital formats.
- ➤ Review the overall balance of the plot to confirm that each layer is clear and distinguishable.

This checklist provides a comprehensive guide for designing effective ridgeline plots. By following these steps, designers can ensure that their plots are clear, accessible, and visually compelling, offering viewers an engaging and insightful way to explore complex distributions. Through careful planning and attention to detail, ridgeline plots can be transformed into powerful tools for comparative analysis and data storytelling.

Suggested Readings

1. Hennig, C.: Ridgeline plot and clusterwise stability as tools for merging gaussian mixture components. p. 109 – 116 (2010). DOI 10.1007/978-3-642-10745-0_11
2. Liu, S., Liu, Y., Li, J., Huang, Y., Shangguan, Y., Deng, Z., Weng, D., Wu, Y.: Ridgebuilder: Interactive authoring of expressive ridgeline plots. In: Proceedings of the 2025 CHI Conference on Human Factors in Computing Systems, CHI '25. Association for Computing Machinery, New York, NY, USA (2025). DOI 10.1145/3706598.3714209. URL https://doi.org/10.1145/3706598.3714209

9 Creating Breathtaking Density Plots

Abstract

This chapter discusses density plots, a versatile visualization tool for communicating patterns in large, complex datasets. Readers are guided through the process of choosing the right type of 2D density plot, including heatmaps, contour plots, and kernel density estimation (KDE) plots, based on data characteristics and visualization goals. The chapter explores essential design principles, such as selecting perceptually uniform color schemes, optimizing transparency to manage overlapping data, and setting appropriate bin sizes and resolution to balance detail and clarity. Advanced topics include layering density plots with other visualizations, using KDE for smoothing data, and adjusting bandwidth and scaling to enhance accuracy and interpretability. Practical tutorials using Matplotlib, Seaborn, and Plotly illustrate step-by-step techniques for creating and customizing 2D Density Plots.

Aims

After reading this chapter, you should be able to:

- ➤ Understand the purpose and applications of 2D Density Plots, including their use in identifying patterns, correlations, and clusters in data.

- ➤ Apply design principles to select appropriate color schemes, transparency levels, and gradients that enhance clarity and accessibility.

- ➤ Construct and customize contour plots, adjusting contour levels, line styles, and color fills to highlight density variations.

- ➤ Design heatmaps that effectively represent data density with color intensity, using appropriate scales, bin sizes, and color gradients.

- ➤ Use advanced techniques to enhance density plots.

- ➤ Apply accessibility best practices, such as choosing color-blind friendly palettes and ensuring readable legends and labels.

- ➤ Identify and avoid common pitfalls in 2D Density Plot design.

© The Author(s), under exclusive license to Springer Nature Switzerland AG 2026

R. Damaševičius, *Human-Centred Scientific Data Visualisation*,
Undergraduate Topics in Computer Science,
https://doi.org/10.1007/978-3-032-01606-5_9

9.1 Introduction to 2D Density Plots

2D Density Plots are indispensable tools in data visualization, providing insight into the distribution, concentration, and relational patterns within a dataset that spans two dimensions. As datasets grow increasingly large and complex, conventional scatter plots and other simplistic visualizations can become cluttered and difficult to interpret, particularly when there is significant overlap between data points. By visualizing the density of data points rather than each individual point, 2D Density Plots offer a refined view that highlights areas of high and low concentration. This approach not only preserves the interpretability of the plot in high-density regions but also reveals underlying structures, patterns, and relationships within the data that may not be immediately evident through raw data visualization. Such clarity is critical for effective decision-making and data-driven analysis, especially in fields where identifying clusters, correlations, and outliers is essential.

9.1.1 Purpose and Applications

The primary purpose of 2D Density Plots is to offer a clear, structured representation of data distribution across two continuous variables. By presenting the data in terms of density, these plots allow viewers to assess where data points are most concentrated, where they are sparse, and how they relate across dimensions. For example, in a dataset with two variables, x and y, representing, respectively, a person's age and income, a 2D Density Plot can reveal income distributions across different age groups, identifying patterns like higher concentrations in certain income brackets within specific age ranges. Such information is invaluable for statistical analysis and decision-making processes as it provides a more nuanced view of data structure beyond raw point data.

One of the most significant advantages of 2D Density Plots is their utility in identifying correlations and trends in large datasets. In scenarios where two variables are expected to correlate, a 2D Density Plot can illustrate the strength and nature of that correlation. In finance, for instance, analysts use 2D Density Plots to understand the relationship between investment risk and return across different asset classes. By observing density patterns, they can identify the typical risk-return profiles of various asset types, which aids in portfolio optimization. Similarly, in the field of ecology, researchers employ 2D Density Plots to explore the relationships between environmental factors, such as temperature and vegetation density, or to identify regions where certain species populations are concentrated. These insights are critical for conservation planning, as they help ecologists pinpoint high-density population zones, enabling targeted interventions to protect biodiversity.

2D Density Plots are also highly valuable in social sciences, where researchers are often tasked with identifying clusters within large demographic datasets. By plotting variables such as income and educational attainment, a 2D Density Plot can reveal socio-economic clusters within a population. For instance, it may show distinct density peaks that correspond to different socio-economic groups, providing insights into educational disparities or economic stratification. This type of analysis is essential for policy-making, as it can guide targeted initiatives to address social inequality or improve educational access for specific demographic groups. Beyond these fields, the applications of 2D Density Plots extend to various disciplines where large datasets are analyzed to uncover meaningful trends, assess associations between variables, and identify clustering phenomena.

9.1.2 Types of 2D Density Plots

There are several types of 2D Density Plots, each with unique strengths and applications, making them suitable for different types of data and visualization goals. The most common types of 2D Density Plots are contour plots, heatmaps, and kernel density estimation (KDE) plots, each offering a distinct perspective on density distribution.

Contour Plots are one of the most popular forms of 2D density visualizations, particularly well-suited for representing continuous data with natural gradients. Contour plots display density through lines that connect points of equal density, creating a topographical effect that resembles elevation maps. Each contour line corresponds to a specific density level, making it easy to discern areas of high and low concentration. By adjusting the number of contour levels, the plot can be fine-tuned to highlight specific density thresholds or gradual changes in data distribution. Contour plots are commonly used in fields like meteorology and geology, where understanding gradual transitions, such as temperature changes over a geographic region or elevation gradients, is crucial. Contour plots provide a clean, minimalist view of density that highlights overall patterns while maintaining interpretability, even in complex datasets. Fig. 9.1 depicts the structure of a typical contour plot, including filled and lined contour levels, scalar legends, and axis annotations.

Heatmaps are another widely used type of 2D density plot, particularly effective for data with discrete or categorical variables in addition to continuous data. In a heatmap, color intensity represents the density of data points, with darker or more intense colors indicating areas of higher density. Heatmaps are highly versatile and can be used across a variety of fields, from biology to business analytics. For example, in web analytics, a heatmap can display the density of clicks or user interactions across a webpage, highlighting the most frequently accessed areas. Heatmaps are also used in genomics to show expression levels of genes across different conditions or sample groups, with color intensity revealing gene activity patterns. The primary strength of heatmaps lies in their visual impact; the use of color intensity allows for an immediate, intuitive understanding of data density, which is especially useful for quickly assessing hotspots or areas of interest within a dataset. As shown in Fig. 9.2, a heatmap consists of color-coded matrix cells, axis labels, and a color scale, all essential for interpreting dense, grid-structured data.

Kernel Density Estimation (KDE) Density Plots represent a non-parametric method for estimating the probability density function of a random variable, making them ideal for smoothing out data and providing a clear density profile across

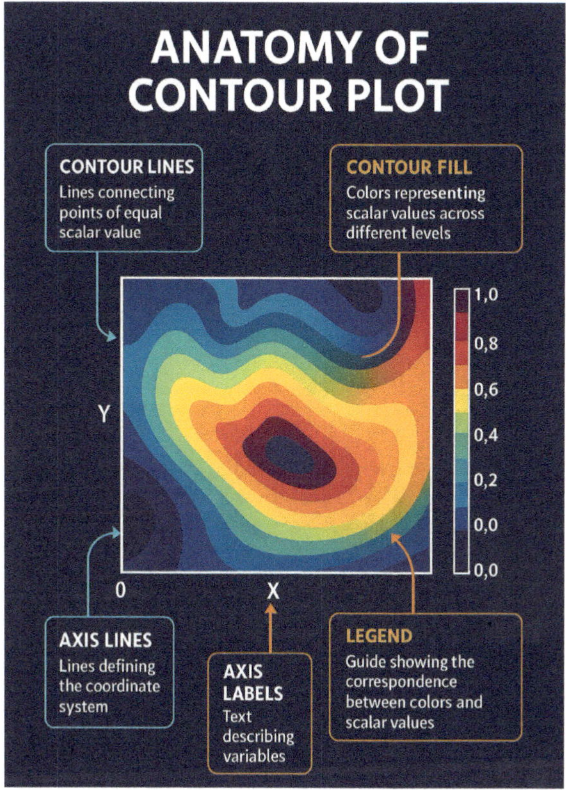

Fig. 9.1: Anatomy of a Basic Contour Plot showing key components of a contour plot, including contour lines, contour fill, axis lines, axis labels, and the legend that maps scalar values to color levels.

two dimensions. Unlike heatmaps or contour plots, which are based on binning or thresholding, KDE plots use a kernel function to estimate density over a continuous area, resulting in a smooth representation of the underlying distribution. KDE plots are advantageous when visualizing datasets with high variability, as the smoothing effect minimizes noise and emphasizes overall trends. In the financial sector, KDE plots are commonly used to understand return distributions for different asset classes, helping analysts identify regions of concentrated or dispersed returns. KDE plots are also valuable in quality control and manufacturing, where they help identify patterns or deviations in product dimensions or performance metrics, thereby aiding in identifying potential defects or process improvements. The key visual components of a KDE-based density plot are illustrated in Fig. 9.3, where the density curve, bandwidth control, and baseline are clearly annotated.

Each type of 2D Density Plot offers unique benefits and is suited to specific analytical needs. While contour plots provide a structured view with clear density

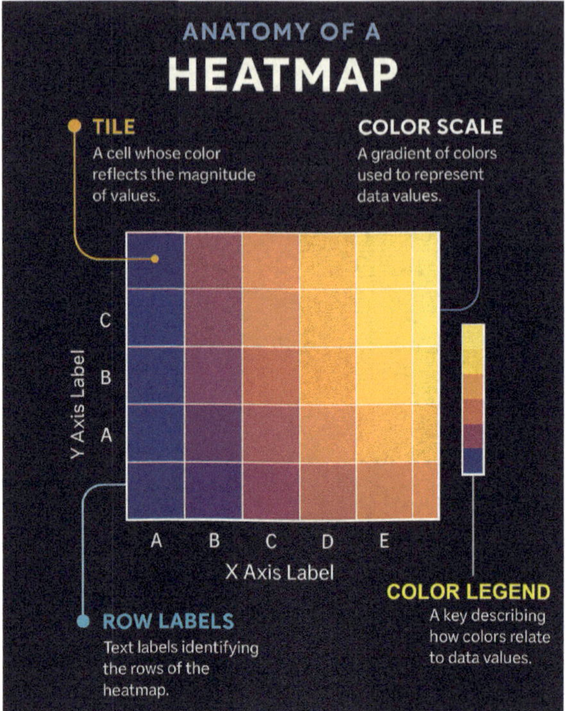

Fig. 9.2: Anatomy of a Basic Heatmap highlighting the key components of a heatmap, including color-coded tiles, axis labels, row labels, color scale, and legend, all essential for interpreting magnitude and structure in matrix-like data.

thresholds, heatmaps emphasize density intensity through color, and KDE plots offer smooth, continuous density estimates. Choosing the appropriate plot type depends on the data characteristics and the intended insights, as well as the need for clarity and accessibility in the final visualization. By using the strengths of each plot type, analysts can create 2D Density Plots that reveal valuable patterns and facilitate a deeper understanding of complex, multidimensional data. Table 9.1 summarizes the distinguishing features of KDE plots, heatmaps, and contour plots, highlighting their respective visual encodings, use cases, and customization options.

9.2   Design Principles for Effective 2D Density Plots

Creating effective 2D Density Plots requires careful consideration of design elements to ensure the visualization accurately represents the data and enhances interpretability. Key principles include selecting the appropriate plot type for the data and analysis objectives, using color schemes that effectively communicate density levels, and applying transparency to manage dense data points. This section discusses

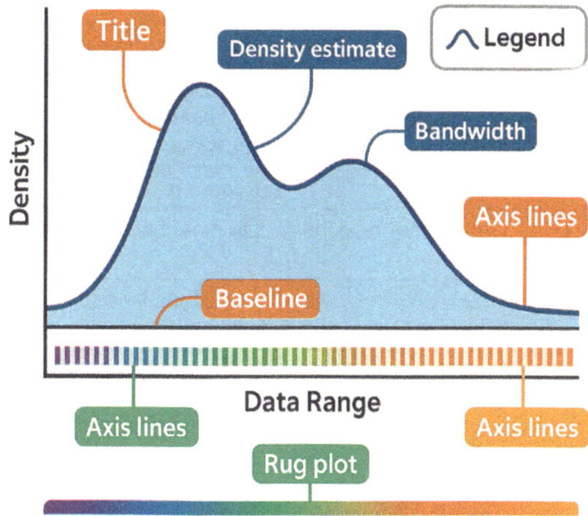

Fig. 9.3: Anatomy of a Basic Density Plot illustrating the key components of a density plot, including the density estimate curve, bandwidth, baseline, rug plot, axis lines, and legend.

these principles in detail, providing guidelines to help create clear, accessible, and visually appealing 2D Density Plots.

9.2.1 Choosing the Right Plot Type

Selecting the most suitable plot type is essential for achieving clarity and conveying meaningful insights. The choice of plot type depends on several factors, including the nature of the data, the specific insights desired, and the audience's familiarity with different visual representations. Each plot type offers unique strengths: contour plots are ideal for showing gradual density changes, heatmaps excel in displaying intensity through color, and KDE plots are effective for smooth density estimation.

For continuous data with smooth transitions, **contour plots** are often the best choice. Contour plots are particularly useful when visualizing data that represents gradients or gradual changes across dimensions, such as temperature distributions or geographic terrain. By displaying data as lines of equal density, contour plots offer a minimalist view that emphasizes the overall structure of the data rather than individual values. This plot type is especially valuable in applications requiring a clean, structured display of density variations, such as meteorological maps or population density studies.

When a visually striking and immediate interpretation of density is required, **heatmaps** provide an intuitive solution. In heatmaps, color intensity represents the

Table 9.1: Comparison of Key Elements in Different Types of 2D Density Plots

Element	KDE Density Plot	Heatmap Plot	Contour Plot
Data Representation	Smoothed probability density	Discrete color-encoded grid	Contour lines or filled regions
Main Feature	Density curve or surface	Color tiles representing value	Lines of equal scalar value
Color Encoding	Gradient for density levels	Gradient based on cell value	Gradient (fill) or line color
Axes Labels	Required	Required	Required
Gridlines / Axis Ticks	Optional for readability	Often included	Usually included
Transparency (`alpha`)	Often used for overlap control	Occasionally used	Optional
Legend / Colorbar	Indicates density values	Indicates value magnitude	Indicates scalar levels
Interactivity Support	Hover, zoom (e.g., Plotly)	Hover, zoom (e.g., Plotly)	Hover, zoom (e.g., Plotly)
Use Case	Smooth distribution overview	Matrix-like or binned data	Highlight continuous gradients
Customization	Bandwidth, color map, levels	Color map, bin size, scale	Levels, line style, fill colors

density of data points within each bin or region, allowing viewers to quickly identify areas of high or low concentration. Heatmaps are well-suited to datasets with discrete or categorical variables, as well as to scenarios where rapid identification of density hotspots is the goal. They are commonly used in fields like web analytics (e.g., to display click density) and genomics (e.g., to visualize gene expression levels). When using heatmaps, it is essential to choose an appropriate bin size; too large a bin size may obscure details, while too small a bin size can lead to excessive visual noise.

> **⚠ Warning**
>
> Too small a bin size can introduce noise and make the plot unreadable; too large a bin size can smooth out important patterns. Choose bin sizes carefully to balance granularity with clarity.

For data requiring a smooth, continuous representation of density, **kernel density estimation (KDE) plots** are ideal. KDE plots use a kernel function to estimate the probability density across a continuous space, resulting in a smooth visualization that reveals underlying patterns without the noise that can arise from raw data points. KDE plots are highly effective for exploratory analysis, where the goal is to understand the general distribution of data rather than precise values. This type of plot is useful in scenarios such as risk analysis in finance or error analysis in manufacturing, where understanding the distribution profile rather than exact values is key.

Ultimately, selecting the right plot type requires aligning the plot's characteristics with the data's attributes and the visualization's purpose. By choosing the appropriate plot type, analysts can create 2D Density Plots that effectively convey the desired insights while maintaining visual clarity and interpretability.

9.2.2 Color Schemes and Gradients

Color is a critical element in 2D Density Plots, as it communicates information about density levels, helping viewers interpret areas of high and low concentration. The choice of color scheme and gradient significantly impacts the plot's readability, accuracy, and accessibility, making it essential to select a color scheme that is both perceptually uniform and accessible to color-blind users.

A **perceptually uniform** color scheme ensures that changes in color intensity correspond consistently to changes in density, preventing visual distortions that might mislead viewers. Non-uniform color schemes, where certain hues appear more intense or saturated than others, can exaggerate density in certain areas, resulting in inaccurate interpretations. Color maps such as `viridis`, `cividis`, and `plasma` are popular choices because they offer perceptually uniform gradients that are suitable for conveying density changes. For example, the `viridis` color map transitions smoothly from dark blue to yellow, providing a consistent representation of density across different levels.

Color accessibility is equally important, as it ensures that the plot is interpretable by all viewers, including those with color vision deficiencies. Color-blind friendly palettes, such as `cividis`, are specifically designed to accommodate color-blind viewers, providing clear and distinct hues without relying on red-green or blue-yellow contrasts. It is also beneficial to use a gradient that varies in both hue and brightness, as this makes the plot readable even for viewers with reduced color sensitivity. Adding color intensity adjustments, such as increasing brightness for higher densities, can improve readability on various screens and under different lighting conditions.

Another effective approach is to use **sequential or diverging color schemes** depending on the data characteristics. Sequential color schemes, which progress from light to dark or from one color to another, are ideal for density plots where higher values should appear more intense. Diverging color schemes, which use contrasting colors for low and high values with a neutral midpoint, can be useful in plots where mid-range densities hold particular significance, such as in plots that highlight deviations from a median density level. In summary, choosing an appropriate color scheme is essential for accurate density interpretation, accessibility, and visual appeal.

9.2.3 Transparency and Overlapping Layers

Transparency is a powerful tool in 2D Density Plots, especially when dealing with dense data or multiple overlapping datasets. By adjusting transparency levels, designers can reduce visual clutter and prevent overplotting , allowing underlying density patterns to emerge clearly. Transparency not only helps balance the visibility of high-density regions but also ensures that low-density areas remain discernible, even when overlaid with dense data points.

In single-layer 2D Density Plots, **applying transparency** to dense regions prevents the plot from appearing overly saturated. A high opacity in densely populated areas may obscure details in adjacent regions, resulting in visual imbalance and loss of information. By setting a lower opacity for dense regions, analysts can highlight data concentration without overshadowing less dense areas. This approach is particularly effective in heatmaps and KDE plots, where transparency can reveal subtle density differences that might otherwise go unnoticed.

When visualizing **multiple datasets on the same plot**, transparency becomes essential for distinguishing between layers without obscuring data. In multi-layered KDE plots or heatmaps, each dataset can be assigned a unique color with adjusted transparency levels to indicate overlap while retaining each dataset's visibility. For example, if analyzing customer demographics across different product segments, each segment can be represented by a unique color, with transparency allowing viewers to see where segments intersect or diverge. This multi-layered approach is valuable in comparative analysis, as it reveals both shared and unique patterns across datasets.

It is important to balance transparency with readability, as excessive transparency can make data difficult to interpret, especially in low-density regions. For best results, start with moderate transparency settings (e.g., an `alpha` value of 0.5) and adjust as necessary based on the density of the data and the complexity of the plot. By using transparency thoughtfully, designers can enhance the clarity and depth of 2D Density Plots, ensuring that density patterns are clearly visible while minimizing visual clutter.

In summary, effective 2D Density Plot design requires selecting the right plot type, applying perceptually uniform and accessible color schemes, and using transparency to handle dense data and overlapping layers. These principles ensure that 2D Density Plots are not only visually appealing but also communicate information clearly and accurately, supporting informed decision-making and analysis.

9.3 Creating 2D Density Plots in Python with `Matplotlib` and `Seaborn`

Python offers a powerful toolkit for creating 2D Density Plots, primarily through libraries like `Matplotlib` and `Seaborn` These libraries provide versatile functions for visualizing data density and allow for a wide range of customization options to enhance clarity, readability, and aesthetic appeal. This section provides a step-by-step guide to creating a basic 2D Density Plot in Python, followed by techniques to customize color scales, axis labels, and gridlines.

9.3.1 Step-by-Step Guide

A basic 2D Density Plot can be created in Python using `Seaborn`, which provides a straightforward interface for visualizing data distributions and density levels. The following example demonstrates how to create a 2D Density Plot using sample data generated from a bivariate normal distribution. This approach highlights the central concentration of data and reveals the spread of values across two variables.

```python
import numpy as np
import seaborn as sns
import matplotlib.pyplot as plt

# Generate sample data
np.random.seed(42)
x = np.random.normal(size=1000)
y = x * 0.5 + np.random.normal(size=1000)

# Create a 2D density plot
plt.figure(figsize=(8, 6))
sns.kdeplot(x=x, y=y, fill=True, cmap='viridis', levels=20)
plt.title('Basic 2D Density Plot')
plt.xlabel('X-axis Label')
plt.ylabel('Y-axis Label')
plt.show()
```

In this example, we start by generating synthetic data that follows a bivariate normal distribution. The `kdeplot` function from `Seaborn` is used to create a kernel density estimate (KDE) plot, where the density of data points is represented by varying shades of color. The `fill=True` parameter fills the density contours with color, while `cmap='viridis'` specifies the color map. Here, the `viridis` color map is chosen for its accessibility, as it is color-blind friendly and perceptually uniform. The plot title and axis labels provide context for interpreting the plot, making this a simple yet informative density plot.

9.3.2 Customization Techniques

To create a more polished and visually appealing 2D density plot, `Matplotlib` and `Seaborn` offer a variety of customization options. These options include adjusting color scales, modifying axis labels, and adding gridlines, each of which enhances readability and aesthetic appeal.

```python
plt.figure(figsize=(8, 6))
sns.kdeplot(x=x, y=y, fill=True, cmap='plasma', levels=15, thresh=0.05)
plt.title('Customized 2D Density Plot with Plasma Color Scale')
plt.xlabel('X-axis Label')
plt.ylabel('Y-axis Label')
plt.show()
```

In this customization, the color map is set to `plasma`, which transitions from dark purple to yellow, providing a clear indication of density levels from low to high. The `levels` parameter is reduced to 15 to simplify the plot and enhance readability, while `thresh=0.05` specifies the minimum contour level to display, removing low-density areas that may clutter the plot.

```python
plt.figure(figsize=(8, 6))
sns.kdeplot(x=x, y=y, fill=True, cmap='cividis', levels=20)
plt.title('Customized 2D Density Plot', fontsize=16, fontweight='bold')
plt.xlabel('Custom X-axis Label', fontsize=14)
plt.ylabel('Custom Y-axis Label', fontsize=14)
plt.xticks(fontsize=12)
plt.yticks(fontsize=12)
plt.show()
```

Here, the title font size is increased and made bold to capture attention, while axis labels are customized with descriptive text and larger font sizes. The `xticks` and `yticks` functions set font sizes for tick labels, ensuring consistent readability across all plot elements.

```python
plt.figure(figsize=(8, 6))
sns.kdeplot(x=x, y=y, fill=True, cmap='viridis', levels=20)
plt.title('2D Density Plot with Gridlines')
plt.xlabel('X-axis Label')
plt.ylabel('Y-axis Label')

# Customize gridlines
plt.grid(color='grey', linestyle='--', linewidth=0.5, alpha=0.7)

plt.show()
```

In this example, dashed gridlines are added with a lighter grey color, ensuring they do not overpower the density information. The `alpha` parameter is set to 0.7 for slight transparency, allowing the gridlines to remain unobtrusive yet functional. Gridlines provide additional context for the viewer, particularly in high-density areas, without cluttering the visualization.

By applying these customization techniques–adjusting color scales, enhancing labels, and adding gridlines–2D Density Plots become more informative and visually appealing. These small adjustments significantly improve the plot's interpretability and ensure it is accessible and engaging for diverse audiences.

9.4   Enhancing 2D Density Plots with Interactive Elements Using `Plotly`

The `Plotly` library in Python offers robust tools for creating interactive, web-based visualizations, including 2D Density Plots. Unlike static visualizations, `Plotly` allows for interactive features such as hover effects, zooming, and responsive scaling, enhancing the user's ability to explore data in depth. This interactivity is particularly beneficial for complex or dense datasets, where interactive elements can provide clarity by allowing users to focus on specific regions or retrieve detailed information with minimal clutter.

9.4.1 Introduction to `Plotly` for Interactivity

Using `Plotly` to create interactive 2D Density Plots provides several advantages, especially when the plots are intended for web applications or presentations where viewers may benefit from exploratory analysis. `Plotly` allows users to zoom into particular regions, pan across the plot to focus on specific areas, and utilize hover tooltips to display precise values or other metadata associated with data points. These interactive features enable users to gain insights beyond what static plots can offer, providing an engaging experience that encourages deeper data exploration.

One of the key benefits of `Plotly` is its adaptability to various screen sizes and devices. Responsive design ensures that plots remain clear and readable on different devices, from desktops to mobile screens. This adaptability makes `Plotly` an ideal choice for creating data visualizations that are accessible to a broad audience, regardless of their viewing platform. Additionally, `Plotly` supports a variety of color scales and customization options, enabling the creation of visually compelling and accessible density plots.

9.4.2 Step-by-Step Guide to Interactive 2D Density Plots

Creating an interactive 2D Density Plot with `Plotly` is straightforward, and the following example provides a step-by-step guide to achieving this. The example also demonstrates how to customize hover tooltips and enable responsive scaling for an optimal viewing experience.

```python
import numpy as np
import plotly.figure_factory as ff

# Generate sample data
np.random.seed(42)
x = np.random.normal(size=1000)
y = x * 0.5 + np.random.normal(size=1000)

# Create a 2D Density Plot\index{2D Density Plot} with Plotly
fig = ff.create_2d_density(x, y, colorscale='Viridis', hist_color='white',
    point_size=2)

# Customize layout for interactivity
fig.update_layout(
    title='Interactive 2D Density Plot with Plotly',
    xaxis_title='X-axis Label',
    yaxis_title='Y-axis Label',
    template='plotly_white',
    hovermode='closest'
)
```

```
20
21  fig.show()
```

In this example, we first generate synthetic data that follows a bivariate normal distribution. The `create_2d_density` function from `Plotly.figure_factory` is used to create a basic 2D density plot, with `colorscale='Viridis'` to provide a perceptually uniform color gradient. The `hist_color='white'` option adds a white histogram overlay, enhancing the contrast between density levels and creating a cleaner look. Finally, the `hovermode='closest'` setting enables precise hover interactions, displaying data values and density information directly at the cursor location.

```
1   fig = ff.create_2d_density(x, y, colorscale='Plasma', hist_color='white',
    ↪   point_size=3)
2
3   # Customize hover tooltips and layout
4   fig.update_traces(hoverinfo='x+y+z',
    ↪   hovertemplate='X: %{x}<br>Y: %{y}<br>Density: %{z}')
5   fig.update_layout(
6       title='2D Density Plot with Customized Hover Tooltips',
7       xaxis_title='Custom X-axis Label',
8       yaxis_title='Custom Y-axis Label'
9   )
10
11  fig.show()
```

In this example, the `hovertemplate` parameter is customized to display values of x, y, and the density level (denoted by z) in the hover tooltip. The tooltip shows the exact coordinates and density at each point, providing a precise, interactive experience that aids in data interpretation. The `hoverinfo` parameter controls which data fields are displayed, allowing further customization based on the information needs of the audience.

```
1   fig = ff.create_2d_density(x, y, colorscale='Cividis', hist_color='white',
    ↪   point_size=2)
2
3   # Enable responsive layout
4   fig.update_layout(
5       title='Responsive 2D Density Plot',
6       xaxis_title='X-axis Label',
7       yaxis_title='Y-axis Label',
8       autosize=True,
9       margin=dict(l=50, r=50, t=50, b=50)
10  )
11
12  fig.show()
```

In this example, `autosize=True` enables responsive scaling, allowing the plot to automatically adjust to the available screen space. The `margin` parameter sets padding around the plot area, ensuring that plot elements do not get cut off on smaller screens. This responsive layout is ideal for web-based applications, as it maintains plot integrity and readability across different viewing devices.

By incorporating interactive features such as hover tooltips, zooming, and responsive scaling, `Plotly` elevates the functionality of 2D Density Plots, transforming them into dynamic tools for data exploration. These enhancements allow users to engage with the data actively, uncovering detailed insights and patterns that static plots cannot provide. Using `Plotly` for interactive density plots thus opens up new possibilities for data visualization, making complex datasets accessible, insightful, and visually compelling.

9.5 Creating and Customizing Contour Plots

Contour plots are an essential visualization tool for representing density levels or gradients across two dimensions. They are particularly valuable in fields requiring a structured and precise visualization of continuous data, such as geographical analysis, statistics, and environmental modeling. This section provides a comprehensive overview of contour plots, including their structure, purpose, and practical tips for customization to enhance clarity and aesthetic appeal. It also explores various applications of contour plots, demonstrating their versatility across disciplines.

9.5.1 Understanding Contour Plots

Contour plots are graphical representations of three-dimensional data in two dimensions, using contour lines to connect points of equal value. Each line, or contour, represents a specific density or intensity level, creating a map-like structure that allows viewers to observe changes and trends across a continuous surface. This type of plot is particularly useful for displaying smooth gradients, showing how values vary spatially without the need for an explicit third dimension.

The structure of a contour plot relies on isolines–lines that connect points sharing the same data value. These contour lines form "bands" that indicate regions of similar density or intensity, with areas of higher density often marked by closely spaced lines and regions of lower density by widely spaced lines. In essence, contour plots provide a topographical view of data, revealing "peaks" and "valleys" of density in a visually accessible manner. For instance, in a contour plot of temperature across a geographic region, contour lines close to one another would indicate a steep temperature gradient, while widely spaced lines would suggest a gradual change in temperature. This clear, structured view allows analysts to assess variations in data, identify regions of interest, and observe density patterns that might be obscured in other types of 2D density visualizations.

Contour plots also support color shading between contour lines, which adds another layer of information by assigning color gradients to different density levels.

This color shading provides viewers with immediate insights into areas of higher and lower density, enabling a more intuitive understanding of the data's structure and distribution.

9.5.2   Design and Customization Tips

To create effective contour plots, attention to design elements such as contour levels, line styles, and color fills is essential. Proper customization enhances the readability and aesthetic appeal of the plot, ensuring that data patterns are clear and interpretable.

```python
import numpy as np
import matplotlib.pyplot as plt

# Generate sample data
x = np.linspace(-3.0, 3.0, 100)
y = np.linspace(-3.0, 3.0, 100)
X, Y = np.meshgrid(x, y)
Z = np.exp(-X**2 - Y**2)

# Create a contour plot\index{Contour Plot} with custom levels
plt.figure(figsize=(8, 6))
contour = plt.contour(X, Y, Z, levels=8, cmap='viridis')
plt.clabel(contour, inline=True, fontsize=10)
plt.title('Contour Plot with Custom Levels')
plt.xlabel('X-axis')
plt.ylabel('Y-axis')
plt.colorbar(contour, label='Density Level')
plt.show()
```

In this example, the `levels` parameter specifies the number of contour levels, set to eight for a moderate level of detail. A color bar is added to indicate density levels, enhancing interpretability.

```python
plt.figure(figsize=(8, 6))
contour = plt.contour(X, Y, Z, levels=10, colors='black',
    ↪ linestyles='dashed', linewidths=0.8)
plt.title('Contour Plot with Customized Line Styles')
plt.xlabel('X-axis')
plt.ylabel('Y-axis')
plt.show()
```

Here, `linestyles='dashed'` applies a dashed style to the contour lines, which may be preferred for plots with multiple datasets, as it enhances differentiation between overlapping lines.

```python
plt.figure(figsize=(8, 6))
plt.contourf(X, Y, Z, levels=10, cmap='plasma')
```

```
3   plt.colorbar(label='Density Level')
4   plt.title('Contour Plot with Color Fills')
5   plt.xlabel('X-axis')
6   plt.ylabel('Y-axis')
7   plt.show()
```

The `contourf` function creates filled contours, providing a visually continuous representation of density levels. Adding a color bar further clarifies the meaning of color intensities, enhancing interpretability for viewers.

9.5.3 Applications of Contour Plots

Contour plots have diverse applications across various fields, where they are used to represent gradients, density distributions, and surface characteristics.

In **geographical data representation**, contour plots are essential for visualizing elevation or terrain. Geographic Information Systems (GIS) frequently use contour plots to map topographic features, showing elevation changes through contour lines. This application is invaluable for urban planning, environmental monitoring, and geological surveys, as it provides a detailed view of landscape features and elevation gradients.

In **statistical density estimation**, contour plots help illustrate probability density functions in multivariate distributions. For example, contour plots are used in statistical analysis to represent confidence intervals in regression models or to visualize bivariate normal distributions in probability theory. By displaying areas of high probability density, these plots enable statisticians to identify patterns, relationships, and central tendencies within datasets.

Contour plots are also widely used in **terrain mapping** and environmental modeling, where they illustrate variations in parameters like temperature, precipitation, or pollution levels. Meteorologists, for instance, employ contour plots to display atmospheric pressure and temperature distributions, enabling the study of weather patterns. Similarly, ecologists use contour plots to examine spatial distributions of vegetation, biodiversity, or soil composition, aiding in the identification of environmentally significant regions.

In summary, contour plots provide a structured, intuitive way of visualizing continuous data distributions across two dimensions. Their application in fields like geography, statistics, and environmental science demonstrates their versatility and effectiveness in communicating complex data patterns. With appropriate customization and attention to design, contour plots can reveal nuanced insights, supporting both exploratory data analysis and professional decision-making.

9.6 Heatmaps for Density Visualization

Heatmaps are a widely-used visualization technique for representing data density, where color intensity indicates the frequency, concentration, or magnitude of values across a two-dimensional grid. This visualization style is particularly valuable for displaying large datasets, as it condenses complex information into an accessible, easily interpretable format. Heatmaps have applications in numerous fields, including biology, finance, web analytics, and climatology, due to their ability to capture both categorical and continuous data. This section introduces heatmaps, outlines essential design principles, and provides examples with customizations to enhance readability and visual appeal.

A heatmap represents data density through varying color intensities, allowing users to quickly identify high and low-density areas within a dataset. In a heatmap, each cell in a grid corresponds to a specific value or count, with color intensity reflecting the magnitude or frequency of that value. Typically, darker or more intense colors signify higher density or values, while lighter colors indicate lower densities. By transforming numerical data into color gradients, heatmaps create an intuitive, visual representation that enables viewers to grasp patterns, distributions, and anomalies at a glance.

Heatmaps are highly versatile, making them suitable for both **categorical and continuous data** For categorical data, heatmaps can represent counts or frequencies across categories, such as user clicks on different sections of a webpage or gene expression levels across different samples in genomics. Continuous data, on the other hand, may represent values that vary smoothly across a spatial or temporal dimension, such as temperature across a geographic region or stock returns over time. The flexibility of heatmaps allows them to be applied to diverse datasets, providing a clear and effective method for density visualization.

Creating effective heatmaps involves careful consideration of color schemes, bin sizes, and scaling, each of which significantly influences the readability and interpretability of the plot.

The following examples demonstrate how to create and customize heatmaps using Python libraries such as `Seaborn` and `Matplotlib` These examples illustrate how color gradients, legends, and axis labels can be tailored to improve both readability and aesthetic appeal.

```python
import numpy as np
import seaborn as sns
import matplotlib.pyplot as plt

# Generate sample data
data = np.random.rand(10, 12)

# Create a basic heatmap
plt.figure(figsize=(8, 6))
sns.heatmap(data, cmap='viridis', annot=True, cbar=True)
plt.title('Basic Heatmap')
plt.xlabel('X-axis Label')
plt.ylabel('Y-axis Label')
plt.show()
```

In this example, the `cmap='viridis'` parameter applies the viridis color map, which is both perceptually uniform and color-blind friendly. The `annot=True` option displays the values in each cell, enhancing interpretability, while the color bar (`cbar=True`) indicates the data scale.

```python
plt.figure(figsize=(8, 6))
sns.heatmap(data, cmap='plasma', annot=False, cbar_kws={'label':
    'Density Level', 'orientation': 'vertical'})
plt.title('Heatmap with Custom Color Gradient and Legend')
plt.xlabel('X-axis Label')
plt.ylabel('Y-axis Label')
plt.show()
```

Here, the `cmap='plasma'` parameter changes the color gradient to plasma, which provides a vibrant color scheme progressing from dark purple to yellow. The `cbar_kws` argument customizes the color bar, adding a label and setting it to a vertical orientation for better readability.

```python
data_with_outliers = np.random.randn(10, 12) * 3

plt.figure(figsize=(8, 6))
sns.heatmap(data_with_outliers, cmap='coolwarm', annot=False, vmin=-2,
    vmax=2, cbar_kws={'label': 'Intensity'})
plt.title('Heatmap with Adjusted Scale for Outliers')
plt.xlabel('X-axis Label')
plt.ylabel('Y-axis Label')
plt.show()
```

In this example, `vmin` and `vmax` are set to -2 and 2, limiting the data range to mitigate the impact of extreme outliers and provide a clearer representation of mid-range values. The `coolwarm` color scheme is a diverging palette that emphasizes both low and high values around a central midpoint, ideal for data with positive and negative deviations.

By applying these customization techniques–choosing an appropriate color scheme, adjusting bin sizes, and setting suitable scales–heatmaps become more effective tools for visualizing data density. Each adjustment enhances the readability and interpretability of the plot, making heatmaps adaptable for various datasets and analytical contexts. Heatmaps are thus powerful, versatile visualizations that, with thoughtful design, provide clear insights into both categorical and continuous data distributions.

9.7 Advanced Techniques for 2D Density Plots

Advanced techniques for 2D Density Plots can significantly enhance data visualization by adding depth and clarity to complex datasets. By layering different plot types, applying smoothing methods like Kernel Density Estimation (KDE), and adjusting bandwidth and resolution, analysts can create multi-dimensional visualizations that reveal intricate patterns and trends. This section discusses these techniques in detail, providing practical guidelines and examples to optimize 2D Density Plots for clarity and accuracy.

9.7.1   Layering Density Plots with Other Plot Types

Overlaying density plots with other plot types, such as scatter plots, contour lines, or histograms, is a powerful way to incorporate multiple data dimensions within a single visualization. This layering technique enhances interpretability by combining different perspectives on the data, allowing viewers to observe overall density while identifying specific data points, trends, or distributions.

```python
import numpy as np
import seaborn as sns
import matplotlib.pyplot as plt

# Generate sample data
np.random.seed(42)
x = np.random.normal(size=1000)
y = x * 0.5 + np.random.normal(size=1000)

# Create 2D Density Plot\index{2D Density Plot} with KDE \index{Kernel
    Density Estimation} and scatter overlay
plt.figure(figsize=(8, 6))
sns.kdeplot(x=x, y=y, fill=True, cmap='viridis', levels=20, alpha=0.7)
plt.scatter(x, y, color='blue', s=10, alpha=0.3)
plt.title('2D Density Plot with Scatter Plot Overlay')
plt.xlabel('X-axis Label')
plt.ylabel('Y-axis Label')
plt.show()
```

In this example, a KDE density plot is created with Seaborn's kdeplot function, and individual data points are overlaid using plt.scatter The scatter plot points are given partial transparency (alpha=0.3) to prevent them from obscuring the underlying density patterns.

```python
plt.figure(figsize=(8, 6))
sns.kdeplot(x=x, y=y, fill=True, cmap='plasma', levels=20)
sns.kdeplot(x=x, y=y, levels=10, color='black', linewidths=0.8)
plt.title('2D Density Plot with Contour Line Overlay')
plt.xlabel('X-axis Label')
plt.ylabel('Y-axis Label')
plt.show()
```

In this example, the contour lines are added using `kdeplot` with `levels=10` and a black color to contrast with the color-filled KDE plot. The combination of filled density and contour lines provides both gradient and structural views of data density.

```python
from matplotlib.gridspec import GridSpec

# Create figure and grid
fig = plt.figure(figsize=(8, 8))
gs = GridSpec(4, 4, figure=fig)
ax_joint = fig.add_subplot(gs[1:4, 0:3])
ax_marg_x = fig.add_subplot(gs[0, 0:3], sharex=ax_joint)
ax_marg_y = fig.add_subplot(gs[1:4, 3], sharey=ax_joint)

# Joint KDE plot
sns.kdeplot(x=x, y=y, ax=ax_joint, fill=True, cmap='cividis')
ax_joint.set_xlabel('X-axis Label')
ax_joint.set_ylabel('Y-axis Label')

# Marginal histograms
sns.histplot(x, ax=ax_marg_x, color='gray', bins=20)
sns.histplot(y, ax=ax_marg_y, color='gray', bins=20,
↪    orientation='horizontal')

plt.suptitle('2D Density Plot with Histogram Marginals')
plt.show()
```

In this example, histograms are added along the top and right axes to display the marginal distributions of x and y, respectively. The `GridSpec` function divides the plot into a grid layout, positioning the density plot in the main area and the histograms on the margins.

9.7.2 Using Smoothing and Kernel Density Estimation (KDE)

Kernel Density Estimation (KDE) is a non-parametric method used for smoothing data in density plots. KDE estimates the probability density function of a dataset, generating a smooth curve that represents the underlying distribution. This technique is valuable for enhancing visual appeal and simplifying complex data by minimizing noise and emphasizing general patterns.

KDE works by placing a kernel (usually a Gaussian function) over each data point and summing the contributions of all kernels. This smoothing process creates a continuous surface that represents the density distribution. KDE is particularly useful for identifying clusters, trends, and anomalies, as it reveals general patterns without being affected by individual data points.

In Python, KDE is implemented in `Seaborn` with the `kdeplot` function. By default, `Seaborn` applies KDE smoothing to create a visually appealing density plot that highlights the overall shape of the data.

```
1  plt.figure(figsize=(8, 6))
2  sns.kdeplot(x=x, y=y, fill=True, cmap='magma', levels=30)
3  plt.title('2D Density Plot with KDE Smoothing')
4  plt.xlabel('X-axis Label')
5  plt.ylabel('Y-axis Label')
6  plt.show()
```

In this example, KDE smoothing is applied automatically, creating a smooth and continuous surface that makes the data distribution more readable. The `levels=30` parameter increases the contour detail, providing a more granular view of density variations.

9.7.3 Adjusting Bandwidth and Resolution

The bandwidth of a KDE plot controls the width of the kernels, which affects the degree of smoothing. A smaller bandwidth produces a plot with more detail, while a larger bandwidth results in greater smoothing, reducing the influence of minor fluctuations. Choosing an appropriate bandwidth is crucial for optimizing plot clarity and accuracy, as it determines the balance between detail and smoothness.

> **⚠ Warning**
>
> Incorrect bandwidth settings in Kernel Density Estimation can lead to over-smoothing or excessive detail. Always experiment with different `bw_adjust` values and validate your results visually.

The following example demonstrates how to adjust bandwidth in a KDE plot to fine-tune the smoothing effect.

```
1  plt.figure(figsize=(8, 6))
2  sns.kdeplot(x=x, y=y, fill=True, cmap='inferno', bw_adjust=0.5, levels=20)
3  plt.title('2D Density Plot with Adjusted Bandwidth (bw_adjust=0.5)')
4  plt.xlabel('X-axis Label')
5  plt.ylabel('Y-axis Label')
6  plt.show()
```

In this example, `bw_adjust=0.5` decreases the bandwidth, resulting in a plot with finer detail. Conversely, increasing the `bw_adjust` parameter would produce a smoother plot. Testing different bandwidth values can help identify the optimal setting based on the dataset's complexity and the desired level of detail.

```
1  plt.figure(figsize=(8, 6))
2  sns.kdeplot(x=x, y=y, fill=True, cmap='cividis', gridsize=100)
3  plt.title('2D Density Plot with Increased Resolution (gridsize=100)')
4  plt.xlabel('X-axis Label')
```

```
5  plt.ylabel('Y-axis Label')
6  plt.show()
```

Here, `gridsize=100` increases the resolution, creating a more detailed density surface. Adjusting resolution can enhance plot accuracy, especially when examining subtle patterns in high-density regions.

By employing advanced techniques such as layering with other plot types, using KDE for smoothing, and fine-tuning bandwidth and resolution, 2D Density Plots can be customized to reveal complex data patterns effectively. These techniques provide flexibility in presentation, allowing analysts to adapt plots to various datasets and analytical needs.

9.8 Accessibility Considerations for 2D Density Plots

Accessibility is a critical aspect of data visualization, ensuring that visual content is interpretable by a diverse audience, including individuals with visual impairments or color vision deficiencies. When designing 2D Density Plots, incorporating accessibility features like color-blind friendly palettes, appropriate contrast, and clear labeling can enhance readability and inclusivity. This section covers key accessibility considerations for 2D Density Plots, focusing on color accessibility and readability of density levels.

9.8.1 Color Accessibility

Color choice plays a significant role in the effectiveness of density plots, as it directly impacts how viewers perceive density variations. For individuals with color blindness or other visual impairments, certain colors may be indistinguishable, making it challenging to interpret the data accurately. To ensure color accessibility, it is essential to choose color palettes that accommodate color vision deficiencies and apply contrast adjustments to make density distinctions clear.

```
1  import numpy as np
2  import seaborn as sns
3  import matplotlib.pyplot as plt
4
5  # Generate sample data
6  np.random.seed(42)
7  x = np.random.normal(size=1000)
8  y = x * 0.5 + np.random.normal(size=1000)
9
10 # Create a density plot with a color-blind friendly palette
11 plt.figure(figsize=(8, 6))
12 sns.kdeplot(x=x, y=y, fill=True, cmap='viridis', levels=20)
13 plt.title('Density Plot with Color-Blind Friendly Palette')
14 plt.xlabel('X-axis Label')
```

```
15  plt.ylabel('Y-axis Label')
16  plt.show()
```

In this example, the `viridis` palette is used to ensure color accessibility, providing a smooth gradient from dark blue to yellow that can be interpreted by color-blind viewers.

```
1  plt.figure(figsize=(8, 6))
2  sns.kdeplot(x=x, y=y, fill=True, cmap='cividis', levels=15, alpha=0.9)
3  plt.title('Density Plot with Enhanced Contrast')
4  plt.xlabel('X-axis Label')
5  plt.ylabel('Y-axis Label')
6  plt.show()
```

Here, the `cividis` color map is selected for its accessibility, and the `alpha=0.9` setting increases contrast, making density distinctions clearer. The `levels=15` parameter provides a moderate level of contour detail, striking a balance between readability and density granularity.

9.8.2 Readability of Density Levels

Readability is another critical component of accessible density plots. Effective label placement, font size, and clear legends contribute to a viewer's ability to interpret data accurately. By optimizing these elements, designers can ensure that density plots remain interpretable, even for individuals with visual impairments.

```
1  plt.figure(figsize=(8, 6))
2  sns.kdeplot(x=x, y=y, fill=True, cmap='magma', levels=20)
3  plt.title('Accessible Density Plot', fontsize=16, fontweight='bold')
4  plt.xlabel('X-axis Label', fontsize=14)
5  plt.ylabel('Y-axis Label', fontsize=14)
6  plt.xticks(fontsize=12)
7  plt.yticks(fontsize=12)
8  plt.show()
```

In this example, the title and axis labels are set to larger font sizes (`fontsize=16` for the title, `fontsize=14` for axis labels), enhancing readability. The use of bold text for the title (`fontweight='bold'`) ensures visibility, especially on smaller screens or in presentations.

```
1  plt.figure(figsize=(8, 6))
2  sns.kdeplot(x=x, y=y, fill=True, cmap='inferno', levels=15)
3  plt.colorbar(label='Density Level')
4  plt.title('Density Plot with Accessible Legend')
5  plt.xlabel('X-axis Label')
```

```python
6  plt.ylabel('Y-axis Label')
7  plt.show()
```

In this example, the `colorbar` is positioned to the right of the plot, labeled clearly with "Density Level" to guide interpretation. The use of the `inferno` color map, which transitions from dark purple to yellow, further ensures that density levels are easy to differentiate.

By incorporating color-blind friendly palettes, enhancing contrast, and optimizing label placement and font size, designers can create 2D Density Plots that are accessible to a broad audience. These adjustments not only improve readability for color-blind or visually impaired viewers but also enhance the overall clarity and effectiveness of the plot, supporting accurate and inclusive data interpretation.

9.9 Common Pitfalls in 2D Density Plot Design and How to Avoid Them

Designing effective 2D Density Plots requires careful consideration of various visual elements, as common pitfalls can obscure data patterns, mislead viewers, or reduce readability. This section addresses some of the most frequent issues encountered in 2D Density Plot design and offers practical guidelines to avoid them, ensuring that density plots remain clear, accurate, and interpretable.

9.9.1 Overplotting and Clutter

Overplotting occurs when too many data points or layers are included in a plot, resulting in visual clutter that makes it difficult for viewers to discern meaningful patterns. In 2D Density Plots, overplotting often arises when dealing with dense datasets, multiple overlapping datasets, or an excessive number of contour levels. To reduce clutter, designers can adjust transparency, optimize bin size, and apply data filtering techniques.

> **⚠ Warning**
>
> When visualizing large datasets, especially with scatter plot overlays, high point density can lead to overplotting . Always apply appropriate transparency using the `alpha` parameter to avoid obscuring meaningful patterns.

```python
1  import numpy as np
2  import seaborn as sns
3  import matplotlib.pyplot as plt
4
5  # Generate sample data
6  np.random.seed(42)
7  x = np.random.normal(size=1000)
8  y = x * 0.5 + np.random.normal(size=1000)
```

```python
# Create 2D Density Plot with reduced opacity
plt.figure(figsize=(8, 6))
sns.kdeplot(x=x, y=y, fill=True, cmap='viridis', levels=20, alpha=0.6)
plt.title('2D Density Plot with Reduced Transparency')
plt.xlabel('X-axis Label')
plt.ylabel('Y-axis Label')
plt.show()
```

In this example, setting `alpha=0.6` reduces opacity, minimizing overlap and enhancing readability by preventing high-density areas from dominating the plot.

9.9.2 Misleading Color Gradients

Color gradients play a crucial role in communicating density variations, but poor color choices can distort interpretation. Misleading gradients, such as non-perceptually uniform color schemes, may exaggerate certain density levels, creating an inaccurate impression of the data distribution. To avoid this, designers should use perceptually accurate color gradients that provide a consistent progression of intensity.

```python
plt.figure(figsize=(8, 6))
sns.kdeplot(x=x, y=y, fill=True, cmap='plasma', levels=20)
plt.title('Density Plot with Perceptually Uniform Color Scheme')
plt.xlabel('X-axis Label')
plt.ylabel('Y-axis Label')
plt.show()
```

In this example, the `plasma` color map provides a smooth, perceptually uniform gradient from dark purple to yellow, ensuring consistent interpretation of density levels across the plot.

> **⚠ Warning**
>
> Avoid non-uniform color maps such as `jet`, which can distort perception of density differences. Use perceptually uniform color schemes like `viridis`, `plasma`, or `cividis` to ensure accurate interpretation of density gradients.

9.9.3 Axis Scaling and Range Issues

Improper axis scaling and range settings can distort the appearance of density patterns, leading to misleading interpretations. When axis ranges are too restrictive or too expansive, the relative density of regions may appear exaggerated or minimized. To prevent these issues, designers should carefully set axis scales and ranges that accurately represent the data distribution without distorting visual density.

```python
plt.figure(figsize=(8, 6))
sns.kdeplot(x=x, y=y, fill=True, cmap='viridis', levels=20)
plt.xlim(-3, 3)
plt.ylim(-3, 3)
plt.title('Density Plot with Adjusted Axis Ranges')
plt.xlabel('X-axis Label')
plt.ylabel('Y-axis Label')
plt.show()
```

In this example, the x-axis and y-axis ranges are set to (-3, 3), centering the plot on the main data cluster and providing a balanced view of density distribution.

> **⚠ Warning**
>
> Setting axis limits arbitrarily or inconsistently across multiple plots may lead to misleading comparisons. Ensure axis ranges reflect the data's actual spread and maintain consistency across plots when comparing multiple datasets.

By addressing common pitfalls such as overplotting , misleading color gradients, and improper axis scaling, designers can create 2D Density Plots that are both clear and accurate. These best practices help avoid visual distortions, maintain readability, and ensure that density plots communicate data patterns effectively to a wide audience.

9.10 Accessibility Considerations for 2D Density Plots

Accessibility improvements in 2D Density Plots ensure that the information is accessible to a diverse audience, including individuals with visual impairments or color vision deficiencies. By choosing color-blind friendly palettes, increasing contrast, and enhancing labels and legends, designers can create plots that are both visually appealing and interpretable by all users. This section provides examples of accessibility improvements with Python code, focusing on color accessibility and label readability.

9.10.1 Color Accessibility

Using color-blind friendly palettes is essential to ensure that density plots are interpretable for viewers with color vision deficiencies. Perceptually uniform color maps, like `viridis`, `plasma`, `cividis`, and `inferno`, are designed to be easily distinguishable by all viewers, regardless of color vision ability. These color maps avoid problematic color pairings (e.g., red-green) and offer a continuous gradient that accurately represents density levels.

```python
import numpy as np
import seaborn as sns
import matplotlib.pyplot as plt
```

```
4
5   # Generate sample data
6   np.random.seed(42)
7   x = np.random.normal(size=1000)
8   y = x * 0.5 + np.random.normal(size=1000)
9
10  # Create a density plot with color-blind friendly palette 'cividis'
11  plt.figure(figsize=(8, 6))
12  sns.kdeplot(x=x, y=y, fill=True, cmap='cividis', levels=20)
13  plt.title('Accessible Density Plot with Cividis Palette')
14  plt.xlabel('X-axis Label')
15  plt.ylabel('Y-axis Label')
16  plt.show()
```

In this example, the `cividis` palette is chosen because it is color-blind friendly, providing a perceptually uniform gradient from dark blue to yellow. This color scheme makes it easier for viewers with color vision deficiencies to interpret density variations, as the gradient is distinguishable without relying on red-green contrasts.

```
1   plt.figure(figsize=(8, 6))
2   sns.kdeplot(x=x, y=y, fill=True, cmap='inferno', levels=15, alpha=0.85)
3   plt.title('Density Plot with Increased Contrast')
4   plt.xlabel('X-axis Label')
5   plt.ylabel('Y-axis Label')
6   plt.show()
```

In this example, the `inferno` color map, which transitions from dark purple to yellow, is selected for its strong contrast. Additionally, setting `alpha=0.85` controls transparency, allowing different density levels to remain distinct and accessible to color-blind and low-vision viewers.

9.10.2   Readability of Density Levels

Readable labels, titles, and legends are essential for effective 2D Density Plots, as they guide viewers in interpreting density levels and plot context. Increasing font size, adjusting label positions, and ensuring high-contrast legend text are important considerations for accessibility.

```
1   plt.figure(figsize=(8, 6))
2   sns.kdeplot(x=x, y=y, fill=True, cmap='magma', levels=20)
3   plt.title('Accessible Density Plot with Enhanced Labels', fontsize=16,
    ↪  fontweight='bold')
4   plt.xlabel('X-axis Label', fontsize=14)
5   plt.ylabel('Y-axis Label', fontsize=14)
6   plt.xticks(fontsize=12)
7   plt.yticks(fontsize=12)
8   plt.show()
```

In this example, the title font size is set to `fontsize=16` with bold styling (`fontweight='bold'`) to increase visibility. Axis labels are given a font size of 14, while tick labels are set to 12, enhancing readability and ensuring that label text is accessible even on smaller screens.

```python
plt.figure(figsize=(8, 6))
kde = sns.kdeplot(x=x, y=y, fill=True, cmap='viridis', levels=20)
colorbar = plt.colorbar(kde, label='Density Level')
colorbar.ax.tick_params(labelsize=12)  # Increase font size for legend
    labels
plt.title('Density Plot with Accessible Legend')
plt.xlabel('X-axis Label')
plt.ylabel('Y-axis Label')
plt.show()
```

In this example, a color bar is added to the density plot to display density levels, with the label "Density Level" to clarify the meaning of color intensities. The tick labels on the color bar are increased in size for improved readability, and the legend is positioned to the side of the plot, preventing it from obscuring density information.

By implementing these accessibility improvements–using color-blind friendly palettes, enhancing contrast, and optimizing label and legend readability–designers can create 2D Density Plots that are inclusive, clear, and visually effective. These adjustments ensure that viewers with diverse visual abilities can accurately interpret density variations, making the plot accessible to a broader audience.

9.11 Summary and Best Practices

Designing effective 2D Density Plots requires careful attention to visualization principles, accessibility, and customization options. This section provides a summary of key takeaways for creating clear, accessible, and visually engaging 2D Density Plots, followed by a practical checklist to guide readers in producing high-quality visualizations.

9.11.1 Key Takeaways

Creating impactful 2D Density Plots involves a balance of clarity, accessibility, and aesthetic appeal. Here are the essential principles:

1. Choose the Right Plot Type Selecting the appropriate type of 2D density plot–whether it's a heatmap, contour plot, or kernel density estimate (KDE) plot–depends on the data characteristics and visualization goals. Each plot type offers unique strengths; for instance, contour plots provide structural clarity with contour lines, while KDE plots offer smooth, continuous density estimates.

2. Use Accessible Color Palettes Color choice is essential for density plot effectiveness and accessibility. Perceptually uniform color maps like `viridis`, `plasma`, and `cividis` ensure that density levels are interpreted accurately by all viewers, including those with color vision deficiencies. Avoid problematic color combinations, such as red-green, that may obscure information for color-blind users.

3. Optimize Transparency and Contrast Adjusting transparency and contrast helps prevent overplotting , especially in dense data regions. Setting appropriate transparency levels allows viewers to distinguish between overlapping areas while maintaining data clarity. Similarly, high-contrast color schemes improve readability for all users.

4. Ensure Readability with Clear Labels and Legends Effective label placement, font size, and legend design are crucial for guiding viewers in interpreting density plots. Titles, axis labels, and legends should be strategically positioned and use sufficiently large font sizes to ensure readability. Legends should provide clear information about color intensities and density levels.

5. Customize Plot Elements for Clarity Customization options, such as adjusting bin size, contour levels, and grid resolution, help tailor density plots to specific data characteristics. Fine-tuning these elements allows viewers to observe detailed patterns without overwhelming the visualization with excessive detail.

6. Use Interactivity for Exploration When sharing density plots in digital formats, interactivity features like hover tooltips, zooming, and responsive scaling enhance the user experience, allowing viewers to explore specific data regions and retrieve detailed information on demand.

7. Apply Scaling and Range Adjustments Carefully Proper axis scaling and range settings prevent visual distortion of density patterns. Use appropriate ranges to represent the full data spread without exaggerating or minimizing certain areas, and consider non-linear scaling for datasets with extreme values or skewed distributions.

9.11.2   Checklist for Effective 2D Density Plots

The following checklist covers essential considerations for designing, customizing, and optimizing 2D Density Plots. This guide helps ensure that density plots are clear, accurate, and accessible to a wide audience.

- ➤ **Plot Type Selection**: Choose a plot type that best suits the data characteristics and visualization goals (e.g., heatmap, contour plot, KDE plot).

- ➤ **Color Accessibility**: Use a color-blind friendly palette (e.g., `viridis`, `cividis`, `inferno`) to ensure accessibility for all viewers.

- ➤ **Transparency Adjustment**: Set appropriate transparency levels (`alpha` parameter) to avoid overplotting and clarify overlapping regions.

- ➤ **Contrast Optimization**: Apply high-contrast color schemes and adjust brightness to enhance visual separation of density levels.

- ➤ **Label Placement and Font Size**: Position labels and legends outside the primary plot area, use large, readable font sizes, and apply bold styling where necessary.

- ➤ **Legend Design**: Add a clear, high-contrast legend that accurately explains color intensities and density levels without overlapping the plot area.

➢ **Customization of Plot Elements**: Adjust bin size, contour levels, and resolution to balance detail and clarity based on the data.

➢ **Interactivity Features**: Enable hover tooltips, zooming, and responsive scaling for digital or web-based density plots to enhance user engagement.

➢ **Axis Scaling and Range**: Set axis ranges that reflect the data spread without distorting density patterns, and consider non-linear scaling for skewed data.

➢ **Testing and Accessibility Checks**: Test the plot's readability across different devices and ensure that color and text are accessible for color-blind or low-vision users.

By following these best practices and the checklist (Table 9.2), designers can create 2D Density Plots that are not only visually engaging but also accessible and insightful. This comprehensive approach to 2D Density Plot design helps ensure that visualizations effectively communicate data patterns to a broad audience, supporting informed analysis and decision-making.

Table 9.2: Checklist for Effective 2D Density Plot Design

Checklist Item	Status
Choose appropriate plot type (heatmap, contour, KDE) based on data characteristics and goals.	
Select a color-blind friendly palette (e.g., `viridis`, `cividis`) to ensure accessibility.	
Set transparency (`alpha`) to manage overplotting and layer visibility.	
Optimize contrast to differentiate density levels effectively.	
Use large, clear labels and place them outside the plot area; set font sizes for readability.	
Add a high-contrast legend outside the plot area to explain density levels.	
Customize bin size, contour levels, and resolution to balance detail and readability.	
Enable interactivity (hover tooltips, zooming, responsive scaling) for digital plots.	
Adjust axis ranges and scaling to accurately represent data spread without distortion.	
Test readability across devices and ensure color and text are accessible for all viewers.	

Suggested Readings

1. Chen, X., Wang, Y., Bao, H., Lu, K., Jo, J., Fu, C.W., Fekete, J.D.: Visualization-driven illumination for density plots. IEEE Transactions on Visualization and Computer Graphics **31**(2), 1631 – 1644 (2025). DOI 10.1109/TVCG.2024.3495695
2. Chen, Y., Lin, X., Zhao, Y., Sun, Y., Zhang, X.: Sunmap: an associated hierarchical data visualization method based on heatmap and sunburst. Jisuanji Fuzhu Sheji Yu Tuxingxue Xuebao/Journal of Computer-Aided Design and Computer Graphics **28**(7), 1075 – 1083 (2016)

3. Kurzyna, M., Kwapinski, T.: Braid plot—a mixed palette plotting method as an extension of contour plot. IEEE Computer Graphics and Applications **42**(1), 95 – 104 (2022). DOI 10.1109/MCG.2021.3057924

4. Zhang, F., Yuan, Z., Xiao, F., You, K., Wang, Z.: Research on heatmap for big data based on spark. Jisuanji Fuzhu Sheji Yu Tuxingxue Xuebao/Journal of Computer-Aided Design and Computer Graphics **28**(11), 1881 – 1886 (2016)

10 Designing Other Beautiful Charts

Abstract

This chapter explores a collection of additional powerful visualization methods such as Parallel Coordinates, Treemaps, Network Graphs, Sankey Diagrams, Chord Diagrams, highlighting their unique capabilities and applications across diverse scientific domains. Through detailed explanations, practical Python code snippets, and visually compelling examples, the chapter provides readers with the tools necessary to implement these advanced visualizations effectively. Case studies demonstrate how these techniques facilitate the exploration of high-dimensional data, reveal intricate relationships, and enhance the communication of complex information. Best practices for designing and deploying these visualizations are discussed, ensuring that they not only present data accurately but also engage and inform the target audience.

Aims

After reading this chapter, you should be able to:

- ➤ Understand the purpose and appropriate use cases for advanced and non-traditional visualization techniques.

- ➤ Implement Treemaps, Network Graphs, Sankey Diagrams, and other advanced charts using Python visualization libraries.

- ➤ Evaluate each chart type in conveying complex, hierarchical, spatial, or high-dimensional data.

- ➤ Apply best practices to ensure clarity, accessibility, interactivity, and comparative power in complex data visualizations.

- ➤ Integrate interactive features into visualizations to enhance user engagement and exploratory capabilities.

- ➤ Effectively combine visual and narrative elements to engage audiences through compelling data storytelling.

© The Author(s), under exclusive license to Springer Nature Switzerland AG 2026 301
R. Damaševičius, *Human-Centred Scientific Data Visualisation*,
Undergraduate Topics in Computer Science,
https://doi.org/10.1007/978-3-032-01606-5_10

10.1 Outline of Advanced Charts

This chapter explores a wide variety of advanced data visualization techniques that go beyond traditional charts like bar graphs and line plots. These advanced charts are particularly valuable when dealing with complex, high-dimensional, hierarchical, or relational datasets, offering specialized ways to uncover insights and communicate findings effectively. Each technique introduces a unique perspective on data, accommodating different analytical goals such as comparison, distribution analysis, relational mapping, and spatial interpretation.

Treemaps provide a compact and space-efficient method to visualize hierarchical data using nested rectangles, where area and color encode quantitative attributes. They are ideal for scenarios requiring proportional comparisons across categories and subcategories.

Network graphs , on the other hand, are well-suited for illustrating relationships or interactions between entities. These graphs use nodes and edges to represent the structure of networks–such as social connections, communication systems, or biological pathways–and reveal patterns like clustering and centrality.

Sankey Diagrams visualize the flow of quantities between different stages or components in a system, with the width of each link proportionally indicating the volume of flow. These diagrams are widely used in energy flow modeling, resource allocation, and financial auditing.

Streamgraphs, a stylized form of stacked area charts, depict temporal evolution across multiple categories in a visually organic manner. They are commonly applied in areas like cultural analytics, user engagement tracking, or media consumption patterns.

Hexbin Plots help address the problem of overplotting in large datasets by aggregating data points into hexagonal bins and using color to indicate density. This facilitates exploration of two-dimensional distributions, especially where scatter plots fall short.

Radial Bar Charts reimagine traditional bar charts in a circular format, making them especially effective for representing cyclical data like monthly sales or seasonal behaviors. They provide an engaging and symmetrical way to compare values around a full circle, though careful design is necessary to avoid misinterpretation due to curvature.

Table 10.1 provides a comparative overview of all advanced chart types discussed in this chapter, highlighting their primary applications, best-fit data formats, and commonly used Python libraries.

To evaluate how interactive each chart type is for users, Table 10.2 summarizes the typical interaction features available across different visualization techniques.

Designing effective and clear visualizations requires following well-established design principles. Table 10.3 outlines the most important best practices specific to each advanced chart type.

Table 10.1: Overview of Advanced Chart Types

Chart Type	Primary Purpose	Typical Data Type
Treemap	Visualize hierarchical proportions	Hierarchical/categorical
Network Graph	Show relationships and interactions	Graph/network structure
Sankey Diagram	Illustrate flow of quantities between entities	Flow-based/tabular
Chord Diagram	Visualize interconnections between categories	Matrix/relational
Streamgraph	Show time-series data with multiple flows	Temporal/categorical
Hexbin Plot	Show data density in 2D space	Large point clouds
Radial Bar Chart	Represent cyclical patterns and categories	Categorical/cyclical

Table 10.2: Interactivity Levels Across Visualization Techniques

Chart Type	Interactivity Level	Typical Interactive Features
Treemap	Medium	Zooming, tooltips, hover highlights
Network Graph	High	Node dragging, filtering, neighborhood highlighting
Sankey Diagram	Medium	Hover effects, highlighting flows
Chord Diagram	Medium	Hover highlights, filter relationships
Streamgraph	Medium	Hover tooltips, stream isolation
Hexbin Plot	Low–Medium	Color bar, hover count per bin
Radial Bar Chart	Low	Tooltip values, animated rotation (rare)

Table 10.3: Design Best Practices Across Visualization Types

Chart Type	Key Best Practices
Treemap	Maintain proportional sizing, limit depth levels, use intuitive color schemes
Network Graph	Avoid clutter, highlight key nodes, use meaningful edge widths
Sankey Diagram	Ensure flow proportionality, minimize crossings, clear node labeling
Chord Diagram	Optimize layout, minimize overlap, use consistent color mapping
Streamgraph	Use smooth curves, apply color separation, keep stream count manageable
Hexbin Plot	Adjust grid size, use readable color gradients, add legend
Radial Bar Chart	Ensure axis symmetry, use consistent scales, avoid angle distortion

These advanced charts and techniques empower analysts and designers to transform complex datasets into compelling visual representations, supporting better analysis, communication, and decision-making in diverse scientific and applied domains.

10.2   Treemaps

10.2.1   Concept of Treemaps

Treemaps are a visualization technique designed to represent hierarchical data using nested rectangles. Each level of the hierarchy is depicted by a set of nested rectangles, where the size and color of each rectangle correspond to specific data attributes. The primary advantage of treemaps is their ability to display large amounts of hierarchical information in a compact and space-efficient manner, making them ideal for visualizing proportions and distributions within complex datasets. By organizing data into a mosaic-like structure, treemaps facilitate the comparison of different categories and subcategories, allowing users to quickly identify patterns, trends, and outliers within the hierarchy. This technique is particularly useful in scenarios where space is limited, and a clear overview of the data hierarchy is essential for effective analysis and decision-making.

10.2.2   Design Principles for Treemaps

Creating effective treemaps involves adhering to several key design principles to ensure clarity, accuracy, and usability:

Proportional Sizing: The area of each rectangle should accurately represent the magnitude of the corresponding data attribute. Ensuring proportional sizing is crucial for maintaining the integrity of the visualization and enabling accurate comparisons between different categories.

Consistent Hierarchical Layout: Maintain a consistent hierarchical structure throughout the treemap. Each level of the hierarchy should be clearly defined, with subcategories nested within their parent categories to reflect the data's inherent organization.

Color Encoding: Utilize color to encode additional data dimensions or to differentiate between categories. A well-chosen color scheme can enhance the visual appeal of the treemap and aid in the quick identification of patterns and anomalies.

Minimalist Design: Avoid clutter by limiting the number of categories and subcategories displayed simultaneously. A minimalist approach helps prevent information overload and ensures that the most important data stands out.

Interactivity: Incorporate interactive elements such as tooltips, zooming, and clicking to explore specific data points in greater detail. Interactivity enhances user engagement and allows for deeper data exploration without overwhelming the primary visualization.

By following these design principles and guidelines (Table 10.4), treemaps can effectively communicate hierarchical data, providing users with a clear and intuitive understanding of complex information structures.

Table 10.4: Best Practices and UX Principles Applied in Treemap Visualization

Design Practice or Feature	Explanation and UX Benefit	Reference / Principle
Sorted Data Input	Sorting values descending improves interpretability and area comparison accuracy.	Cleveland & McGill (1984), Shneiderman's Information-Seeking Mantra
Colorblind-Friendly Palette	Selected colors ensure accessibility for users with color vision deficiency.	WCAG 2.1, Color Universal Design (CUD) Guidelines
Large Font with High Contrast	Bold white text over vivid colors increases legibility of labels in small regions.	Nielsen's Usability Heuristics: Visibility of System Status
Data Labels with Values	Embedding both category name and percentage within the block reduces cognitive load.	Tufte's Principle of Data-Ink Maximization
Legend with High Readability	Clearly labeled legend supports category decoding without re-scanning treemap blocks.	Gestalt Principle of Proximity and Labeling Guidelines
Minimalism: Axis Removed	Axes are redundant in this layout; removing them reduces visual clutter.	Tufte's Principle of Chartjunk Elimination
Semantic Title with Padding	Descriptive title explains context and improves comprehension; padding ensures spatial clarity.	Cognitive Load Theory, UX Writing Standards
Consistent Padding in Blocks	Uniform spacing improves separation and alignment, enhancing visual balance.	Aesthetic-Usability Effect, White Space Usage Principles
Bold Font in Legend Title	Enhances visual hierarchy and directs attention appropriately.	Fitts's Law (guiding attention via prominence)
Tight Layout Management	`tight_layout` prevents overlap and enhances visual compactness on all screen sizes.	Responsive Design Guidelines

10.2.3   Facilitating Comparative Analysis

Treemaps excel in facilitating comparative analysis by allowing users to simultaneously view multiple categories and their subcategories within a single, cohesive layout. The proportional sizing of rectangles enables quick visual comparisons of different data segments, making it easy to identify which categories dominate the dataset and which are less significant. The hierarchical nesting of rectangles allows for the examination of relationships between parent and child categories, highlighting how subcategories contribute to overall trends. Color encoding further enhances comparative analysis by enabling users to differentiate between categories or represent additional data dimensions, such as performance metrics or growth rates. This multi-faceted approach empowers users to conduct comprehensive comparisons across various levels of the data hierarchy, uncovering insights that inform strategic decision-making and data-driven analysis.

10.2.4 Implementation Techniques

Implementing treemaps effectively involves utilizing appropriate tools and libraries that support hierarchical data representation and offer customization options for design and interactivity. Below is a Python code snippet using the `Squarify` and `Matplotlib` libraries to create a basic treemap.

```python
# Sample data (sorted for better visual comparison)
data = {
    'Company': ['Company A', 'Company B', 'Company C', 'Company D',
    ↪ 'Company E'],
    'Market Share': [40, 35, 15, 10, 5]
}
df = pd.DataFrame(data).sort_values(by='Market Share', ascending=False)

# Color palette: colorblind-friendly
colors = ['#1b9e77', '#d95f02', '#7570b3', '#e7298a', '#66a61e']

# Plot configuration
fig, ax = plt.subplots(figsize=(10, 6))
squarify.plot(
    sizes=df['Market Share'],
    label=[f"{label}\n{value}%" for label, value in zip(df['Company'],
    ↪ df['Market Share'])],
    color=colors,
    alpha=0.9,
    pad=True,
    text_kwargs={'fontsize': 14, 'color': 'white', 'weight': 'bold'}
)

# Add a legend formatted for accessibility)
legend_patches = [
    mpatches.Patch(color=color, label=label)
    for color, label in zip(colors, df['Company'])
]
plt.legend(handles=legend_patches, title="Companies", loc='upper left',
↪ bbox_to_anchor=(1, 1), fontsize=12,
↪ title_fontproperties={'weight':'bold', 'size':14})

# Remove axes\index{Axes} and set title
plt.axis('off')
plt.title('Market Share Distribution by Company', fontsize=18,
↪ fontweight='bold', pad=20)

# Display
```

```
34  plt.tight_layout()
35  plt.show()
```

In Figure 10.1, the treemap visualizes the market share of five companies. Each rectangle's size corresponds to the company's market share, and distinct colors differentiate between the companies. This clear representation allows for an immediate understanding of the relative market positions, highlighting the dominance of Company A and the minimal presence of Company E.

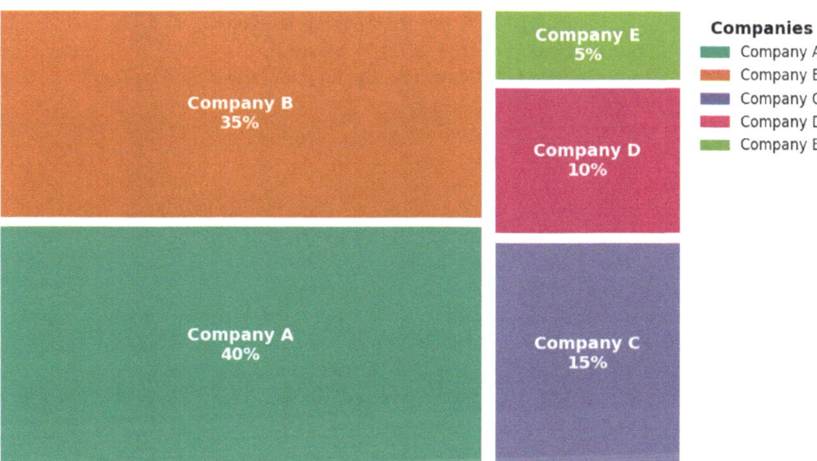

Fig. 10.1: Treemap Diagram Example Representing Market Share by Company

10.2.5 Example

An illustrative example is the use of treemaps in `Plotly` to create interactive and dynamic treemap visualizations. Below is a Python code snippet demonstrating how to implement a treemap using Plotly's `express` module.

```python
# Sample data
data = {
    'Region': ['North', 'North', 'South', 'South', 'East', 'East', 'West',
        'West'],
    'Country': ['USA', 'Canada', 'Brazil', 'Argentina', 'China', 'Japan',
        'Germany', 'France'],
    'Sales': [300, 200, 400, 150, 500, 350, 250, 180],
    'Profit': [50, 30, 80, 20, 100, 70, 40, 25]
}
```

```python
8    df = pd.DataFrame(data)
9
10   # Create treemap
11   fig = px.treemap(
12       df,
13       path=['Region', 'Country'],
14       values='Sales',
15       color='Profit',
16       hover_data={
17           'Sales': True,
18           'Profit': True,
19           'Region': False,
20           'Country': False
21       },
22       color_continuous_scale=px.colors.diverging.Tealrose,
23       color_continuous_midpoint=df['Profit'].mean(),
24       title='<b>Sales and Profit Overview by Region and Country</b>'
25   )
26
27   # Update layout with larger bold fonts
28   fig.update_layout(
29       title_font=dict(size=28, color='black', family='Arial'),
30       font=dict(size=18, family='Arial', color='black'),
31       margin=dict(t=80, l=10, r=10, b=10),
32       coloraxis_colorbar=dict(
33           title='<b>Profit ($)</b>',
34           ticksuffix='',
35           titlefont=dict(size=18, family='Arial', color='black'),
36           tickfont=dict(size=16, family='Arial')
37       )
38   )
39
40   # Make labels inside boxes larger and bold
41   fig.update_traces(
42       texttemplate='<b>%{label}</b><br><b>Sales:</b> %{value}',
43       textfont=dict(size=20, family='Arial', color='white'),
44       marker=dict(line=dict(width=0.7, color='black'))
45   )
46
47   # Show figure
48   fig.show()
```

In this example (Figure 10.2), the treemap displays sales and profit data across different regions and countries. The hierarchical structure allows users to drill down from regions to individual countries, while color encoding represents profit levels, providing a comprehensive view of sales performance and profitability.

These examples demonstrate the versatility and effectiveness of treemaps in scientific visualization, enabling researchers and analysts to present hierarchical data in an intuitive and informative manner.

Table 10.5 presents a curated set of UX design practices that were deliberately applied in the construction of a high-quality treemap visualization. Each design

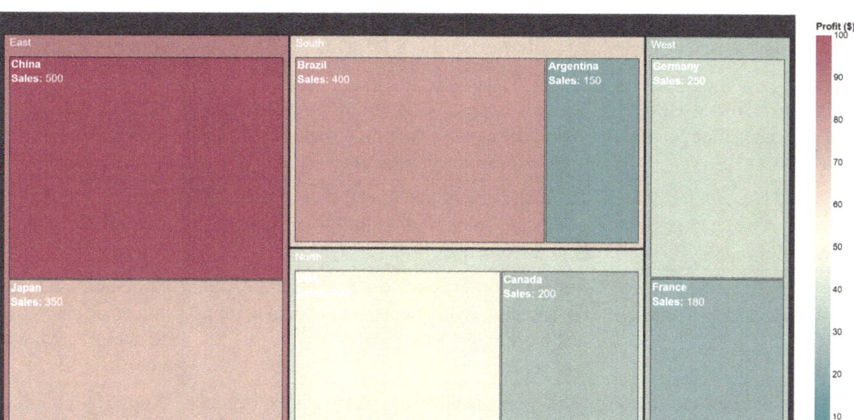

Fig. 10.2: Interactive Treemap Representing Sales and Profit by Region and Country

element contributes to improving accessibility, interpretability, and cognitive usability, aligning with well-established human-centered design principles and visualization best practices. For example, the use of a *Color Universal Design* palette with high contrast (e.g., Tealrose) ensures the treemap is interpretable by users with color vision deficiencies. This aligns with both WCAG 2.1 standards and CUD guidelines [15, 19]. Similarly, embedding bold white labels directly within the treemap tiles maximizes data-ink ratio, as advocated by Tufte [18], and enhances in-situ comprehension of data values. The inclusion of an intuitive drill-down hierarchy (Region $\rightarrow$ Country) facilitates progressive exploration while respecting the user's mental model, an embodiment of Shneiderman's information-seeking mantra [16]. Interactive hover states further support Nielsen's heuristic of maintaining system status visibility without visual clutter [13]. Spatial grouping by geographical region, use of white space, and a consistent typographic scale follow cognitive psychology principles, such as Gestalt proximity and cognitive load theory [20]. These decisions guide the user's eye and reduce interpretation time.

10.2.6 Best Practices for Treemaps

To maximize the effectiveness of treemaps, adhere to the following best practices:

Limit Hierarchy Depth: Avoid excessive levels of hierarchy, as deep nesting can make the treemap cluttered and difficult to interpret. Typically, two to three levels of hierarchy are sufficient for most applications.

Maintain Proportionality: Ensure that the size of each rectangle accurately reflects the underlying data value. Proportional sizing is crucial for enabling accurate comparisons between different categories and subcategories.

Table 10.5: UX Design Practices Applied in Treemap Visualization

Design Practice or Feature	Explanation and UX Benefit	Reference / Principle
Color Universal Design palette and color contrast (Tealrose)	Ensures accessibility for users with color vision deficiency; diverging scale highlights high and low profit effectively	WCAG 2.1, CUD Guidelines [15, 19]
Embedded data labels in each box using large bold white font	Improves clarity and readability of data values inside colored areas, following best text-contrast practices	Tufte's Data-Ink Maximization [18]
Hierarchical drill-down path: Region → Country	Supports progressive disclosure and exploration; aligns with mental model of geographical hierarchy	Shneiderman's Information-Seeking Mantra [16]
Interactive hover data with Profit and Sales values	Maintains visibility of system status without clutter; reduces cognitive overload	Nielsen's Usability Heuristics [13]
Logical grouping by Region and visual proximity of countries	Enhances recognition and reduces cognitive effort through spatial organization of related data	Gestalt Proximity Principle [20]
White space via padding and margins	Supports scannability and visual separation, using aesthetic-usability effect and responsiveness	Aesthetic-Usability Effect [13]
Consistent typography and size hierarchy	Improves legibility and comprehension across devices; respects cognitive load thresholds	Cognitive Load Theory, Responsive Design [17, 10]
Color legend with labeled axis and value bar	Ensures accurate interpretation of value encoding; supports accessibility and semantic mapping	Chartjunk Elimination [18]

Use Distinct Colors: Apply a consistent and meaningful color scheme to differentiate between categories or to represent additional data dimensions. Avoid using too many colors, which can overwhelm the viewer and reduce clarity.

Provide Clear Labels: Label rectangles clearly to identify the represented categories. For larger rectangles, full labels can be displayed, while smaller rectangles may require abbreviated labels or hover tooltips to prevent overlap and maintain readability.

Incorporate Interactivity: Utilize interactive features such as tooltips, zooming, and clickable areas to enhance user engagement and facilitate deeper exploration of the data without cluttering the visualization.

Optimize Layout: Arrange rectangles in a manner that minimizes wasted space and maintains a balanced layout. Efficient use of space ensures that all categories are visible and easily comparable.

By following these best practices, treemaps can effectively convey hierarchical data, providing users with clear and actionable insights. Figure 10.3 exemplifies best practices in treemap design, showcasing a balanced layout with clear labels, consistent color schemes, and interactive elements that enhance data exploration and interpretation.

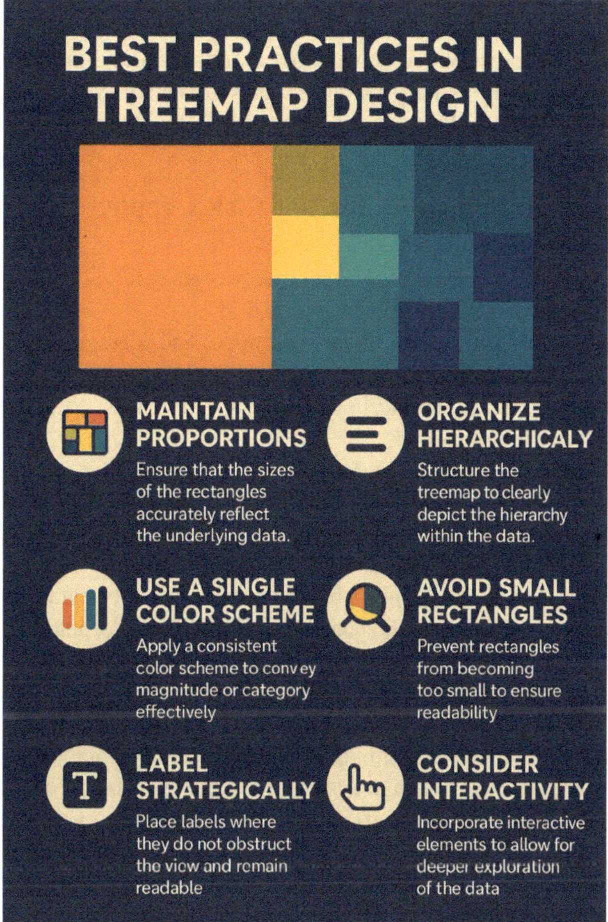

Fig. 10.3: Best Practices in Treemap Design

10.3 Network Graphs

10.3.1 Concept of Network Graphs

Network graphs are a visualization technique used to represent relationships and interactions between entities, known as nodes, connected by edges. These graphs are instrumental in illustrating complex networks such as social interactions, biological systems, communication networks, and transportation systems. Each node represents an individual entity, while the edges denote the connections or relationships between them. The structure of a network graph can reveal important insights into the underlying patterns, such as clusters, central nodes, and hierarchical relationships. By visualizing these connections, network graphs facilitate the understanding of the dynamics and complexities inherent in interconnected systems, making them a powerful tool for both exploratory data analysis and the communication of intricate relational data.

10.3.2 Design Principles for Network Graphs

Designing effective Network graphs involves several key principles to ensure clarity, readability, and meaningful interpretation:

Clarity of Nodes and Edges: Ensure that nodes and edges are clearly distinguishable. Nodes should be appropriately sized and colored to represent different categories or attributes, while edges should be styled (e.g., thickness, color) to indicate the strength or type of relationship.

Layout Optimization: Utilize layout algorithms (e.g., force-directed, circular, hierarchical) that enhance the readability of the graph by minimizing overlapping edges and evenly distributing nodes. The chosen layout should align with the nature of the data and the insights being sought.

Scalability: Design the graph to handle varying sizes of datasets without compromising readability. Techniques such as clustering, filtering, and zooming can help manage larger networks by focusing on relevant subsets of data.

Interactive Features: Incorporate interactivity, such as hover tooltips, clickable nodes, and dynamic filtering, to allow users to explore the graph in greater detail. Interactive elements enhance user engagement and facilitate deeper data exploration.

Consistent Encoding: Maintain consistency in the encoding of visual attributes (e.g., color, size) across different graphs to enable accurate comparisons and pattern recognition.

By adhering to these design principles, Network graphs can effectively convey the structure and dynamics of complex relationships, providing users with clear and actionable insights.

10.3.3 Implementation Techniques

Implementing Network Graphs effectively involves utilizing appropriate tools and libraries that offer robust functionalities for graph creation, layout optimization, and interactivity. Below is a Python code snippet using `NetworkX` and `Plotly` to create an interactive Network Graph.

```python
# Create a random geometric graph
G = nx.random_geometric_graph(40, 0.2, seed=1111)
pos = nx.get_node_attributes(G, 'pos')
```

```python
 4
 5  # Find the node closest to the center (0.5, 0.5)
 6  dmin = 1
 7  ncenter = 0
 8  for n in pos:
 9      x, y = pos[n]
10      d = (x - 0.5) ** 2 + (y - 0.5) ** 2
11      if d < dmin:
12          ncenter = n
13          dmin = d
14
15  # Compute shortest paths from the center
16  p = nx.single_source_shortest_path_length(G, ncenter)
17
18  # Build edge trace
19  edge_x = []
20  edge_y = []
21
22  for edge in G.edges():
23      x0, y0 = pos[edge[0]]
24      x1, y1 = pos[edge[1]]
25      edge_x += [x0, x1, None]
26      edge_y += [y0, y1, None]
27
28  edge_trace = go.Scatter(
29      x=edge_x,
30      y=edge_y,
31      line=dict(width=3, color='#888'),
32      hoverinfo='none',
33      mode='lines')
34
35  # Build node trace
36  node_x = []
37  node_y = []
38  node_color = []
39  node_text = []
40
41  for node in G.nodes():
42      x, y = pos[node]
43      node_x.append(x)
44      node_y.append(y)
45      num_connections = len(list(G.adj[node]))
46      node_color.append(num_connections)
47      node_text.append(f'# of connections: {num_connections}')
48
49  node_trace = go.Scatter(
50      x=node_x,
51      y=node_y,
52      mode='markers',
53      hoverinfo='text',
54      text=node_text,
55      marker=dict(
56          showscale=True,
57          colorscale='YlGnBu',
```

```
58          reversescale=True,
59          color=node_color,
60          size=30,
61          colorbar=dict(
62              thickness=25,
63              title='Node Connections',
64              xanchor='left',
65              titleside='right'),
66          line=dict(width=4)
67      )
68  )
69
70  # Create figure
71  fig = go.Figure(
72      data=[edge_trace, node_trace],
73      layout=go.Layout(
74          title='<b>Geometric Network Graph</b>',
75          titlefont_size=24,
76          showlegend=False,
77          hovermode='closest',
78          margin=dict(b=20, l=20, r=20, t=40),
79          xaxis=dict(showgrid=False, zeroline=False, showticklabels=False),
80          yaxis=dict(showgrid=False, zeroline=False, showticklabels=False))
81  )
82  fig.show()
```

In Figure 10.4, the interactive Network Graph visualizes social connections between individuals. Nodes represent people, with size and color indicating different attributes, while edges represent the strength of their relationships. Hovering over a node displays additional information, enhancing the interactivity and informativeness of the visualization.

10.3.4   Best Practices for Network Graphs

To maximize the effectiveness of Network Graphs, adhere to the following best practices:

Simplify Complexity: Avoid overcrowding the graph with too many nodes and edges. Focus on the most relevant connections to maintain clarity and prevent visual overload.

Use Meaningful Attributes: Encode node and edge attributes with meaningful visual properties such as color, size, and thickness to convey additional layers of information effectively.

Optimize Layout: Choose layout algorithms that enhance readability and reveal underlying structures. Force-directed layouts are popular for their ability to minimize edge crossings and distribute nodes evenly.

Incorporate Interactivity: Implement interactive features like zooming, panning, and tooltips to allow users to explore the graph in detail without overwhelming the initial view.

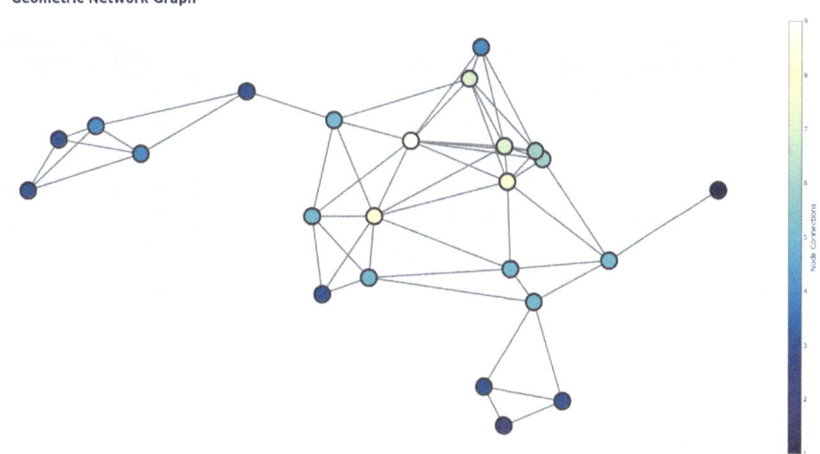

Fig. 10.4: Interactive Network Graph Representing Social Connections

Highlight Key Nodes: Emphasize important nodes or clusters through distinct colors, sizes, or annotations to draw attention to significant elements within the network.

Maintain Consistency: Ensure consistency in the representation of similar nodes and edges across multiple visualizations to facilitate accurate comparisons and pattern recognition.

> **⚠ Warning**
>
> **Network graphs become unreadable when too many nodes and edges are included.** Always apply filtering, clustering, or zoom-based interactions to reduce clutter and improve interpretability, especially in large-scale graphs.

By following these best practices (Table 10.6), Network graphs can effectively represent complex relationships, providing users with clear and actionable insights into the underlying data structures.

Figure 10.5 exemplifies best practices in Network Graph design, showcasing a clear layout with highlighted key nodes, consistent color encoding, and interactive elements that enhance user engagement and data exploration.

Table 10.6: UX Design and Visualization Principles Applied in the Network Graph

Design Feature or Strategy	Purpose and UX Impact	Guideline / Principle
Color encoding of node connectivity (degree) via 'Yl-GnBu' scale	Reveals structural patterns, encourages exploration of hubs; colorblind-friendly	Color Universal Design, WCAG 2.1 [15, 19]
Centering graph on most central node (closest to (0.5, 0.5))	Enhances layout symmetry and navigational clarity for users	Fitts's Law, Visual Attention Focus [5]
Large circular nodes with colorbar for connection count	Embeds data within elements, minimizes external legend lookup	Tufte's Data-Ink Maximization [18]
Hover tooltip with descriptive text ('# of connections')	Makes hidden structure visible on demand, supports visibility of system status	Nielsen's Heuristics [13]
Edge widths, spacing, and grid-less layout	Minimizes clutter and enables easier pattern recognition	Tufte's Chartjunk Elimination [18]
Exclusion of small disconnected components	Reduces visual noise, prevents distraction, and focuses on main structure	Cognitive Load Theory [17]
Interactive visual encoding with hover, zoom and pan support	Empowers user control and interactive engagement	Shneiderman's Information-Seeking Mantra [16]
Margin and axis suppression for borderless framing	Achieves aesthetic balance and emphasizes structure over scaffolding	Aesthetic-Usability Effect [13]
Thick node outlines and bold title	Enhances contrast and readability for accessibility	Responsive Typography & Layout Design [10]

10.4   Sankey Diagrams

10.4.1   Concept of Sankey Diagrams

Sankey Diagrams are a specialized visualization technique used to depict the flow of quantities between different stages, categories, or entities. They are particularly effective in illustrating the magnitude of transfers or transformations within a system, with the width of the arrows proportional to the flow quantities they represent. This visual proportionality allows for an immediate understanding of the relative significance of different flows, making Sankey Diagrams ideal for analyzing energy transfers, financial transactions, supply chains, and resource allocations. By clearly showing how inputs are distributed across various outputs, Sankey Diagrams provide

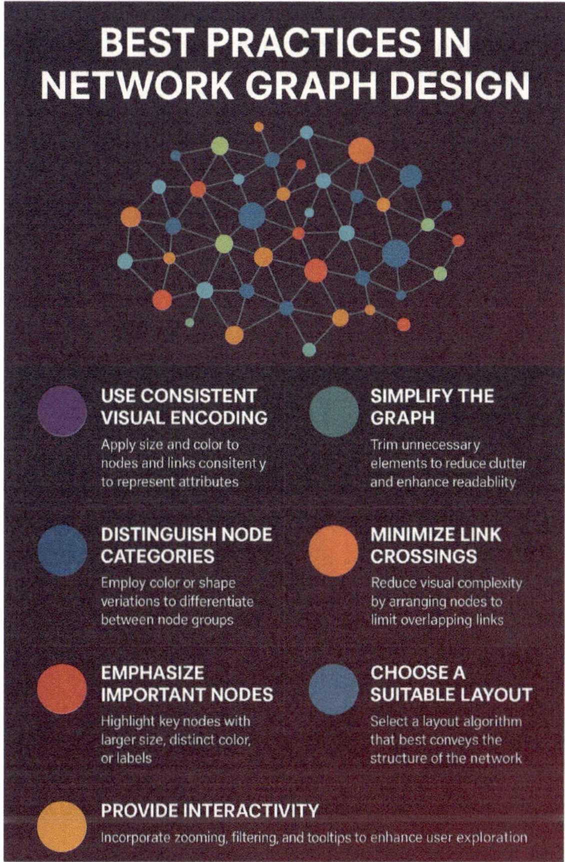

Fig. 10.5: Best Practices in Network Graph Design

a comprehensive overview of the dynamics within complex systems, facilitating the identification of inefficiencies, major contributors, and critical pathways.

10.4.2   Design Principles for Sankey Diagrams

Creating effective Sankey Diagrams involves adhering to several key design principles to ensure clarity, accuracy, and meaningful interpretation:

Proportional Flow Representation: Ensure that the width of the flows accurately represents the magnitude of the quantities being depicted. This proportionality is essential for maintaining the integrity of the data and enabling accurate comparisons.

Clear Labeling: Label all nodes and flows clearly to provide context and facilitate understanding. Labels should be concise and positioned to avoid clutter and overlap with other elements.

Consistent Color Coding: Use consistent color schemes to differentiate between various categories or entities. Consistent colors help in distinguishing different flows and maintaining visual coherence.

Minimal Overlapping: Design the layout to minimize overlapping flows and ensure that all flows are clearly visible. This can be achieved through thoughtful positioning of nodes and optimization of flow paths.

Scalability: Design the diagram to handle varying amounts of data without losing readability. Techniques such as grouping similar flows or using interactive features can help manage larger datasets effectively.

By following these design principles, Sankey Diagrams can effectively communicate the flow dynamics within a system, providing users with clear and actionable insights into the distribution and transformation of quantities.

10.4.3 Facilitating Comparative Analysis

Sankey Diagrams facilitate comparative analysis by allowing users to easily compare the magnitude and direction of flows across different categories or stages. The proportional width of the flows provides a visual cue for comparing the relative significance of different transfers or transformations within the system. For example, in an energy flow analysis, users can quickly identify which energy sources contribute the most to overall consumption and which sectors are the largest consumers. Similarly, in financial Sankey Diagrams, stakeholders can compare the distribution of funds across various departments or projects, highlighting areas of significant investment and potential inefficiencies. By presenting multiple flows within a single diagram, Sankey Diagrams enable users to conduct comprehensive comparisons, uncovering patterns and relationships that inform strategic decision-making and optimization efforts.

10.4.4 Implementation Techniques

Implementing Sankey Diagrams effectively involves utilizing specialized tools and libraries that support flow-based visualizations. Below is a Python code snippet using the `Plotly` library to create an interactive Sankey Diagram.

```python
# Nodes
node_labels = [
    "Fossil Fuels",        # 0
    "Renewable Energy",    # 1
    "Electricity",         # 2
    "Heat",                # 3
    "Industrial",          # 4
    "Residential",         # 5
    "Transport"            # 6
]

# Assign colors for nodes
node_colors = [
```

```python
14        "#d62728",   # Fossil Fuels
15        "#2ca02c",   # Renewable Energy
16        "#1f77b4",   # Electricity
17        "#ff7f0e",   # Heat
18        "#8c564b",   # Industrial
19        "#7f7f7f",   # Residential
20        "#9467bd"    # Transport
21    ]
22
23    # Define links (sources and targets with updated indices)
24    link_sources = [0, 0, 0, 0, 1, 1]    # Fossil Fuels and Renewable Energy
25    link_targets = [2, 3, 4, 5, 2, 3]    # Electricity, Heat, etc.
26    link_values  = [120, 150, 100, 80, 180, 120]  # Values of flow
27    link_colors  = [
28        "rgba(214, 39, 40, 0.6)",   # Fossil
29        "rgba(214, 39, 40, 0.6)",
30        "rgba(214, 39, 40, 0.6)",
31        "rgba(214, 39, 40, 0.6)",
32        "rgba(44, 160, 44, 0.6)",   # Renewable
33        "rgba(44, 160, 44, 0.6)"]
34
35    # Create Sankey diagram
36    fig = go.Figure(data=[go.Sankey(
37        arrangement="snap",
38        node=dict(
39            pad=30,
40            thickness=35,
41            line=dict(color="black", width=2),
42            label=node_labels,
43            color=node_colors,
44            hovertemplate='%{label}<extra></extra>'
45        ),
46        link=dict(
47            source=link_sources,
48            target=link_targets,
49            value=link_values,
50            color=link_colors,
51            hovertemplate='Flow: %{value} units<br>From %{source.label}
52            to %{target.label}<extra></extra>'
53        )
54    )])
55
56    # Update layout
57    fig.update_layout(
58        title=dict(
59            text="<b>Energy Flow Distribution</b>",
60            font=dict(size=22),
61            x=0.5,
62            xanchor="center"),
63        font=dict(size=16, weight='bold'),
64        margin=dict(l=50, r=50, t=80, b=40),
65        height=400,
66        width=600,
67        paper_bgcolor="white"
```

```
68  )
69  fig.show()
```

In Figure 10.6, the Sankey Diagram visualizes the flow of energy from the input source to various end-use sectors. The width of each flow corresponds to the quantity of energy transferred, providing a clear and immediate understanding of how energy is distributed across different applications.

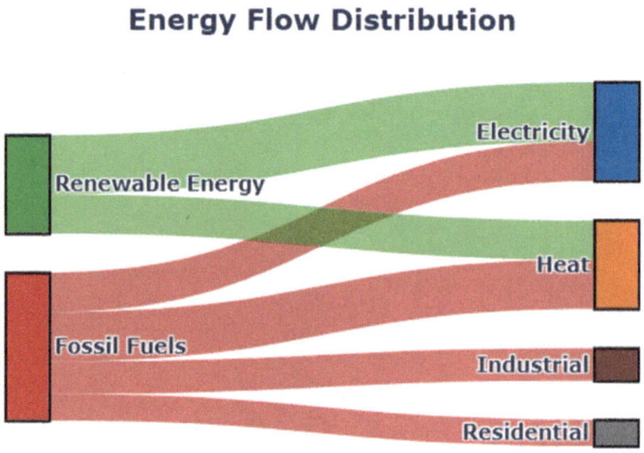

Fig. 10.6: Interactive Sankey Diagram Representing Energy Flow

10.4.5 Best Practices for Sankey Diagrams

To ensure that Sankey Diagrams are effective and informative, adhere to the following best practices:

Limit the Number of Nodes and Flows: Avoid overcrowding the diagram with too many nodes and flows. Focus on the most significant flows to maintain clarity and prevent visual clutter.

Use Meaningful Colors: Apply colors strategically to differentiate between various categories or to represent additional data dimensions. Consistent color schemes enhance readability and facilitate easier interpretation of the flows.

Provide Clear Labels: Ensure that all nodes and flows are clearly labeled with descriptive text. Labels should be legible and positioned to avoid overlapping with other elements in the diagram.

Maintain Proportionality: Accurately represent the magnitude of flows through the proportional width of the arrows. This proportionality is essential for conveying the relative significance of different flows effectively.

Optimize Layout: Arrange nodes and flows in a logical and organized manner to minimize overlapping and ensure that the diagram remains easy to follow. Thoughtful layout design enhances the overall readability and aesthetic appeal of the Sankey Diagram.

Incorporate Interactivity: Where possible, add interactive features such as hover tooltips, clickable nodes, and dynamic filtering to allow users to explore the diagram in greater detail. Interactivity enhances user engagement and facilitates deeper data exploration.

By following these best practices (see Table 10.7), Sankey Diagrams can effectively communicate the flow dynamics within a system, providing users with clear and actionable insights into the distribution and transformation of quantities.

Table 10.7: UX Design Principles and Guidelines in Sankey Diagram Visualization

Design Element	UX Purpose and Rationale	Principle / Reference
Alphabetical node ordering (left to right)	Improves cognitive parsing and visual search efficiency	Shneiderman's Information-Seeking Mantra [16]
Color coding for energy sources (Fossil: red, Renewable: green)	Encodes categorical differences clearly with intuitive color metaphors (e.g., red = fossil, green = renewable)	WCAG 2.1, Color Universal Design [15, 19]
Link colors aligned with energy source	Enables traceability of flows from origin to destination with minimal cognitive load	Gestalt Principle of Common Fate [20]
Bold, centered and large title with strong font hierarchy	Supports immediate user orientation and comprehension of visualization intent	Aesthetic-Usability Effect, Responsive Typography [13, 10]
Thick node boxes with padding and border outlines	Ensures element distinguishability, aids target acquisition, improves accessibility	Fitts's Law, Accessible UI Guidelines [5]
Hover tooltips with structured labels and flow values	Encourages exploration, supports visibility of information on demand	Nielsen's Usability Heuristics [13]
Consistent margins and white background	Reduces visual clutter, ensures clean and readable layout	Tufte's Principle of Chartjunk Elimination [18]
Fixed node positioning ('arrangement="snap"')	Maintains structural consistency and prevents disorientation on interaction	Cognitive Load Theory [17]

Figure 10.7 illustrates best practices in Sankey Diagram design, showcasing a clear layout with proportional flow representation, consistent color coding, and well-placed labels that enhance data interpretation and user engagement.

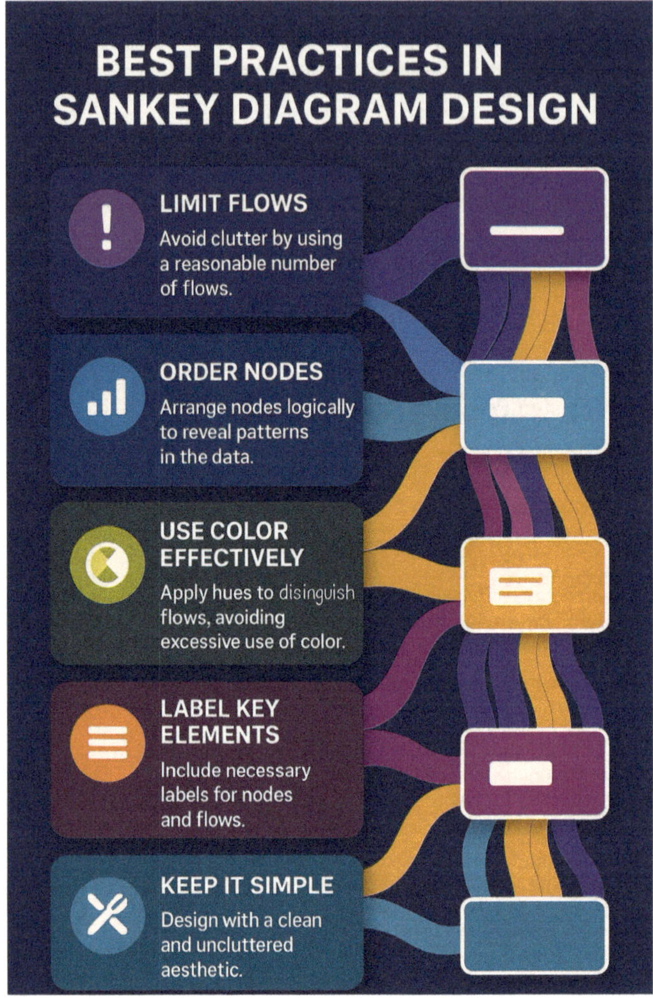

Fig. 10.7: Best Practices in Sankey Diagram Design

10.5 Chord Diagrams

10.5.1 Concept of Chord Diagrams

Chord Diagrams are a visualzation technique used to represent the interrelationships and flows between different entities within a dataset. They are particularly effective in illustrating complex connections, such as trade relationships between countries, collaborations between organizations, or interactions within biological systems. In a Chord Diagram, entities are arranged in a circular layout, and the relationships between them are depicted as arcs or ribbons connecting the corresponding segments on the circle. The width or color of the ribbons can represent the magnitude or type of the relationship, providing a clear and intuitive overview of the network's structure and dynamics. By displaying connections in a circular format, Chord Diagrams facilitate the visualization of multiple relationships simultaneously, making it easier to identify patterns, clusters, and key players within the network.

10.5.2 Design Principles for Chord Diagrams

Creating effective Chord Diagrams involves adhering to several key design principles to ensure clarity, accuracy, and meaningful interpretation:

Circular Layout: Arrange entities in a circular layout to provide a balanced and organized structure for depicting relationships. This layout minimizes edge crossings and distributes connections evenly around the circle.

Proportional Encoding: Encode the magnitude or significance of relationships through the width or color of the chords. This proportionality ensures that the visual representation accurately reflects the underlying data.

Clear Labeling: Label each segment clearly to identify the entities being represented. Labels should be positioned to avoid overlap and maintain readability, possibly using interactive tooltips for detailed information.

Minimize Overlapping Chords: Design the diagram to minimize overlapping or tangled chords, which can obscure relationships and hinder interpretation. Techniques such as chord ordering algorithms can help in reducing clutter.

Consistent Color Schemes: Use consistent color schemes to differentiate between entities or to represent different types of relationships. Consistent coloring enhances the visual coherence and facilitates easier interpretation of the diagram.

Interactive Features: Incorporate interactive elements such as hover effects, clickable segments, and dynamic filtering to allow users to explore specific relationships in greater detail without overwhelming the overall visualization.

By following these design principles, Chord Diagrams can effectively communicate the complex interrelationships within a dataset, providing users with clear and actionable insights into the network's structure and dynamics.

10.5.3 Facilitating Comparative Analysis

Chord Diagrams facilitate comparative analysis by allowing users to simultaneously view and compare multiple relationships within a network. The circular layout enables the visualization of connections between all pairs of entities, making it easy to identify which entities have the strongest or most frequent interactions. By encoding the magnitude or type of relationships through chord properties such as width or color, users can quickly assess the relative significance of different connections. This comprehensive view supports the identification of key players, clusters, and outliers within the network, enhancing the ability to compare and contrast different aspects of the data. Interactive features such as highlighting specific chords or filtering certain entities allow users to focus on particular segments of the network, facilitating more targeted and detailed comparative analyses.

10.5.4 Implementation Techniques

Implementing Chord Diagrams effectively involves utilizing specialized tools and libraries that support the creation and customization of flow-based visualizations. Below is a Python code snippet using the `Plotly` library to create an interactive Chord Diagram.

```python
# Sorted country labels
labels = sorted(["USA", "China", "Germany", "India", "Brazil"])

# Trade matrix matching sorted labels
matrix = [
    [0, 80, 250, 150, 300],    # Brazil
    [80, 0, 100, 120, 400],    # China
    [250, 100, 0, 70, 200],    # Germany
    [150, 120, 70, 0, 100],    # India
    [300, 400, 200, 100, 0],   # USA
]

# Generate edge list from matrix
edges = []
for i, source in enumerate(labels):
    for j, target in enumerate(labels):
        if matrix[i][j] > 0:
            edges.append((labels[i], labels[j], matrix[i][j]))

# Convert to DataFrame
edge_df = pd.DataFrame(edges, columns=["source", "target", "value"])

# Create node DataFrame (needed for coloring/styling)
nodes = pd.DataFrame({'name': labels})

# Create Chord object
chord = hv.Chord((edge_df, hv.Dataset(nodes, 'name')))

# Apply best-practice visual styling
chord.opts(
    opts.Chord(
        labels='name',
        node_color='name',        # refers to node dataset column
        edge_color='source',      # refers to edge_df column
        cmap='Category10',
        edge_cmap='Category20',
        edge_alpha=0.75,
        node_size=18,
        width=700,
        height=700,
```

```
41        tools=['hover'],
42        fontscale=1.5,
43        label_text_font_size='14pt',
44        show_frame=False,
45        bgcolor='lightgray',
46        hooks=[rotate_label])
47 )
48
49 # Render the Chord diagram as a Bokeh object
50 bokeh_obj = renderer('bokeh').get_plot(chord).state
51
52 # Display it in a browser window
53 show(bokeh_obj)
```

In Figure 10.8, the Chord Diagram visualizes the trade relationships between five countries. Each segment of the circle represents a country, and the chords connecting them indicate the volume of trade between each pair. The width of the chords corresponds to the magnitude of trade flows, providing a clear and immediate understanding of the most significant trade relationships.

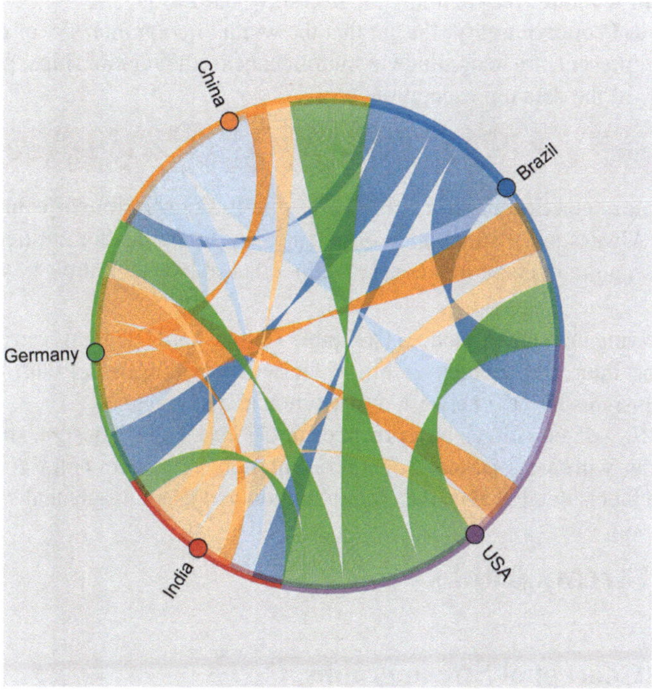

Fig. 10.8: Interactive Chord Diagram Representing Trade Relationships

10.5.5   Best Practices for Chord Diagrams

To ensure that Chord Diagrams are effective and informative, adhere to the following best practices:

Limit the Number of Entities: Avoid including too many entities in a single Chord Diagram, as this can lead to cluttered and unreadable visuals. Focus on the most significant entities to maintain clarity.

Use Meaningful Color Coding: Apply color schemes that distinguish between different types of relationships or categories. Consistent color usage helps in interpreting the diagram accurately.

Provide Clear Labels: Ensure that all segments and chords are clearly labeled to identify the entities and the nature of their relationships. Consider using interactive tooltips for additional information to prevent label overlap.

Optimize Layout: Arrange the segments to minimize overlapping chords and enhance the readability of the diagram. Utilizing chord ordering algorithms can help in reducing clutter and improving visual coherence.

Incorporate Interactivity: Add interactive features such as hover effects, clickable segments, and dynamic filtering to allow users to explore specific relationships in greater detail without overwhelming the overall visualization.

Maintain Proportionality: Ensure that the width or color intensity of the chords accurately represents the magnitude or significance of the relationships, preserving the integrity of the data representation.

> **⚠ Warning**
>
> **Poor color selection in chord diagrams can make relationships hard to follow.** Always use distinct and consistent color mappings for entities, and avoid using gradients or low-contrast colors that reduce legibility.

By following these best practices (see Table 10.8), Chord Diagrams can effectively illustrate the interrelationships within a dataset, providing users with clear and actionable insights into the network's structure and dynamics.

Figure 10.9 demonstrates best practices in Chord Diagram design, showcasing a clear layout with proportional flow representation, consistent color coding, and well-placed labels that enhance data interpretation and user engagement.

10.6   Streamgraphs

10.6.1   Concept of Streamgraphs

Streamgraphs are a variation of stacked area charts that display data streams flowing over time, creating an organic and flowing visual representation. Unlike traditional area charts, streamgraphs eliminate the rigid baseline, allowing the streams

Table 10.8: UX Design Principles and Best Practices Applied in Chord Diagram

Principle / Practice	Explanation	Reference
Consistent Label Orientation	Dynamically adjusted text rotation and alignment prevent upside-down labels, improving readability.	[24]
Sorted Node Ordering	Alphabetically sorting countries enhances logical flow and cognitive ease of finding elements.	[18]
Color Differentiation by Category	Distinct colors for nodes and links enhance category distinction and reduce cognitive load.	[6]
Semantic Edge Coloring	Coloring links by source entity provides immediate semantic encoding of data flow origin.	[20]
Reduced Visual Clutter	'edge_alpha=0.75' softens connections and emphasizes nodes without overwhelming users.	[9]
Hover Tooltips	Interactivity via hover shows detailed information on demand, aligning with progressive disclosure.	[16]
High Font Contrast	Font scale and background contrast improve accessibility and legibility for all users.	[23]
Frame Suppression	'show_frame=False' creates a cleaner and more modern aesthetic, avoiding unnecessary borders.	[13]

to curve and shift, which enhances the aesthetic appeal and makes them more engaging. Each stream in the graph represents a different category or variable, with the height of the stream corresponding to the magnitude of the data at any given time. This fluid layout enables users to visualize the evolution of multiple categories simultaneously, making it easier to identify trends, patterns, and shifts in the data over time. Streamgraphs are particularly useful for displaying time-series data with multiple overlapping categories, such as music genres over decades, website traffic sources, or resource allocations in projects.

10.6.2 Design Principles for Streamgraphs

Creating effective Streamgraphs involves adhering to several key design principles to ensure clarity, readability, and meaningful data representation:

Flow Continuity: Ensure that each stream flows smoothly over time without abrupt jumps or breaks. Smooth transitions enhance the visual appeal and make the data easier to follow.

Consistent Color Coding: Use a consistent color scheme to differentiate between streams. Assign distinct colors to each category to facilitate easy identification and comparison.

Avoid Overlapping Streams: Design the stream layout to minimize overlapping or excessive layering of streams, which can obscure data and reduce readability.

Scalable Layout: Ensure that the Streamgraph scales appropriately with the data, maintaining proportionality and preventing distortion as data magnitudes change over time.

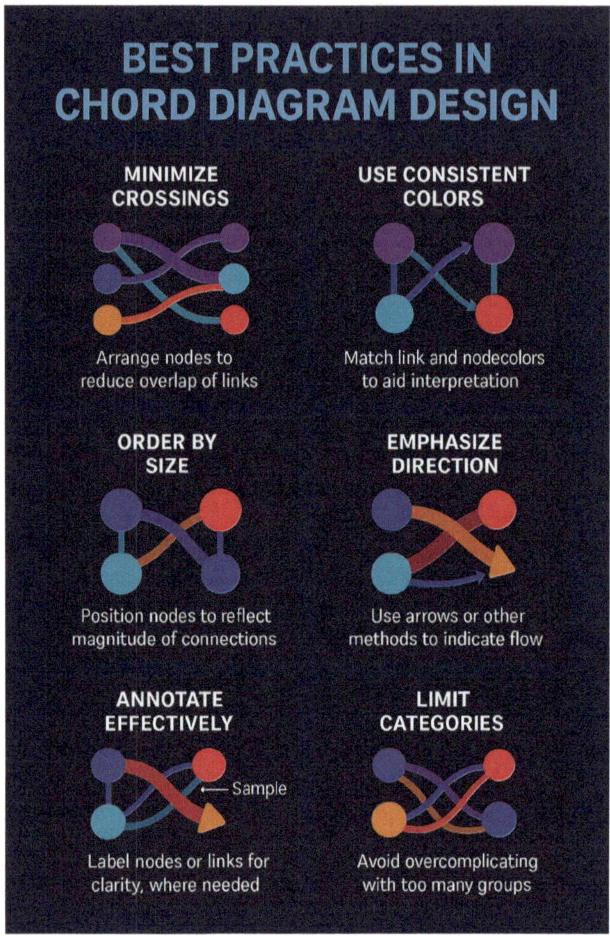

Fig. 10.9: Best Practices in Chord Diagram Design

Clear Labeling: Incorporate clear labels or legends to identify each stream, ensuring that users can easily associate colors with their corresponding categories.

Interactive Features: Where possible, add interactive elements such as tooltips, highlighting, and dynamic filtering to allow users to explore specific streams in greater detail without cluttering the overall visualization.

By following these design principles, Streamgraphs can effectively represent the dynamic flow of multiple categories over time, providing users with a clear and engaging means of exploring and analyzing time-series data.

10.6.3 Facilitating Comparative Analysis

Streamgraphs facilitate comparative analysis by allowing users to view and compare multiple data streams simultaneously within a single visualization. The flowing layout of the streams makes it easy to identify trends and patterns across different categories over time, highlighting periods of growth, decline, and stability. By maintaining consistent color coding and proportional scaling, users can quickly compare the relative magnitudes of different streams, making it easier to discern which categories dominate at various points in time and how their contributions evolve. The organic nature of streamgraphs enables the visualization of subtle shifts and interactions between streams, providing deeper insights into the dynamics of the data. This simultaneous display of multiple streams enhances the ability to conduct comprehensive comparative analyses, uncovering relationships and trends that inform data-driven decision-making.

10.6.4 Implementation Techniques

Implementing Streamgraphs effectively involves utilizing specialized tools and libraries that support the creation and customization of flowing, stacked area charts. Below is a Python code snippet using the `Plotly` library to create an interactive Streamgraph.

```python
# Define time range and genres
years = list(range(2000, 2021))
genres = ['Rock', 'Pop', 'Hip-Hop', 'Jazz', 'Classical']

# Generate sample data
data = {genre: np.random.randint(20, 100, len(years)) for genre in genres}
df = pd.DataFrame(data, index=years)

# Create Streamgraph \index{Streamgraph} with improved design
fig = go.Figure()

# Define color palette
palette = ['#1f77b4', '#ff7f0e', '#2ca02c', '#d62728', '#9467bd']

for genre, color in zip(genres, palette):
    fig.add_trace(go.Scatter(
        x=df.index,
        y=df[genre],
        mode='lines',
        stackgroup='one',        # Enables Streamgraph \index{Streamgraph}
            ↪   stacking
        name=genre,
        line=dict(width=2.5),
        fill='tonexty',
        hoverinfo='x+y+name',
```

```python
25          marker=dict(color=color)
26      ))
27
28  # Apply best practice layout updates
29  fig.update_layout(
30      title={
31          'text': 'Music Genre Popularity (2000{2020)',
32          'x': 0.5,
33          'xanchor': 'center',
34          'font': dict(size=24, family='Arial Black')
35      },
36      xaxis=dict(
37          title='Year',
38          tickmode='linear',
39          tick0=2000,
40          dtick=2,
41          titlefont=dict(size=18,weight='bold'),
42          tickfont=dict(size=14,weight='bold'),
43          gridcolor='lightgrey',
44      ),
45      yaxis=dict(
46          title='Popularity Index',
47          titlefont=dict(size=18,weight='bold'),
48          tickfont=dict(size=14,weight='bold'),
49          gridcolor='lightgrey',
50      ),
51      legend=dict(
52          orientation="h",
53          yanchor="bottom",
54          y=1.02,
55          xanchor="center",
56          x=0.5,
57          font=dict(size=14)
58      ),
59      plot_bgcolor='white',
60      margin=dict(t=80, l=50, r=50, b=50)
61  )
62
63  # Show the interactive plot
64  fig.update_layout(
65      autosize=False,
66      width=800,
67      height=600,
68  )
69  fig.show()
```

In Figure 10.10, the Streamgraph visualizes the popularity of different music genres from 2000 to 2020. Each stream represents a genre, with the flow indicating changes in popularity over time. The stacked layout allows for easy comparison of genre trends and the identification of dominant and emerging genres throughout the years.

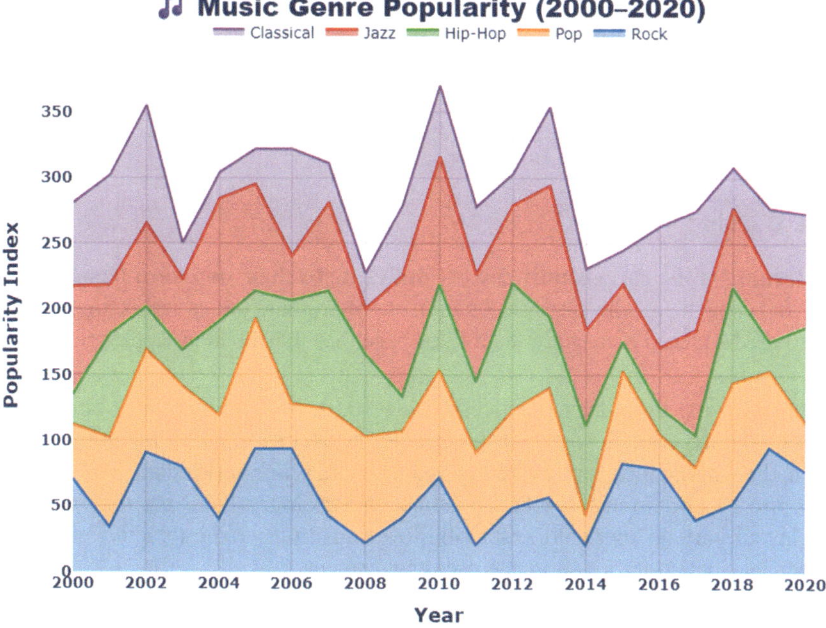

Fig. 10.10: Interactive Streamgraph Representing Music Genre Popularity Over Time

10.6.5   Best Practices for Streamgraphs

To ensure that Streamgraphs are effective and informative, adhere to the following best practices:

Limit the Number of Streams: Avoid including too many streams in a single Streamgraph, as this can lead to visual clutter and make it difficult to interpret individual trends. Focus on the most significant categories to maintain clarity.

Use Distinct Colors: Assign distinct and consistent colors to each stream to facilitate easy identification and comparison. Avoid using overly similar colors that can cause confusion.

Ensure Proportionality: Maintain proportional scaling of the streams to accurately represent the underlying data values. Avoid distortions that can mislead interpretation.

Provide Clear Legends: Include clear legends that map colors to their corresponding categories, ensuring that users can easily understand the representation of each stream.

Enhance Interactivity: Incorporate interactive features such as tooltips, high-lighting, and dynamic filtering to allow users to explore specific streams in greater detail without overwhelming the overall visualization.

Maintain Smooth Transitions: Design the stream layout to ensure smooth and continuous flows, enhancing the visual appeal and making it easier to follow trends over time.

> **⚠ Warning**
>
> **Streamgraphs can visually distort data due to their non-zero baseline.** Avoid using them when precise comparisons of absolute values across time are required–stacked area charts with fixed baselines may be more appropriate.

By following these best practices (Table 10.9), Streamgraphs can effectively represent the dynamic flow of multiple categories over time, providing users with a clear and engaging means of exploring and analyzing time-series data.

Figure 10.11 demonstrates best practices in Streamgraph design, showcasing a clear layout with distinct colors, smooth flow transitions, and interactive elements that enhance data interpretation and user engagement.

10.7 Hexbin Plots

10.7.1 Concept of Hexbin Plots

Hexbin Plots are a visualization technique used to represent the density of data points in a two-dimensional space by aggregating them into hexagonal bins. Unlike traditional scatter plots, which can become cluttered and difficult to interpret with large datasets, Hexbin Plots provide a clear and concise representation of data density by summarizing the number of points within each hexagonal bin. The color or shading of each hexagon indicates the concentration of data points, allowing users to quickly identify areas of high and low density. Hexbin Plots are particularly useful for visualizing large datasets with overlapping points, such as geographical distributions, sensor data, and performance metrics, enabling efficient exploration of data density patterns and underlying trends.

10.7.2 Design Principles for Hexbin Plots

Creating effective Hexbin Plots involves adhering to several key design principles to ensure clarity, accuracy, and meaningful data representation:

Hexagon Size and Orientation: Choose an appropriate size and orientation for the hexagons to balance detail and readability. Smaller hexagons provide finer granularity, while larger ones offer a broader overview of data density.

Table 10.9: UX Design Principles and Best Practices in Streamgraph Visualization

Design Feature	UX Purpose and Justification	Guideline / Reference
`stackgroup='one'` (Streamgraph layering)	Preserves temporal relationships and allows flow-based area comparison over time	Byron & Wattenberg (2008); Heer et al. (2010) [1, 7]
Ordered genre layering with consistent palette	Ensures category recognition and supports color traceability for users	Color Universal Design; Ware (2012) [15, 20]
Custom hover text: `hoverinfo ='x+y+name'`	Enhances interpretability and interaction without cluttering visual space	Shneiderman's Visual Information-Seeking Mantra [16]
Bold axis titles and ticks	Improves readability for users with visual impairments and supports accessibility	WCAG 2.1 Guidelines; Nielsen (1994) [19, 13]
Centralized and bold plot title	Draws attention and improves cognitive load by clearly stating chart purpose	Tufte (2001); Visual Hierarchy Principles [18]
Use of evenly spaced ticks (`dtick=2`)	Enhances temporal trend comparison and maintains scale clarity	Cleveland & McGill (1984) [3]
Gridlines in light grey	Provides reference without overwhelming the visual focus	Few (2009); Tufte's Minimal Ink Principle [4, 18]
White background with strong line contrast	Ensures contrast compliance and improves visual clarity for low-vision users	WCAG 2.1 Contrast Guidelines [19]
Horizontal legend on top	Optimizes space, keeps categories visible, and improves scanning efficiency	Gestalt Proximity and Alignment Principles [20]
Responsive size and margins	Prevents clutter while adapting to various display sizes and screen types	Responsive Visualization Design; UX Flexibility Guidelines [21]

Color Encoding: Use a clear and intuitive color scheme to represent data density, with distinct gradients or color variations that accurately reflect the concentration of data points within each bin.

Consistent Scaling: Maintain consistent scaling across the axes to ensure that the spatial representation accurately reflects the underlying data distribution.

Legibility: Ensure that the hexagons are clearly visible and distinguishable, avoiding excessive overlap or blending that can obscure data density patterns.

Interactive Features: Incorporate interactive elements such as tooltips, zooming, and panning to allow users to explore specific areas of the plot in greater detail without losing the overall context.

Data Normalization: Apply appropriate normalization techniques to account for varying bin sizes or data distributions, ensuring that the color encoding accurately represents data density.

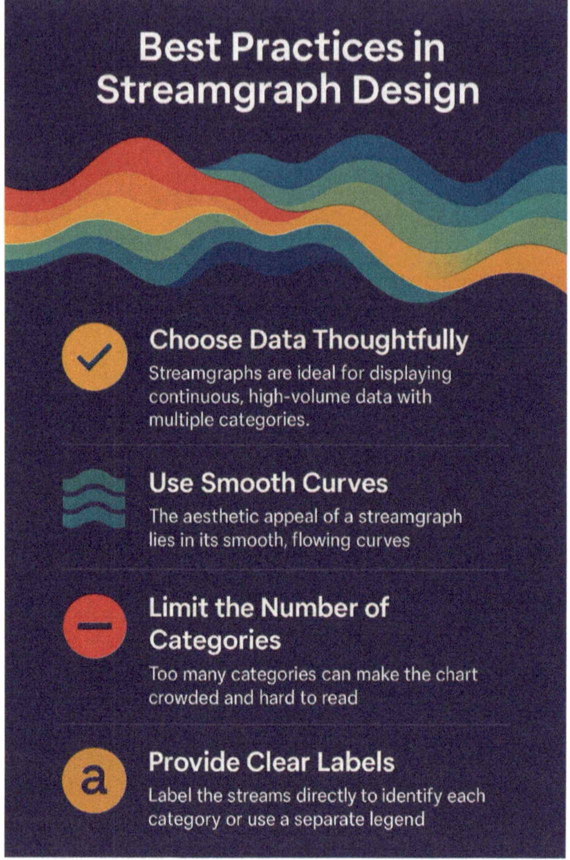

Fig. 10.11: Best Practices in Streamgraph Design

By following these design principles, Hexbin Plots can effectively convey data density and distribution, providing users with a clear and intuitive means of exploring large and complex datasets.

10.7.3 Facilitating Comparative Analysis

Hexbin Plots facilitate comparative analysis by allowing users to assess data density across different regions of the plot simultaneously. The aggregated representation of data points within hexagonal bins makes it easier to compare the concentration of points across various sections of the dataset. For example, in geographical data analysis, Hexbin Plots can compare population densities across different cities or regions, highlighting areas of high and low concentration. Similarly, in performance metrics analysis, Hexbin Plots can compare the distribution of results across different parameters, identifying trends and anomalies that inform optimization efforts. By providing a clear and concise summary of data density, Hexbin Plots enable users to conduct efficient comparative analyses, uncovering patterns and insights that drive data-driven decision-making.

10.7.4   Implementation Techniques

Implementing Hexbin Plots effectively involves utilizing specialized tools and libraries that support the creation and customization of hexagonal binning. Below is a Python code snippet using `Matplotlib` to create a basic Hexbin Plot.

```python
# Simulated performance metrics
np.random.seed(42)
x = np.random.normal(loc=0.0, scale=1.0, size=10000)  # Simulated metric A
y = np.random.normal(loc=0.0, scale=1.0, size=10000)  # Simulated metric B

# Set up figure and styling
fig, ax = plt.subplots(figsize=(8, 6), dpi=150)

# Create Hexbin plot with clear grid and perceptually uniform color map
hb = ax.hexbin(x, y, gridsize=40, cmap='viridis', mincnt=4, linewidths=0.6,
     edgecolors='white')

# Add colorbar with a meaningful label
cb = fig.colorbar(hb, ax=ax)
cb.set_label('Data Point Density (per bin)', fontsize=12)

# Axis labels with improved readability and font styling
ax.set_xlabel('Normalized Task Score', fontsize=13)
ax.set_ylabel('Normalized Response Time', fontsize=13)

# Set axes\index{Axes} ranges
ax.set_xlim([-3, 3])
ax.set_ylim([-3, 3])

# Grid and layout
ax.grid(visible=True, linestyle='--', linewidth=0.5, alpha=0.6)
ax.set_facecolor('whitesmoke')

# Add contextual annotation\index{Annotation}
ax.annotate('Most \ndense \nregion', xy=(0, 0), xytext=(2, 2),
            arrowprops=dict(facecolor='black', shrink=0.05),
            fontsize=11, fontweight='semibold', color='black')

plt.tight_layout()
plt.show()
```

In Figure 10.12, the Hexbin Plot visualizes the density of randomly distributed data points. Each hexagon's color intensity represents the number of points within that bin, providing a clear overview of data concentration and distribution.

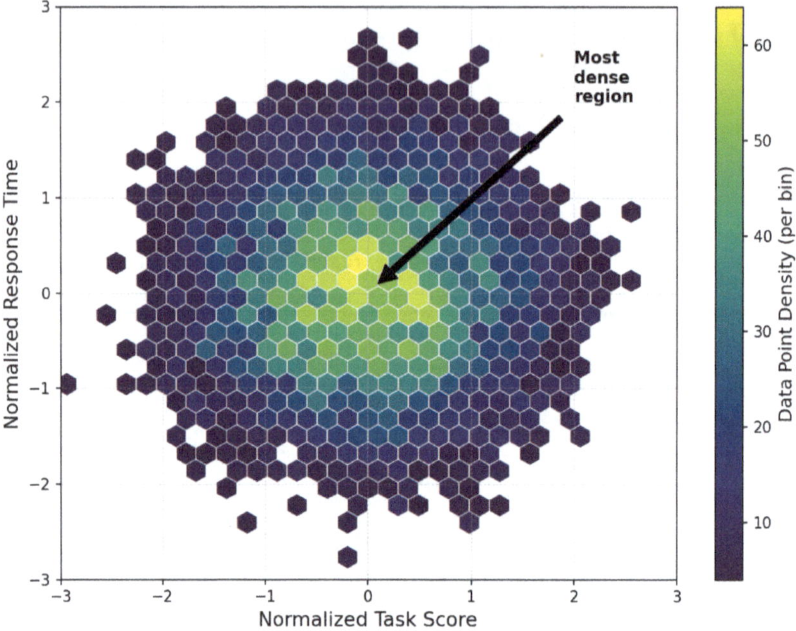

Fig. 10.12: Hexbin Plot Representing Data Density

10.7.5   Best Practices for Hexbin Plots

To ensure that Hexbin Plots are effective and informative, adhere to the following best practices:

Choose Appropriate Grid Size: Select a hexagon size that balances detail and readability. Smaller grids capture finer density variations, while larger grids provide a broader overview of data distribution.

Use Clear Color Schemes: Apply color schemes that clearly differentiate between varying levels of data density, ensuring that color gradients accurately reflect the concentration of data points.

Provide Legends and Labels: Include clear legends and axis labels to help users interpret the plot accurately. Legends should explain the color encoding used for data density.

Incorporate Interactive Elements: Where possible, add interactive features such as tooltips, zooming, and panning to allow users to explore specific areas of the plot in greater detail without losing the overall context.

Avoid Overlapping Bins: Ensure that hexagons are spaced appropriately to prevent excessive overlap or blending, which can obscure data density patterns.

Normalize Data: Apply data normalization techniques if necessary to account for varying bin sizes or data distributions, ensuring that color encoding accurately represents data density.

By following these best practices (Table 10.10), Hexbin Plots can effectively convey data density and distribution, providing users with a clear and intuitive means of exploring large and complex datasets.

Table 10.10: UX Design Principles and Best Practices in Hexbin Visualization

Design Feature	UX Purpose and Justification	Guideline / Reference
`hexbin` aggregation	Represents data density visually, minimizing overplotting in dense distributions	Wilkinson (2005); Cleveland (1985) [22, 2]
`cmap='viridis'` color scale	Perceptually uniform, colorblind-friendly colormap improving accessibility and interpretability	Moreland (2016); Ware (2012) [11, 20]
`linewidths=0.6,` `edgecolors=` `'white'`	Enhances bin separability and clarity, especially in high-density regions	Tufte (2001); Few (2009) [18, 4]
Contextual annotation using `annotate()`	Directs user attention to important regions; reduces cognitive load by guiding focus	Shneiderman (1996); Gestalt Theory [16, 20]
Axis labels with semantic context	Labels ("Normalized Task Score", "Normalized Response Time") support clear and domain-specific interpretability	WCAG 2.1; Norman (2013) [19, 14]
`ax.grid()` with light styling	Reinforces orientation and readability without visual clutter	Few (2009); Tufte's minimal ink principle [4, 18]
Colorbar with precise label	Supports quantitative density interpretation; improves clarity of encoded variable	Cleveland & McGill (1984); Munzner (2014) [3, 12]
`tight_layout()` and DPI=150	Ensures rendering clarity across screen resolutions and print outputs; avoids element clipping	UX Responsiveness Guidelines [21]
`ax.set_facecolor` `('whitesmoke')`	Soft background improves figure contrast and visual aesthetics	WCAG Contrast Guidelines; Ware (2012) [19, 20]

Figure 10.13 illustrates best practices in Hexbin Plot design, showcasing an optimized layout with appropriate grid size, clear color encoding, and informative legends that enhance data interpretation and user engagement.

10.8 Radial Bar Charts

10.8.1 Concept of Radial Bar Charts

Radial Bar Charts are a variation of traditional bar charts where the bars are arranged in a circular layout around a central axis. This radial arrangement allows for the representation of multiple data categories in a compact and visually appealing

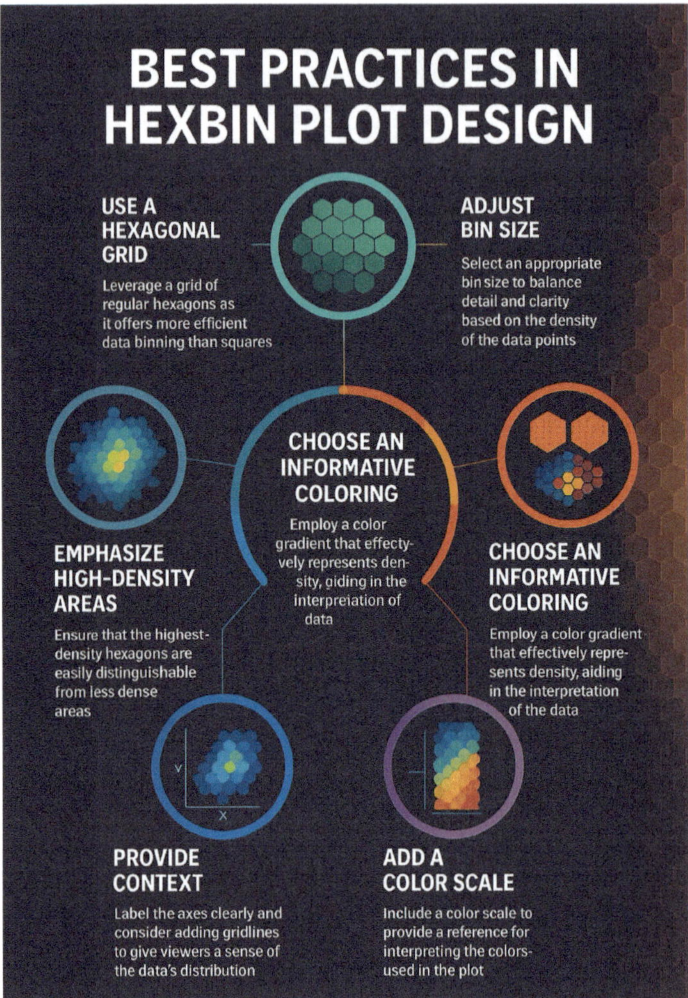

Fig. 10.13: Best Practices in Hexbin Plot Design

format. Each bar's length corresponds to the magnitude of the data attribute it represents, similar to standard bar charts, while its angular position facilitates the comparison of different categories. Radial Bar Charts are particularly useful for displaying cyclical data, such as monthly sales figures or seasonal trends, as the circular layout inherently conveys the cyclical nature of the data. Their aesthetic appeal makes them suitable for presentations and dashboards where visual engagement is essential.

10.8.2 Design Principles for Radial Bar Charts

Creating effective Radial Bar Charts involves adhering to several key design principles to ensure clarity, accuracy, and meaningful data representation:

Consistent Scaling: Ensure that all bars are scaled consistently relative to a common reference point, such as the central axis or a uniform angular spacing, to facilitate accurate comparisons between categories.

Clear Labeling: Label each bar clearly to identify the corresponding data category. Labels should be positioned to avoid overlap and maintain readability, potentially using interactive tooltips for additional information.

Appropriate Color Coding: Use distinct and consistent color schemes to differentiate between categories or to represent different data attributes. Consistent color usage enhances visual coherence and facilitates easier interpretation of the chart.

Minimize Clutter: Avoid overcrowding the radial layout with too many bars or excessive detail. Focus on the most significant categories to maintain clarity and prevent visual confusion.

Use of Highlights: Incorporate highlighting techniques, such as contrasting colors or increased bar thickness, to emphasize key data points or categories, drawing the viewer's attention to important insights.

Interactive Features: Where possible, add interactive elements such as hover effects, clickable bars, and dynamic filtering to allow users to explore specific data points in greater detail without overwhelming the overall visualization.

> ⚠ **Warning**
>
> **Radial bar charts may exaggerate differences due to angular distortion.**
> The circular layout can visually amplify longer bars and understate shorter ones. Use with caution and always clarify the scale to the reader.

By following these design principles, Radial Bar Charts can effectively represent multiple data categories in a visually engaging and interpretable manner, facilitating insightful data analysis and presentation.

10.8.3 Facilitating Comparative Analysis

Radial Bar Charts facilitate comparative analysis by enabling the simultaneous comparison of multiple data categories within a single, cohesive visualization. The circular layout allows for an organized and structured representation of data, making it easier to compare the lengths of bars across different angles. This arrangement helps in identifying patterns, trends, and outliers among the categories, enhancing the ability to discern relative magnitudes and relationships within the dataset. The radial symmetry of the chart supports the comparison of cyclical data, such as seasonal variations or periodic trends, by inherently conveying the cyclical nature through the circular layout. By providing a clear and concise means of comparing multiple data categories, Radial Bar Charts enhance the depth and efficiency of comparative data analysis, leading to more informed and data-driven insights.

10.8.4 Best Practices for Radial Bar Charts

To ensure that Radial Bar Charts are effective and informative, adhere to the following best practices:

Limit the Number of Categories: Avoid including too many categories in a single chart, as this can lead to clutter and make it difficult to interpret individual bars. Focus on the most significant categories to maintain clarity.

Use Distinct Colors: Apply distinct and consistent color schemes to differentiate between categories or to represent different data attributes. Consistent color usage enhances visual coherence and facilitates easier interpretation of the chart.

Ensure Proportional Scaling: Maintain proportional scaling of the bars to accurately represent the underlying data values. This proportionality is crucial for conveying the relative significance of different categories effectively.

Provide Clear Labels: Ensure that all categories are clearly labeled to identify the corresponding data points. Labels should be legible and positioned to avoid overlap with other elements in the chart.

Incorporate Interactivity: Add interactive features such as tooltips, highlighting, and dynamic filtering to allow users to explore specific bars in greater detail without overwhelming the overall visualization.

Maintain Aesthetic Balance: Design the chart to maintain a balanced and visually appealing layout, ensuring that all bars are evenly distributed around the circle and that the overall structure is harmonious.

By following these best practices, Radial Bar Charts can effectively represent multiple data categories in a visually engaging and interpretable manner, facilitating insightful data analysis and presentation.

Figure 10.14 illustrates best practices in Radial Bar Chart design, showcasing a clear layout with distinct colors, proportional scaling, and well-placed labels that enhance data interpretation and user engagement.

10.8.5 Example

Implementing Radial Bar Charts effectively involves utilizing specialized tools and libraries that support circular plotting and customization of bar properties. Below is a Python code snippet using the `Matplotlib` library to create a basic Radial Bar Chart.

```python
# Sample data
months = ['Jan', 'Feb', 'Mar', 'Apr', 'May', 'Jun',
          'Jul', 'Aug', 'Sep', 'Oct', 'Nov', 'Dec']
sales = [150, 200, 180, 220, 240, 210, 230, 250, 190, 220, 260, 300]

# Parameters
```

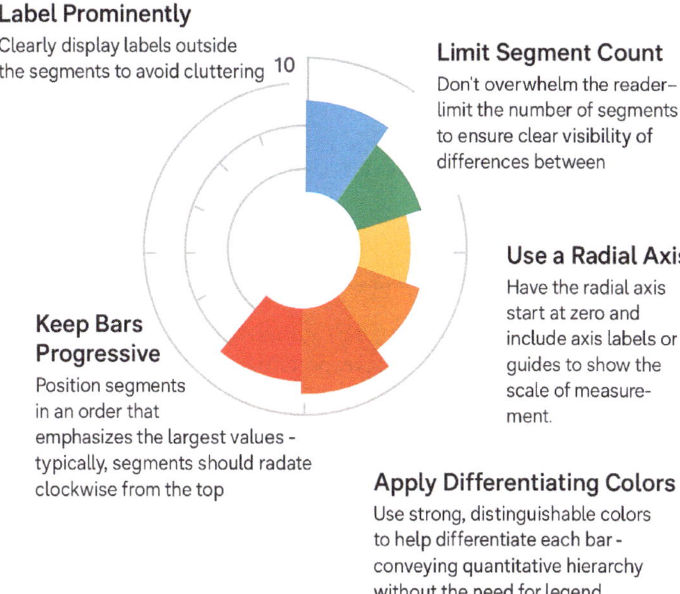

Fig. 10.14: Best Practices in Radial Bar Chart Design

```python
N = len(months)
angles = np.linspace(0, 2 * np.pi, N, endpoint=False).tolist()
angles += angles[:1]
sales += sales[:1]

# Plot setup
fig, ax = plt.subplots(figsize=(5, 5), subplot_kw=dict(polar=True),
↪   dpi=200)
fig.patch.set_facecolor('lightgray')

# Styling the plot line and area
ax.plot(angles, sales, color='#007acc', linewidth=4)
ax.fill(angles, sales, color='#a6dcef', alpha=0.6)

# Set month labels around the circle
ax.set_xticks(angles[:-1])
ax.set_xticklabels(months, fontsize=12, fontweight='bold')

# Radial ticks (y-axis) styling
ax.set_yticklabels([])
ax.set_ylim(0, max(sales) + 50)
ax.grid(color='gray', linestyle='--', linewidth=0.6, alpha=0.4)

# Add title with padding
ax.set_title('Monthly Sales', fontsize=18, fontweight='bold', y=1.08)
```

```
31
32  # Add custom radial labels manually for better readability
33  for i in range(N):
34      angle = angles[i]
35      value = sales[i]
36      ax.text(angle, value + 30, str(sales[i]), ha='center', va='center',
         ↪   fontsize=8, fontweight='semibold', color='black')
37
38  plt.tight_layout()
39  plt.show()
```

In Figure 10.15, the Radial Bar Chart visualizes monthly sales data. Each bar represents a month, with its length corresponding to the sales figures. The circular layout provides an organized and aesthetically pleasing representation, making it easy to compare sales across different months and identify seasonal trends.

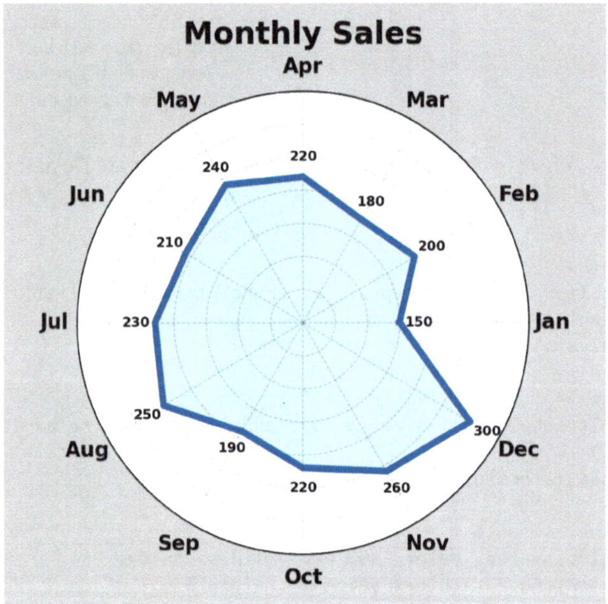

Fig. 10.15: Radial Bar Chart Representing Monthly Sales

To ensure that the radial bar chart visualization effectively communicates cyclical data while maintaining high usability and aesthetic quality, a series of user-centered design principles and visualization best practices were applied. These practices focus on enhancing readability, interpretability, and accessibility for diverse audiences, including users with visual impairments. Table 10.11 summarizes and explains each principle employed in the visualization, its purpose from a UX perspective, and relevant literature or guideline references supporting its implementation.

Table 10.11: UX Design Principles and Best Practices in Radial Bar Chart

Design Feature	UX Purpose and Justification	Reference / Source
`fig.patch.set_faceco`	Soft background improves contrast and enhances visual hierarchy	Ware (2012); WCAG Guidelines [20, 19]
Radial layout with consistent angles	Ensures cyclical data (monthly sales) is represented clearly in a circular format	Cleveland (1985); Tufte (2001) [2, 18]
`ax.plot()` + `ax.fill()`	Uses line+area encoding for dual emphasis on trends and magnitude	Munzner (2014); Few (2009) [12, 4]
Color choice: #007acc, #a6dcef	Visually appealing, high-contrast and colorblind-friendly hues	Color Universal Design (CUD); Moreland (2016) [11]
Bold, evenly spaced `set_xticklabels()`	Enhances legibility of categorical axis (months) in radial layout	Norman (2013); WCAG 2.1 [14, 19]
Removed radial tick labels with custom annotations	Reduces clutter and provides clear data context directly on chart	Tufte's Data-Ink Principle [18]
`ax.text()` for data labels	Supports numeric literacy with direct value annotations, improving precision and interpretability	Shneiderman (1996); Cleveland & McGill (1984) [16, 3]
`tight_layout()` + DPI=200	Guarantees sharp rendering and avoids label clipping for screen and print	Wilke (2019); ISO 9241-210 [21, 8]

10.9 Insights and Takeaways

The exploration of advanced visualization techniques in this chapter–ranging from radial bar charts and hexbin plots to streamgraphs, chord diagrams, 3D scatter plots, treemaps, Sankey diagrams, and network graphs –demonstrates the growing demand for expressive and intuitive data communication. Each chart type serves unique analytical purposes and user interaction needs, particularly when visualizing multidimensional or relational data.

➢ **Design Clarity and Purpose:** Every visualization must serve a clear analytical purpose. Whether representing hierarchical structures in a treemap, flow dynamics in a Sankey diagram, or dense distributions in a hexbin plot, the visual encoding should match the nature of the data and the questions being asked.

➢ **Color and Accessibility:** Color should be used meaningfully and accessibly. Colorblind-safe palettes, adequate contrast, and intuitive gradients (e.g., diverging or sequential) enhance both usability and equity in data interpretation.

➢ **Interactivity Enhances Understanding:** Interactive visualizations (using Plotly, HoloViews, or Bokeh) empower users to explore data dynamically. Hover tooltips, zooming, and drill-down capabilities improve cognitive engagement and user experience.

➢ **Labeling and Annotation:** Effective annotations–including node labels in network graphs , bar values in radial charts, or contextual tooltips–support storytelling and reduce cognitive effort. Always consider which information should be visible immediately and what can be revealed on demand.

➢ **Minimizing Clutter:** Avoid overplotting by aggregating data (as in hexbin plots), stacking layers smartly (as in streamgraphs), or selectively displaying key nodes and edges (as in network graphs). Whitespace and layout spacing are not empty–they serve as navigational aids.

➢ **Responsiveness and Readability:** Visualizations must adapt to different screen sizes and resolutions. Use scalable vector graphics when publishing in digital or print formats, and ensure legibility of all text elements across display contexts.

➢ **Data Context is Crucial:** Charts without context can mislead. All visualizations should be accompanied by descriptive titles, legends, units, and source acknowledgments. Streamgraphs, for instance, must explain what is flowing over time–whether it be sales, population, or attention.

➢ **Tool Selection Matters:** Python libraries such as Plotly, Matplotlib, Holoviews, and NetworkX offer different strengths. The choice of tool should reflect your priorities–whether they are interactivity, customization, computational efficiency, or publication quality.

Suggested Readings

1. Byron, L., Wattenberg, M.: Stacked graphs—geometry & aesthetics. IEEE Transactions on Visualization and Computer Graphics **14**(6), 1245–1252 (2008)
2. Cleveland, W.S.: The Elements of Graphing Data. Wadsworth (1985)
3. Cleveland, W.S., McGill, R.: Graphical perception: Theory, experimentation, and application to the development of graphical methods. Journal of the American Statistical Association **79**(387), 531–554 (1984)
4. Few, S.: Now You See It: Simple Visualization Techniques for Quantitative Analysis. Analytics Press (2009)
5. Fitts, P.M.: The information capacity of the human motor system in controlling the amplitude of movement. Journal of Experimental Psychology **47**(6), 381 (1954)
6. Healey, C.G., Enns, J.T.: Visualizing data with motion. IEEE Computer Graphics and Applications **16**(4), 20–30 (1996)
7. Heer, J., Boyd, D.: A streamgraph: a form of stacked area graph for visualization of thematic changes over time. In: Proc. of InfoVis (2010)
8. International Organization for Standardization: Ergonomics of human-system interaction — part 210: Human-centred design for interactive systems (2010). ISO 9241-210
9. Mackinlay, J.D.: Automating the design of graphical presentations of relational information. In: ACM Transactions on Graphics (TOG), vol. 5, pp. 110–141 (1986)
10. Marcotte, E.: Responsive web design. A Book Apart (2011)
11. Moreland, K.: Diverging color maps for scientific visualization. Proceedings of the 5th International Symposium on Visual Computing (2016)

12. Munzner, T.: Visualization Analysis and Design. CRC Press (2014)
13. Nielsen, J.: Usability Engineering. Morgan Kaufmann (1994)
14. Norman, D.A.: The Design of Everyday Things. Basic Books (2013)
15. Okabe, M., Ito, K.: Color universal design (cud): How to make figures and presentations that are friendly to colorblind people. J. Sci. Commun 3(2) (2008)
16. Shneiderman, B.: The eyes have it: A task by data type taxonomy for information visualizations. In: Proceedings 1996 IEEE Symposium on Visual Languages, pp. 336–343. IEEE (1996)
17. Sweller, J.: Cognitive load during problem solving: Effects on learning. Cognitive science 12(2), 257–285 (1988)
18. Tufte, E.R.: The visual display of quantitative information. Graphics Press (2001)
19. W3C: Web content accessibility guidelines (wcag) 2.1. World Wide Web Consortium (W3C) (2018). https://www.w3.org/TR/WCAG21/
20. Ware, C.: Information Visualization: Perception for Design. Elsevier (2012)
21. Wilke, C.O.: Fundamentals of Data Visualization. O'Reilly Media, Inc. (2019)
22. Wilkinson, L.: The Grammar of Graphics. Springer (2005)
23. Williams, R.: The Non-Designer's Design Book. Peachpit Press (2014)
24. Wong, D.M.: The Wall Street Journal Guide to Information Graphics. W. W. Norton & Company (2011)

11 Advanced Visualization Techniques

Abstract

This chapter focuses on advanced visualization techniques aimed at enhancing charts, thereby making intricate data more accessible and interpretable. Techniques such as Zoom-Out, Panning, Brushing, Drill-Down, Filtering, Linking, Annotation, Animation, Small Multiples, and Focus + Context (Fisheye Views) are systematically explored, each accompanied by clear definitions, benefits, and practical implementation strategies. Through illustrative Python code snippets and visually appealing examples, readers are guided on how to integrate these methods into their data visualization workflows effectively. The chapter also outlines best practices to ensure the optimal use of each technique, emphasizing design principles that enhance clarity, interactivity, and user engagement.

Aims

After reading this chapter, you should be able to:

> ➤ Understand advanced visualization techniques used in data analysis.

> ➤ Implement Zoom-Out functionality to provide broader context and overall trends in data visualizations.

> ➤ Utilize Panning to navigate large datasets without altering the zoom level, enhancing data exploration.

> ➤ Apply Brushing techniques to select and highlight subsets of data points across linked visualizations for multivariate analysis.

> ➤ Employ Drill-Down methods to interactively explore data from summary levels to more detailed layers, facilitating deeper investigation of trends and outliers .

> ➤ Implement effective Filtering strategies to focus on specific data points.

> ➤ Integrate Linking across multiple visualizations to synchronize interactions and enhance comprehensive data analysis.

R. Damaševičius, *Human-Centred Scientific Data Visualisation*,
Undergraduate Topics in Computer Science,
https://doi.org/10.1007/978-3-032-01606-5_11

> ➤ Add Annotations to visualizations to provide context, highlight key insights, and improve the interpretability of data charts.

> ➤ Create and utilize Animations to represent changes in data over time or transitions between different data states, aiding in the understanding of temporal trends.

> ➤ Design and apply Small Multiples to display multiple versions of the same chart for different data subsets, facilitating easy comparison across categories or time periods.

> ➤ Implement Focus + Context (Fisheye Views) to magnify areas of interest within a visualization while maintaining surrounding data context.

> ➤ Integrate multiple advanced visualization techniques to create comprehensive and insightful charts.

11.1 Introduction

11.1.1 Overview of Advanced Visualization Techniques

In the realm of data analysis, the ability to effectively visualize complex datasets is paramount. Traditional static charts and graphs often fall short in conveying the intricate relationships and multifaceted patterns inherent in large-scale data. This chapter focuses on a suite of advanced visualization techniques designed to enhance the accessibility and interpretability of charts. Techniques such as Zoom-Out, Panning, Brushing, Drill-Down, Filtering, Linking, Annotation, Animation, Small Multiples, and Focus + Context (Fisheye Views) are explored in depth. Each of these methods offers unique capabilities for interacting with data, allowing users to navigate, manipulate, and extract meaningful insights from their visualizations. By integrating these advanced techniques, scientists and researchers can transform raw data into intuitive and informative visual representations, facilitating a deeper understanding of underlying trends and patterns.

The chapter begins with an exploration of **Zoom-Out**, a technique that enables users to step back and view the entire dataset, providing a comprehensive overview after detailed examination. **Panning** follows, allowing seamless horizontal or vertical movement across the visualization without altering the zoom level, which is particularly useful for navigating expansive datasets. **Brushing** introduces interactive selection of data subsets, highlighting specific points and often linking them to other visualizations for enhanced multivariate analysis. The **Drill-Down** method offers a hierarchical approach to data exploration, enabling users to shift focus from summary views into more granular details. **Filtering** focuses on isolating specific variables or data points, thereby improving clarity and enabling targeted analysis. **Linking** connects multiple visualizations, ensuring that interactions in one chart dynamically update corresponding elements in others, fostering a cohesive analytical environment. **Annotation** adds explanatory text and markers to highlight key insights directly within the visualization. **Animation** brings data to life by depicting changes over time or transitions between different states, aiding in the comprehension of temporal dynamics. **Small Multiples** facilitate the comparison of different data subsets by displaying multiple versions of the same chart side by side. Finally, **Focus + Context (Fisheye Views)** balances detailed inspection of specific areas with the preservation of surrounding data context, allowing for a nuanced exploration of complex datasets. Together, these advanced visualization techniques empower users to create more dynamic, interactive, and insightful charts.

A broad summary of the advanced visualization techniques covered in this chapter, including their primary purposes, is presented in Table 11.1.

Table 11.1: Overview of Advanced Visualization Techniques

Technique	Primary Purpose
Zoom-Out	Provides a broader view of the dataset to reveal overall patterns and trends.
Panning	Enables horizontal/vertical navigation within a visualization without zooming.
Brushing	Allows selection and highlighting of data subsets within and across charts.
Drill-Down	Enables hierarchical exploration from summary views to detailed layers.
Filtering	Focuses the visualization on specific variables or categories by excluding irrelevant data.
Linking	Synchronizes interactions across multiple visualizations.
Annotation	Adds contextual labels, notes, or highlights to improve interpretability.
Animation	Depicts temporal or state transitions to highlight dynamic patterns.
Small Multiples	Displays multiple charts side by side to support comparative analysis.
Focus + Context	Emphasizes a data region while maintaining overall context (e.g., fisheye views).

The level of interaction complexity for each technique is outlined in Table 11.2, which helps in choosing the appropriate visualization method based on user experience and computational resources.

11.1.2 Importance in Data Analysis

The significance of advanced visualization techniques in data analysis cannot be overstated. As the volume and complexity of data continue to grow, researchers are increasingly challenged to extract meaningful insights efficiently and accurately. Advanced visualization techniques serve as essential tools in this endeavor, bridging the gap between raw data and actionable knowledge. By transforming complex datasets into intuitive visual formats, these techniques enhance the ability to identify

Table 11.2: Interaction Complexity of Visualization Techniques

Technique	Interaction Type	Complexity (Low/Med/High)
Zoom-Out	Scroll, zoom controls	Medium
Panning	Mouse drag, arrow keys	Low
Brushing	Selection with mouse or touch	Medium
Drill-Down	Click to navigate deeper levels	Medium
Filtering	Sliders, dropdowns	Medium
Linking	Dynamic data binding across views	High
Annotation	Static or dynamic labels	Low
Animation	Frame transitions, playback controls	High
Small Multiples	Static layout with shared settings	Low
Focus + Context	Distortion, fisheye interaction	High

patterns, detect anomalies, and uncover relationships that might remain obscured in traditional data representations.

Advanced visualizations facilitate more effective communication of scientific findings. Clear and compelling visual representations make it easier for researchers to convey their discoveries to diverse audiences, including peers, stakeholders, and the general public. This is particularly important in interdisciplinary research, where visualizations can serve as a common language that transcends specialized jargon and methodological differences. Interactive visualization techniques empower users to engage with data in a more hands-on manner, fostering exploratory analysis and encouraging a deeper cognitive engagement with the information presented.

In practical terms, the application of advanced visualization techniques can lead to significant improvements in research outcomes. For instance, in fields such as genomics, climate science, and astrophysics, where datasets are inherently large and multidimensional, advanced visualizations enable scientists to navigate vast information landscapes with ease. Techniques like Drill-Down and Focus + Context allow for the exploration of data at multiple levels of granularity, facilitating the discovery of both broad trends and fine-scale details. Filtering and Linking enhance the ability to isolate and examine specific variables, while Animation and Small Multiples support the analysis of temporal and comparative data. Ultimately, the integration of these advanced visualization methods into scientific workflows not only accelerates the data analysis process but also elevates the quality and depth of scientific insights, driving innovation and advancing our understanding of complex phenomena.

11.2   Zoom-Out

11.2.1   Definition and Purpose

Zoom-Out is a fundamental interaction technique in data visualization that allows users to step back from a detailed view to a broader perspective of the entire dataset. This functionality is crucial for providing context and understanding the overarching trends and patterns that may not be immediately apparent when focusing solely on specific data points or regions. By enabling users to seamlessly transition between different levels of detail, Zoom-Out facilitates a more comprehensive analysis, ensuring that insights derived from detailed examinations are grounded within the larger dataset. This technique is particularly valuable in scenarios where datasets are extensive or multidimensional, as it helps prevent tunnel vision and promotes a holistic understanding of the data landscape.

11.2.2   Benefits of Zooming Out

Implementing Zoom-Out techniques in data visualizations offers several advantages. Firstly, it enhances the user's ability to perceive overall trends and patterns, which might be obscured when viewing only a subset of the data. This macro-level perspective is essential for identifying general behaviors, anomalies, and correlations that inform strategic decisions. Secondly, Zoom-Out improves navigational efficiency by allowing users to quickly shift their focus from specific details to the entire dataset without losing their place or context. This flexibility supports exploratory data analysis, where users iteratively refine their inquiries based on insights gained at different levels of granularity. Zoom-Out contributes to better data comprehension and reduces cognitive load by preventing information overload, as users can control the degree of detail with which they wish to engage at any given moment.

11.2.3   Implementation Strategies

Effectively implementing Zoom-Out in data visualizations involves several strategies to ensure a smooth and intuitive user experience. One common approach is to incorporate interactive controls such as sliders, buttons, or scroll gestures that allow users to adjust the zoom level dynamically. For example, a slider can provide granular control over the zoom scale, enabling precise adjustments. Another strategy is to implement zooming functionality through mouse interactions, such as double-clicking to reset the view or using scroll wheel gestures to zoom in and out. Maintaining the aspect ratio and ensuring that key data points remain visible during zoom transitions are critical for preserving data integrity and preventing distortion. It is also beneficial to provide visual indicators or breadcrumbs that inform users of their current zoom level and position within the dataset, enhancing navigational awareness.

> **⚠ Warning**
>
> Excessive use of the Zoom-Out or Zoom-In functions without clear contextual indicators can disorient users and obscure the broader data landscape. Always provide visual cues, such as overview panels or reset buttons, to maintain spatial awareness.

11.2.4 Best Practices

To maximize the effectiveness of Zoom-In techniques, it is essential to adhere to several best practices. Firstly, ensure that the transition between zoom levels is smooth and visually coherent to prevent disorientation. Animated transitions can enhance the user experience by providing visual continuity. Secondly, maintain the integrity of the data by preserving relative positions and scales during zoom operations, avoiding misleading distortions. Thirdly, provide clear and accessible controls for zooming out, such as prominent buttons or intuitive gestures, to facilitate ease of use. Offer contextual cues or annotations that guide users in understanding the current view and available zoom levels. It is also advisable to implement constraints that prevent users from zooming out excessively, which could lead to overly abstract representations that obscure meaningful insights. Lastly, consider the responsiveness of the visualization, ensuring that performance remains optimal even when handling large datasets during zoom transitions.

Figure 11.1 exemplifies best practices by showcasing a user interface with clearly labeled zoom controls, smooth transition animations, and contextual indicators that enhance navigational clarity and user engagement.

11.2.5 Example

Zoom-In techniques are widely applicable across various domains of data analysis. In genomics, for instance, researchers often deal with vast genomic sequences where Zoom-In allows them to examine the overall genetic landscape before focusing on specific genes or mutations. Similarly, in climate science, meteorologists can use Zoom-In to view global weather patterns before drilling down into localized weather events. An illustrative example is visualizing stock market data, where analysts can observe long-term trends across multiple years and then zoom in to investigate daily or hourly fluctuations. Below is a Python code snippet using Matplotlib that demonstrates a simple Zoom-In functionality by displaying a global view of data points and allowing the user to zoom out to see the entire dataset.

```python
#Implementing Zoom-In in Matplotlib
import matplotlib.pyplot as plt
import numpy as np
from mpl_toolkits.axes_grid1.inset_locator import inset_axes, mark_inset

```

Best Practices for Implementing Zoom-In

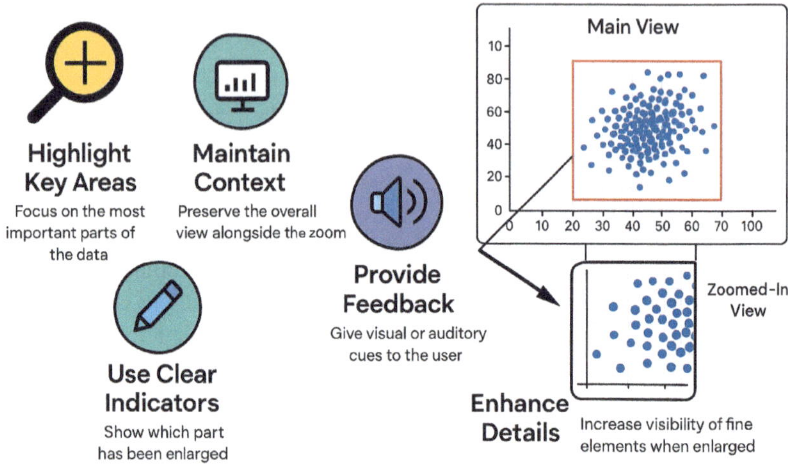

Fig. 11.1: Best Practices for Implementing Zoom-In

```python
# Generate sample data
np.random.seed(0)
x = np.random.normal(50, 15, 200)
y = np.random.normal(50, 15, 200)

# Main figure and axis
fig, ax = plt.subplots(figsize=(6, 6))

scatter_main = ax.scatter(x, y, alpha=0.7, edgecolors='k', linewidth=0.5,
    ↪  s=30, cmap='viridis')
ax.set_title('Main View with Zoomed-In Overlay', fontsize=12,
    ↪  fontweight='bold')
ax.tick_params(labelsize=10)
ax.set_xlim(20, 90)
ax.set_ylim(20, 90)
ax.grid(True, linestyle='--', linewidth=0.5, alpha=0.5)

# Define zoom-in region
x1, x2, y1, y2 = 40, 60, 40, 60

# Create inset axes\index{Axes} for zoomed-in view
axins = inset_axes(ax, width="35%", height="35%", loc='upper right',
                   borderpad=2)

# Plot data in inset
scatter_zoom = axins.scatter(x, y, alpha=0.8, edgecolors='k',
    ↪  linewidth=0.75, s=60)
```

```python
30
31  # change all spines
32  for axis in ['top','bottom','left','right']:
33      axins.spines[axis].set_linewidth(2)
34
35  # set axis limits and ticks
36  axins.set_xlim(x1, x2)
37  axins.set_ylim(y1, y2)
38  axins.set_title('Zoomed-In View', fontsize=10, pad=5,  fontweight='bold')
39  axins.tick_params(axis='both', which='minor', labelsize=8)
40  axins.set_xticks(np.linspace(x1, x2, 5))
41  axins.set_yticks(np.linspace(y1, y2, 5))
42  axins.set_xticks(np.linspace(x1, x2, 25), minor=True)
43  axins.set_yticks(np.linspace(y1, y2, 25), minor=True)
44
45  # And a corresponding grid
46  axins.grid(which='both')
47  axins.grid(which='major', linestyle='--', linewidth=1, alpha=1)
48  axins.grid(which='minor', linestyle='--', linewidth=0.5, alpha=0.5)
49
50  # Highlight zoomed area on main plot
51  mark_inset(ax, axins, loc1=2, loc2=4, fc="none", ec="red", lw=2)
52
53  # Layout adjustment
54  plt.tight_layout()
55  plt.show()
```

In Figure 11.2, the initial scatter plot presents a detailed view of the data points within a specific range. By pressing the 'o' key, the view zooms out to encompass the entire dataset, providing a comprehensive overview of the distribution and clustering of the data.

11.3   Panning

11.3.1   Understanding Panning in Visualization

Panning is an interactive technique in data visualization that allows users to navigate horizontally or vertically across a dataset without altering the current zoom level. This functionality is essential for exploring large or expansive datasets where viewing the entire data range simultaneously is impractical. By enabling seamless movement across different sections of the visualization, panning helps users maintain their context while examining specific areas of interest. For instance, in a geographical map, panning allows users to move from one region to another effortlessly, ensuring that they can explore various locations without losing sight of the overall spatial

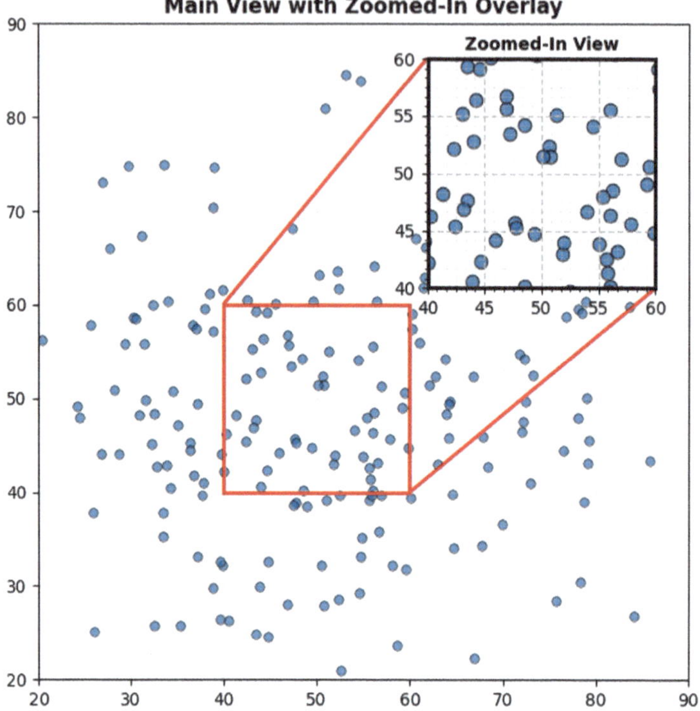

Fig. 11.2: Zoom-In Example in a Scatter Plot

distribution. Panning enhances the exploratory capabilities of visualizations, making it easier to uncover patterns, trends, and anomalies spread across extensive data fields.

11.3.2   Techniques for Effective Panning

Implementing effective panning in visualizations involves several techniques that ensure smooth and intuitive navigation. One common method is to incorporate click-and-drag functionality, where users can click on the visualization and drag it in the desired direction to pan across the data. Another technique involves using scroll bars or arrow keys to facilitate controlled movement within the visualization space. Touch-based interactions, such as swipe gestures on touchscreens, provide a natural and responsive way to pan through data. To enhance user experience, it is crucial to implement inertia or momentum scrolling, which allows the visualization to continue moving briefly after the user has stopped dragging, mimicking real-world motion and making the panning feel more fluid. Providing visual cues, such as a mini-map or overview panel, can help users understand their current position within the larger dataset and navigate more effectively.

11.3.3   Interactive Panning Mechanisms

Interactive panning mechanisms elevate the user experience by making navigation intuitive and responsive. One effective mechanism is the integration of mouse-based panning , where users can click and drag the visualization to move in any direction. This method is particularly useful for desktop environments, providing precise control over the panning process. Another mechanism involves keyboard shortcuts, allowing users to use arrow keys or other designated keys to pan the visualization incrementally. For touch-enabled devices, multi-touch gestures, such as swiping with one or more fingers, offer a natural way to navigate through the data. Incorporating panning buttons on the user interface can provide an alternative means of navigation for users who prefer clicking over dragging. Enhancing these interactive mechanisms with real-time feedback, such as smooth transitions and dynamic loading of data, ensures that panning remains responsive even with large and complex datasets.

> **⚠ Warning**
>
> Unlimited panning can cause users to get lost in the visualization space or interpret out-of-range data as valid. Implement logical boundaries and consider adding a minimap or position indicators to assist orientation.

11.3.4   Design Considerations

When designing panning features for data visualizations, several key factors must be considered to ensure an effective and user-friendly experience. Firstly, the responsiveness of the panning mechanism is paramount; any lag or delay can hinder user interaction and reduce the overall usability of the visualization. Ensuring smooth and instantaneous panning enhances the fluidity of data exploration. Secondly, maintaining visual continuity is essential. As users pan through the data, the visualization should preserve the integrity of the data representation, avoiding any distortions or misalignments that could mislead interpretation. Providing contextual indicators, such as a minimap or current viewport highlight, helps users maintain orientation within the dataset. Accessibility is another crucial consideration; panning controls should be easily discoverable and operable by users with varying levels of technical proficiency. Finally, optimizing performance for large datasets is vital to prevent the visualization from becoming sluggish or unresponsive, which can detract from the user experience. By addressing these design considerations, panning features can significantly enhance the navigational capabilities and overall effectiveness of data visualizations.

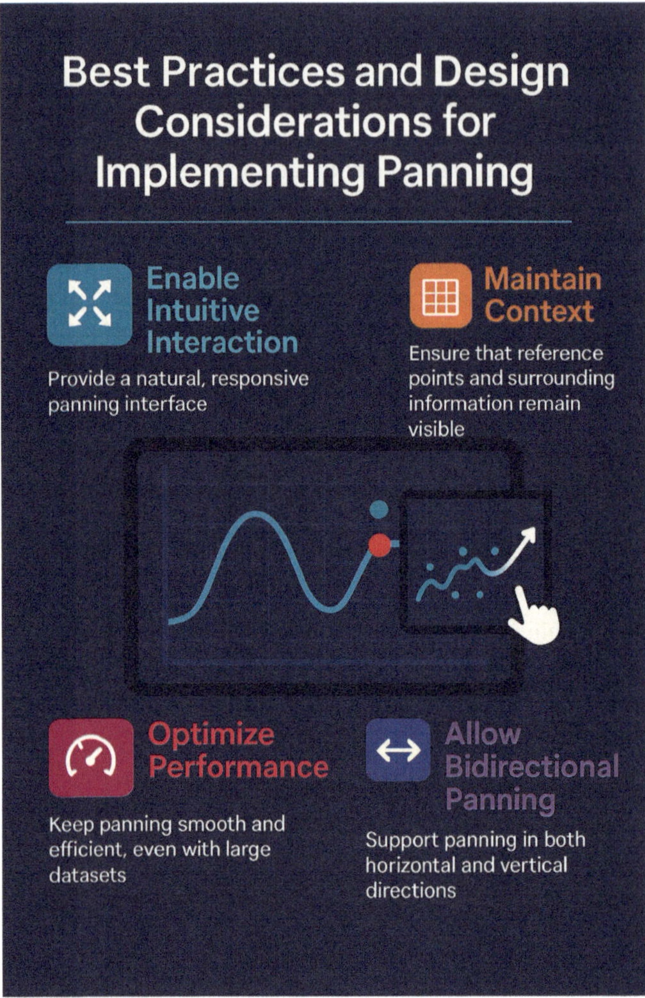

Fig. 11.3: Best Practices and Design Considerations for Implementing Panning

11.3.5 Example

Panning techniques have been effectively applied across various domains to enhance data exploration and analysis. In the field of astronomy, interactive star maps utilize panning to allow astronomers to navigate across different sections of the night sky, facilitating the study of celestial objects in diverse regions. Similarly, in financial analytics, panning enables analysts to explore extensive time-series data, moving through years of stock performance to identify long-term trends and short-term fluctuations. A notable example is the `Plotly` library in Python, which provides built-in panning capabilities for interactive charts. Below is a Python code snippet demonstrating how to implement panning in a scatter plot using Plotly.

```python
#Implementing Panning with Plotly
import plotly.express as px
import pandas as pd

# Sample data
df = pd.DataFrame({
    'X': range(1000),
    'Y': [i + (i**0.5)*10 for i in range(1000)]
})

# Create scatter plot
fig = px.scatter(df, x='X', y='Y',
    title='Interactive Scatter Plot with Panning')

# Update layout for better panning \index{Panning}  experience
fig.update_layout(
    dragmode='pan',
    xaxis=dict(rangeslider=dict(visible=True)),
    yaxis=dict(scaleanchor="x")
)

fig.show()
```

11.4   Brushing

11.4.1   Concept of Brushing

Brushing is an interactive technique in data visualization that allows users to select and highlight specific subsets of data points within a visualization. This selection can be performed using various input methods such as click-and-drag, lasso tools, or rectangular selectors. The primary purpose of brushing is to enable users to focus on particular data segments, making it easier to examine and analyze relevant information without being overwhelmed by the entire dataset. By visually distinguishing the brushed elements, users can quickly identify patterns, correlations, and outliers within the selected subset. Brushing serves as a foundational interaction method that enhances the exploratory data analysis process, allowing for dynamic and intuitive data manipulation directly within the visualization interface.

11.4.2   Linking Brushing with Multiple Views

One of the most powerful aspects of brushing is its ability to link multiple visualizations, creating a cohesive and synchronized analytical environment. When brushing is applied to one view, the corresponding data points in other linked views are automatically highlighted or filtered based on the selection. This interconnectedness facilitates multivariate analysis by allowing users to observe how selected data points behave across different dimensions and representations. For example, brushing a subset of points in a scatter plot can simultaneously highlight the same points in a parallel coordinates plot or a bar chart, revealing relationships and dependencies that might not be evident in a single view. Linking brushing across multiple views enhances the user's ability to draw comprehensive insights from complex and multidimensional datasets, promoting a more integrated and holistic understanding of the data.

> **⚠ Warning**
>
> If brushing is not properly synchronized across linked views, users may draw incorrect conclusions from inconsistent subsets. Ensure all related views are dynamically updated to reflect selection changes.

11.4.3   Enhancing Multivariate Analysis through Brushing

Brushing significantly enhances multivariate analysis by enabling the simultaneous examination of multiple variables within a dataset. By selecting a subset of data points in one visualization, users can observe how these points correlate with other variables in different visual representations. This capability is particularly useful in identifying complex relationships and interactions between variables that may not be apparent through univariate or bivariate analyses alone. For instance, in a dataset containing information on various species of plants, brushing can help identify how specific species cluster based on attributes like height, leaf size, and growth rate across multiple plots or charts. This multidimensional exploration fosters a deeper understanding of the underlying data structure, facilitating the discovery of nuanced patterns and insights that drive informed decision-making and scientific discovery.

> **Design Tip: Highlight Selected Data in Brushing Interactions**
>
> Use distinctive colors or shapes to visually emphasize brushed elements. This immediate feedback improves the clarity of user interactions and supports comparative exploration across linked views.

11.4.4   Implementation Tools and Libraries

Several tools and libraries facilitate the implementation of brushing in data visualizations, each offering unique features and capabilities to support interactive data exploration. `D3.js` is a powerful JavaScript library that provides extensive support for creating dynamic and interactive visualizations, including brushing and linking functionalities. Its flexibility allows developers to customize brushing interactions to suit specific data analysis needs. `Plotly` is another versatile library that offers built-in brushing capabilities, enabling users to create interactive charts with minimal code. For Python users, `Bokeh` stands out as a robust library that supports sophisticated brushing interactions, allowing seamless integration with web-based visualizations. `Matplotlib` in conjunction with `mpld3` can be used to add brushing functionality to static plots, bridging the gap between traditional plotting and interactive exploration. These tools empower developers and data scientists to incorporate brushing into their visualizations, enhancing the interactivity and analytical depth of their data representations. Table 11.3 highlights the Python libraries that support the implementation of various advanced visualization techniques, along with their capabilities for interactive data exploration.

Table 11.3: Python Library Support for Visualization Techniques

Technique	Common Libraries	Interactive Support
Zoom-Out	Matplotlib, Plotly	Key press/scroll events, zoom buttons
Panning	Plotly, Bokeh	Click-and-drag, pan tools
Brushing	Bokeh, Plotly, D3.js	Box select, lasso tool, link views
Drill-Down	Plotly, Dash, D3.js	On-click callbacks, hierarchical updates
Filtering	Matplotlib + widgets, Plotly Dash	Sliders, dropdowns, checkboxes
Linking	Bokeh, Plotly, Tableau	Shared data source, events and callbacks
Annotation	Matplotlib, Plotly, Seaborn	Arrows, labels, tooltips
Animation	Plotly, Matplotlib (FuncAnimation)	Time sliders, animated frames
Small Multiples	Seaborn (FacetGrid), Plotly subplots	Shared scales, consistent layout
Focus + Context	Matplotlib, Plotly, D3.js	Fisheye distortion, insets, draggable views

11.4.5 Best Practices and Design Considerations for Implementing Brushing

Brushing is a core interaction technique in data visualization that empowers users to select, highlight, and investigate specific data subsets. When implemented effectively, brushing facilitates exploratory data analysis, supports multivariate pattern recognition, and enhances the interpretability of linked visualizations. However, poor brushing design can confuse users or lead to misleading insights. Therefore, careful attention must be paid to interaction design, visual clarity, and user experience.

Figure 11.4 presents a visual summary of the most important best practices and design considerations for brushing in scientific visualizations. Key principles include the use of **responsive visual feedback**–where brushed elements are immediately and distinctly highlighted to confirm user interaction. Consistency in **linked views** is crucial; brushing in one visualization should dynamically update all related plots, maintaining analytical coherence.

Another best practice involves offering **multiple selection modes**, such as rectangular, free-form (lasso), or categorical selectors, to accommodate different analytical needs. It is also important to **avoid visual clutter** by choosing distinguishable colors or opacities and to provide **clear legends or tooltips** that explain what the brushing highlights represent. The ability to **reset or undo selections** improves usability and encourages experimentation.

From an accessibility standpoint, brushing controls and highlights should be perceivable by users with color vision deficiencies–this includes using patterns or bold outlines in addition to color changes. Finally, brushing should be performant even on large datasets, meaning that rendering optimizations or simplified visual feedback (e.g., skeleton highlights) may be needed for real-time responsiveness.

Overall, by following these best practices, developers and designers can create brushing interactions that are intuitive, responsive, and highly effective for scientific data exploration.

11.4.6 Example

Brushing has been effectively utilized in various scientific studies to facilitate detailed data exploration and analysis. In genomics, researchers employ brushing techniques to explore large-scale genetic data, enabling the identification of gene clusters and the examination of gene expression patterns across different conditions. For example, in a study analyzing gene expression levels in cancer cells, brushing allows scientists to highlight specific genes of interest and observe their behavior across multiple samples and conditions, uncovering potential biomarkers and therapeutic targets. In environmental science, brushing is used to analyze complex climate data, such as temperature and precipitation patterns, enabling the detection of regional trends and anomalies. An illustrative example is the use of brushing in interactive climate models, where researchers can select specific time periods or geographic regions to examine detailed climate variables and their interdependencies. These applications demonstrate how brushing enhances the ability to manage and interpret vast and intricate datasets, driving advancements in scientific research and discovery.

```python
# Implementing Brushing with Bokeh
from bokeh.io import show, output_file
from bokeh.plotting import figure
from bokeh.models import ColumnDataSource, CategoricalColorMapper,
    CustomJS
from bokeh.layouts import row
from bokeh.palettes import Category10
import pandas as pd
import numpy as np

# Data
data = pd.DataFrame({
    'x': range(100),
    'y': [i**0.5 * 10 for i in range(100)],
    'category': ['A' if i < 50 else 'B' for i in range(100)]
```

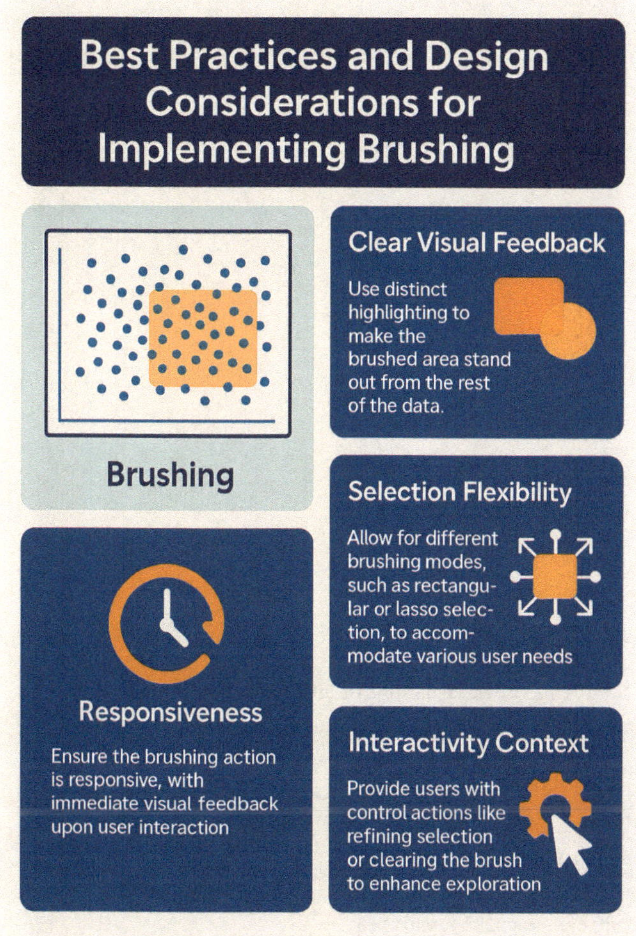

Fig. 11.4: Best Practices and Design Considerations for Implementing Brushing in Scientific Visualizations. The infographic outlines essential principles such as responsive feedback, linked view synchronization, lasso selection, minimal visual clutter, and clear highlighting to ensure effective and user-friendly brushing interactions.

```
15  })
16
17  source = ColumnDataSource(data)
18
19  color_mapper = CategoricalColorMapper(factors=["A", "B"],
    ↪    palette=Category10[3][:2])
20
21  # Scatter plot
22  p1 = figure(title="Scatter Plot with Brushing",
    ↪    tools="pan,wheel_zoom,box_select,reset", width=600, height=600)
```

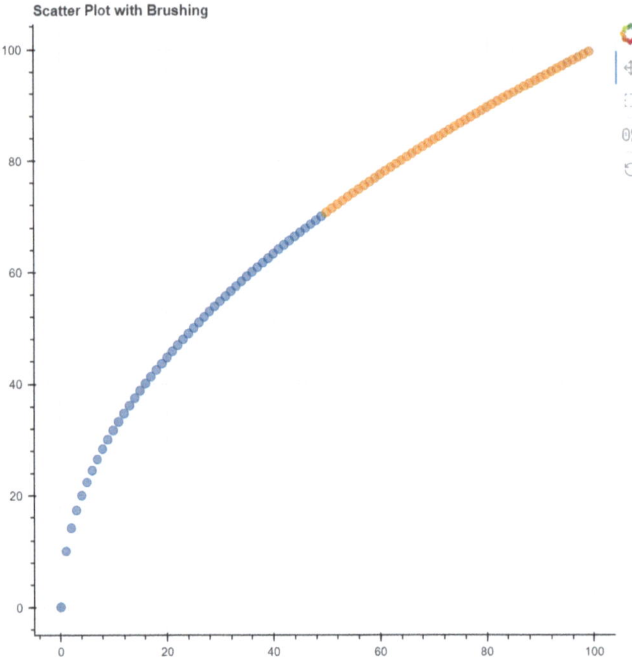

Fig. 11.5: Brushing in an Interactive Scatter Plot

```
23  p1.scatter('x', 'y', source=source, size=8, color={'field': 'category',
    ↪    'transform': color_mapper}, alpha=0.6)
24
25  # Log brushing\index{Brushing} to console
26  callback = CustomJS(args=dict(source=source), code="""
27      const indices = cb_obj.indices;
28      console.log("Selected indices:", indices);
29  """)
30  source.selected.js_on_change('indices', callback)
31
32  show(row(p1))   # opens in browser
```

In Figure 11.5, the interactive scatter plot allows users to select a subset of data points through brushing. The selection is dynamically linked to a corresponding histogram, which updates to reflect the distribution of the brushed subset, demonstrating the power of linking multiple views through brushing.

11.5 Drill-Down

11.5.1 Interactive Exploration with Drill-Down

Drill-Down is an interactive exploration technique in data visualization that allows users to navigate from a high-level summary view to more detailed layers of data. This hierarchical exploration enables users to investigate underlying patterns, trends, and outliers by progressively accessing finer granularities of the dataset. By providing a structured pathway from overview to detail, Drill-Down enhances the user's ability to uncover insights that might remain hidden in aggregated data representations. For instance, in a sales dashboard, a user can start by viewing total sales across different regions and then drill down into specific countries, cities, or even individual stores to analyze performance metrics in greater depth. This methodical approach to data exploration fosters a deeper understanding of complex datasets, facilitating informed decision-making and strategic planning.

11.5.2 Hierarchical Data Structures

Effective implementation of Drill-Down techniques relies on well-organized hierarchical data structures that support multi-level data exploration. Hierarchical data models, such as trees or nested datasets, provide the necessary framework for representing data at varying levels of detail. Each node in the hierarchy corresponds to a specific data aggregation level, allowing seamless transitions between summaries and detailed views. For example, in a geographical hierarchy, the top level might represent continents, the next level countries, followed by states or provinces, and finally cities. Similarly, in a time-based hierarchy, data can be organized from years to quarters, months, weeks, and days. Utilizing hierarchical data structures ensures that Drill-Down interactions are logical and intuitive, enabling users to navigate through data layers without confusion or loss of context. These structures facilitate efficient data retrieval and rendering, which are crucial for maintaining performance and responsiveness in interactive visualizations.

> **⚠ Warning**
>
> Deep hierarchical drill-down interfaces without breadcrumb trails or return paths can confuse users and prevent them from efficiently navigating back to higher-level views. Always provide a clear way to backtrack.

11.5.3 Techniques for Smooth Transitions

Ensuring smooth transitions during Drill-Down interactions is essential for maintaining user engagement and preventing disorientation. Several techniques can be employed to achieve seamless navigation between different data levels. One effective method is the use of animated transitions that visually guide users as the view shifts from a summary to a detailed perspective. These animations can include zooming effects, fading elements in and out, or morphing shapes to represent the change in data granularity. Another technique involves maintaining visual continuity by preserving common visual elements, such as color schemes, axes, and labels, across different drill-down levels. This consistency helps users retain their spatial and contextual understanding of the data. Providing breadcrumb trails or hierarchical navigation menus allows users to track their exploration path and easily return to previous levels. By combining these techniques, visualizations can offer an intuitive and fluid Drill-Down experience that enhances data exploration without disrupting the user's cognitive flow.

> **Design Tip: Support Drill-Down with Breadcrumbs or Tabs**
>
> Help users navigate complex drill-down interfaces by including breadcrumbs or tabbed navigation that reflects the hierarchy. This prevents cognitive overload and maintains a sense of orientation.

11.5.4 Best Practices and Design Considerations for Implementing Drill-Down

Drill-Down is a powerful technique in interactive data visualization that enables users to navigate from a high-level summary to more granular layers of data. When designed effectively, it allows for focused exploration without overwhelming the viewer with unnecessary detail upfront. However, implementing drill-down functionality requires thoughtful user interface design and clear visual guidance to maintain usability and orientation.

Figure 11.6 visually summarizes the best practices and design considerations for implementing drill-down in scientific visualizations. The infographic emphasizes a clean, visual-first approach with minimal text to promote clarity and accessibility.

A key best practice is the use of**clear hierarchical structures** to guide users through the data levels intuitively. Each level should build contextually upon the previous one, maintaining consistent visual metaphors and layout alignment.**Interactive breadcrumbs** or progress indicators help users understand their current location within the data hierarchy and allow easy backtracking.

Animated transitions between levels are recommended to preserve visual continuity and reduce cognitive load. Sudden changes can disorient users, especially in complex visualizations. Smooth zoom-ins, fades, or morphing shapes enhance the drill-down experience and reinforce spatial memory.

Designers should also ensure that**labels and legends remain legible** at all levels and that**focus areas are visually highlighted** when drilling deeper. **Action cues,** such as clickable icons or expandable regions, should be easily discoverable without cluttering the interface.

Finally, drill-down interactions must be optimized for performance. Responsive feedback is essential to keep the user engaged and to support exploration across large or nested datasets.

By following these principles, as illustrated in Figure 11.6, developers can create intuitive and visually coherent drill-down interfaces that support effective scientific discovery.

Fig. 11.6: Best Practices and Design Considerations for Implementing Drill-Down. The infographic emphasizes visual hierarchy, smooth transitions, interactive breadcrumbs, and intuitive navigation cues using minimal text and strong graphical elements.

11.5.5 Applications in Data Analysis

Drill-Down techniques find application across various domains of data analysis, enabling users to explore complex datasets with ease and precision. In business intelligence, Drill-Down is commonly used in sales and financial dashboards to analyze performance metrics at multiple organizational levels, from overall company revenue down to individual product lines or sales representatives. In healthcare, clinicians utilize Drill-Down to examine patient data, starting with general health indicators and delving into specific medical histories, treatment plans, and outcomes. Environmental scientists apply Drill-Down to climate data, enabling the exploration of global weather patterns and the identification of localized anomalies. In academic research, Drill-Down facilitates the analysis of large-scale experimental data, allowing researchers to investigate specific variables or subsets of data that are pertinent to their hypotheses. These diverse applications demonstrate the versatility and critical importance of Drill-Down techniques in enhancing data-driven analysis and fostering deeper insights across multiple fields.

11.5.6 User Experience Considerations

Designing effective Drill-Down interactions requires careful consideration of User Experience (UX) factors to ensure that the navigation process is intuitive, efficient, and satisfying. One key consideration is the clarity of the hierarchical structure, which should be easily understandable and logically organized to prevent user confusion. Visual indicators, such as expandable menus, icons, or labels, can help users comprehend the available drill-down paths and the relationships between different data levels. Another important aspect is the responsiveness of the Drill-Down interactions; delays or lag during navigation can disrupt the user experience and hinder data exploration. Optimizing performance through efficient data loading and rendering techniques is therefore essential. Providing contextual information, such as current drill-down level and available actions, helps users maintain their orientation within the data hierarchy. Accessibility features, including keyboard navigation and support for screen readers, ensure that Drill-Down interactions are usable by all individuals, regardless of their technical proficiency or physical abilities. By prioritizing these UX considerations, Drill-Down techniques can be designed to offer a seamless and engaging data exploration experience that empowers users to uncover meaningful insights with ease.

```python
#Implementing Drill-Down
import pandas as pd
import plotly.graph_objects as go

# Sample detailed dataset
data = {
```

```python
    'Region': ['North', 'North', 'North', 'South', 'South', 'South',
    ↪  'East', 'East', 'West', 'West'],
    'Country': ['USA', 'Canada', 'Mexico', 'Brazil', 'Argentina', 'Chile',
    ↪  'China', 'Japan', 'Germany', 'France'],
    'City': ['New York', 'Toronto', 'Mexico City', 'Sao Paulo',
    ↪  'Buenos Aires', 'Santiago',
            'Beijing', 'Tokyo', 'Berlin', 'Paris'],
    'Sales': [320, 210, 180, 450, 230, 190, 500, 470, 380, 360]
}
df = pd.DataFrame(data)

# Aggregate sales by Region
region_df = df.groupby('Region', as_index=False).agg({'Sales': 'sum'})

# Define initial figure
fig = go.Figure()

# Base regional view
fig.add_trace(go.Bar(
    x=region_df['Region'],
    y=region_df['Sales'],
    marker=dict(color='lightskyblue', line=dict(color='royalblue',
    ↪  width=2)),
    name="Region Sales",
    hovertemplate="<b>%{x}</b><br>Total Sales: %{y}k USD<extra></extra>"
))

# Add drill-up button placeholder
fig.update_layout(
    title="\faGlobe Regional Sales Overview",
    xaxis_title="Region",
    yaxis_title="Sales (in thousands USD)",
    template="plotly_white",
    font=dict(family="Segoe UI", size=14),
    clickmode="event+select",
    margin=dict(t=90, l=60, r=30, b=60),
    updatemenus=[
        {
            "buttons": [
                {
                    "label": "Back to Region View",
                    "method": "restyle",
                    "args": [{"x": [region_df['Region']], "y":
                    ↪  [region_df['Sales']]}]
                }
            ],
            "direction": "left",
            "type": "buttons",
            "x": 0,
            "xanchor": "left",
            "y": 1.1,
            "yanchor": "top"
        }
    ]
```

```
56  )
57  fig.show()
```

In Figure 11.7, the hierarchical bar chart initially displays sales aggregated by region. By clicking on a specific region, the visualization drills down to reveal sales figures for individual countries within that region. This interactive Drill-Down capability allows users to seamlessly transition from a broad overview to detailed insights, facilitating a more comprehensive analysis of the sales data.

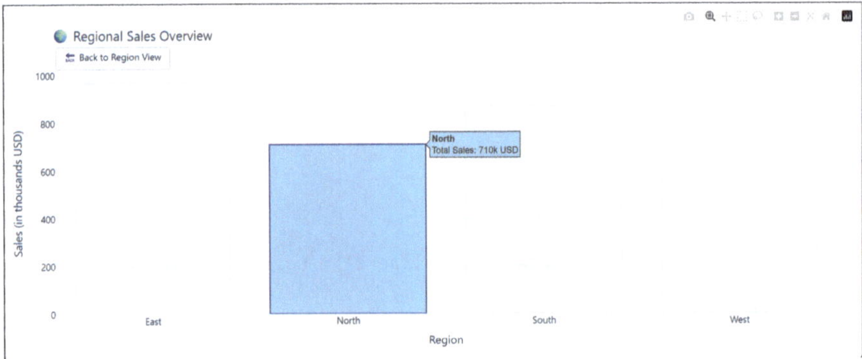

Fig. 11.7: Drill-Down Interaction in a Hierarchical Bar Chart

11.6   Filtering

11.6.1   Purpose of Data Filtering

Data filtering is a critical technique in data visualization that enables users to focus on specific subsets of data by excluding irrelevant or less important information. The primary purpose of filtering is to enhance the clarity and readability of visualizations, making complex datasets more manageable and interpretable. By allowing users to isolate particular variables, ranges, or categories, filtering helps in highlighting significant patterns, trends, and outliers that are pertinent to the analysis at hand. This selective visibility not only reduces cognitive load but also facilitates a more targeted and efficient exploration of the data. In scenarios where datasets are vast and multifaceted, filtering serves as an essential tool for distilling the information into meaningful insights, thereby supporting informed decision-making and deeper analytical understanding.

11.6.2   Types of Filters

There are various types of filters available in data visualization, each serving distinct purposes and catering to different analytical needs.

Range Filters allow users to specify numerical ranges for variables, enabling the focus on data points that fall within the defined boundaries. For example, filtering sales data to include only transactions above a certain value can help identify high-performing regions or products.

Category Filters enable the selection of specific categories or groups within a dataset. This type of filter is particularly useful for comparing different segments, such as filtering customer data by age groups or product types.

Text Filters allow users to search and filter data based on textual input, facilitating the identification of specific entries or keywords within the dataset. This is beneficial in scenarios like filtering articles by keywords or searching for specific entries in a database.

Boolean Filters provide binary filtering options, typically based on true/false conditions. For instance, filtering customer data to show only those who have made purchases within the last month can help in targeting active users.

Date Filters are used to narrow down data based on specific timeframes, such as filtering events that occurred within a particular date range. This is essential in time-series analyses where temporal patterns are of interest.

Each type of filter offers unique capabilities that can be used to tailor visualizations according to the specific requirements of the analysis, thereby enhancing the overall effectiveness of data exploration.

11.6.3 Dynamic vs. Static Filtering

Filtering techniques can be broadly categorized into dynamic and static filtering approaches, each with its own set of advantages and use cases.

Dynamic Filtering refers to interactive filtering methods that allow users to adjust filter parameters in real-time while exploring the visualization. This interactivity provides a flexible and responsive way to manipulate data, enabling users to experiment with different filter settings and immediately observe the resulting changes in the visualization. Dynamic filters are particularly useful in exploratory data analysis, where users need to iteratively refine their focus to uncover insights. For example, using sliders to adjust numerical ranges or checkboxes to select multiple categories dynamically updates the visualization based on user input.

Static Filtering, on the other hand, involves applying predefined filters that do not change during the interaction with the visualization. Static filters are typically set before the visualization is rendered and remain constant throughout the analysis. This approach is suitable for scenarios where specific data subsets need to be consistently displayed, such as generating reports or dashboards with fixed criteria. Static filtering ensures that the visualization remains focused on the intended data subset, providing a stable and reliable representation of the selected information.

Both dynamic and static filtering approaches play vital roles in data visualization, and the choice between them depends on the specific needs of the analysis and the desired level of interactivity.

11.6.4 Implementing Effective Filters

Creating efficient filters in data visualization involves several strategies to ensure that the filtering process is intuitive, responsive, and enhances the overall user experience.

User-Friendly Interface: Designing an intuitive interface for filters is paramount. This includes using clear labels, appropriate input controls (such as sliders, dropdowns, and checkboxes), and logical placement within the visualization layout. A well-designed interface minimizes user confusion and facilitates easy interaction with the filters.

Performance Optimization: Ensuring that the visualization remains responsive during filtering operations is crucial, especially when dealing with large datasets. Implementing efficient data processing techniques and optimizing rendering performance can prevent lag and maintain a smooth user experience.

Visual Feedback: Providing immediate visual feedback when filters are applied helps users understand the impact of their actions. This can be achieved through animations, highlighting filtered data points, or updating related visual elements in real-time.

Multi-Filter Support: Allowing the combination of multiple filters enables users to perform more complex data slicing and dicing. Supporting hierarchical or dependent filters can further enhance the filtering capabilities, allowing for more nuanced data exploration.

Accessibility Considerations: Designing filters that are accessible to all users, including those with disabilities, ensures inclusivity. This involves providing keyboard navigation support, ensuring compatibility with screen readers, and maintaining sufficient contrast and size for filter controls.

> **⚠ Warning**
>
> Filters that remove significant portions of the dataset without indicating what has been excluded may result in misleading interpretations. Always show filtering status and offer a reset option to restore the full dataset view.

By adhering to these strategies, filters can be implemented in a manner that maximizes their effectiveness and usability within data visualizations.

11.6.5 Best Practices and Design Considerations for Implementing Filtering

Filtering plays a crucial role in interactive data visualizations by allowing users to focus on specific subsets of data that are most relevant to their analysis. However, the effectiveness of filtering depends not just on functionality, but also on how intuitively it is designed and visually presented.

Figure 11.8 illustrates best practices and key design considerations for implementing filtering using a clear, visual-forward infographic. The graphic emphasizes the importance of simplicity, user control, and contextual awareness while minimizing textual overload.

One fundamental best practice is the inclusion of **clear, intuitive controls**–such as sliders, checkboxes, dropdown menus, or toggle switches–that allow users to easily define filtering criteria. These controls should be **clearly labeled**, consistently styled, and placed prominently within the visualization interface.

Effective filtering design also involves providing **real-time visual feedback**. As users apply or adjust filters, the visualization should update immediately, reflecting the change while preserving overall layout and structure. This responsiveness reinforces trust and encourages exploratory analysis.

Another important consideration is to **visibly communicate the active filter state**. Indicators such as filter tags, color-coded selections, or summaries of selected ranges ensure that users are always aware of which filters are applied. A **reset or undo option** should be readily accessible to revert the view to its unfiltered state.

In complex visualizations, it is helpful to allow **combinatorial filtering**–enabling users to apply multiple filters across different dimensions. However, designers must ensure this does not lead to confusion or data omission, which is why **contextual counts** (e.g., "showing 128 of 500 items") are also recommended.

From an accessibility perspective, filtering controls should be fully keyboard-navigable, use distinguishable color contrast, and include descriptive tooltips or ARIA labels to support users with varying abilities.

As visualized in Figure 11.8, combining usability, clarity, and performance in filtering design ensures that users can engage with large or complex datasets effectively, revealing meaningful insights without being overwhelmed.

11.6.6 Example

Filtering techniques have been successfully applied in various scientific research studies to enhance data clarity and facilitate targeted analysis.

```python
#Implementing Filtering with Pandas and Matplotlib
import pandas as pd
import matplotlib.pyplot as plt
from matplotlib.widgets import RangeSlider

# Sample data
data = {
    'Age': [23, 45, 12, 36, 27, 52, 19, 33, 41, 60],
    'Income': [50000, 80000, 20000, 60000, 55000, 120000, 30000, 65000,
    ↪ 70000, 150000],
    'Spending Score': [60, 80, 20, 50, 55, 90, 25, 65, 70, 95]
}

df = pd.DataFrame(data)

```

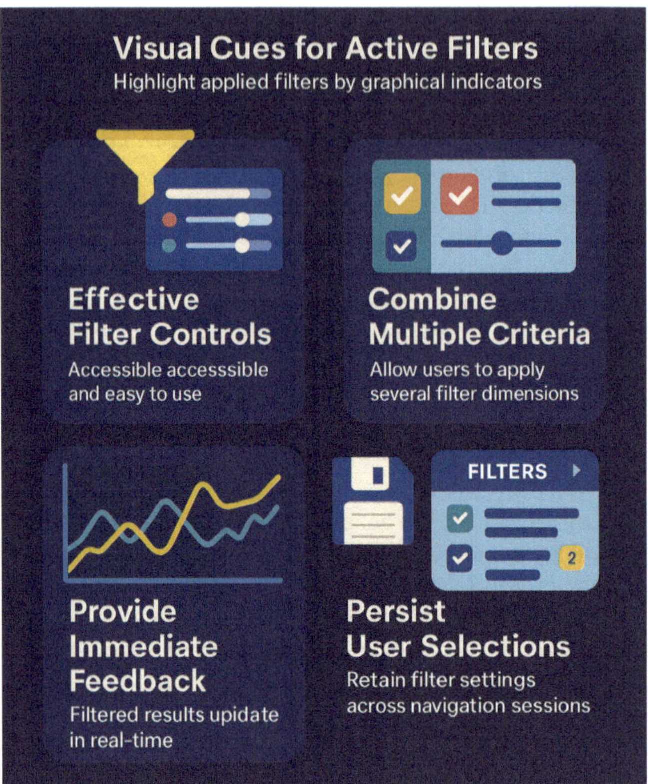

Fig. 11.8: Best Practices and Design Considerations for Implementing Filtering. This infographic highlights essential UX principles including clarity of filter controls, real-time responsiveness, state visibility, multi-filter support, and accessibility features.

```python
fig, ax = plt.subplots()
plt.subplots_adjust(bottom=0.25)

# Initial scatter plot
scatter = ax.scatter(df['Age'], df['Income'], c=df['Spending Score'],
    cmap='viridis')
ax.set_xlabel('Age')
ax.set_ylabel('Income')
ax.set_title('Age vs. Income with Spending Score')

# Slider for Income filtering
ax_income = plt.axes([0.25, 0.1, 0.65, 0.03])
income_slider = RangeSlider(ax_income, 'Income Range', 0, 200000,
    valinit=(20000, 120000))

def update(val):
    income_min, income_max = income_slider.val
```

```
30    filtered_df = df[(df['Income'] >= income_min) & (df['Income'] <=
      ↪  income_max)]
31    ax.clear()
32    scatter = ax.scatter(filtered_df['Age'], filtered_df['Income'],
      ↪  c=filtered_df['Spending Score'], cmap='viridis')
33    ax.set_xlabel('Age')
34    ax.set_ylabel('Income')
35    ax.set_title('Age vs. Income with Spending Score')
36    fig.canvas.draw_idle()
37
38 income_slider.on_changed(update)
39
40 plt.show()
```

In Figure 11.9, an interactive scatter plot demonstrates the application of filtering to focus on specific income ranges. The range slider allows users to dynamically adjust the income boundaries, instantly updating the visualization to display only the data points that fall within the selected range. This interactive filtering enhances data clarity by enabling users to concentrate on relevant subsets, thereby facilitating more precise analysis and insight generation.

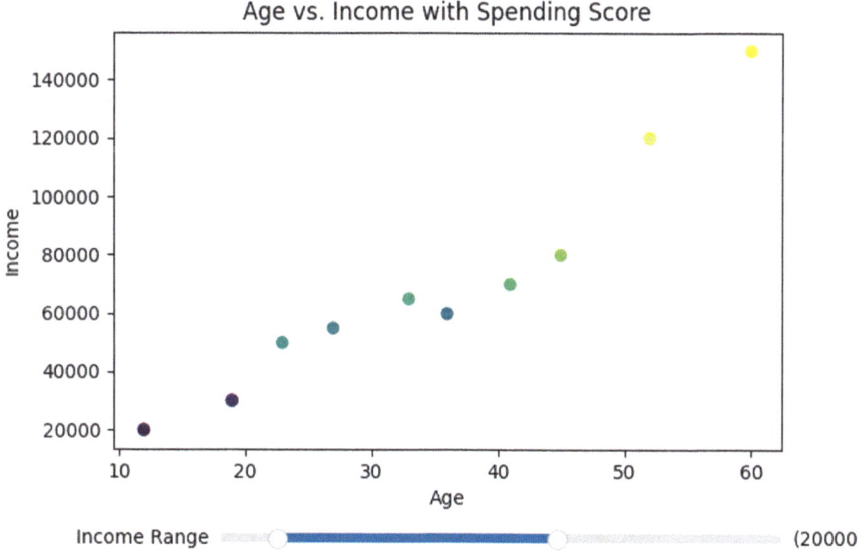

Fig. 11.9: Interactive Filtering in a Multi-Dimensional Scatter Plot

> **Design Tip: Combine Filters with Visual Summaries**
>
> Pair filter controls with charts or counters that indicate the number of visible vs. total data points. This helps users understand the scope of their filtered view and ensures transparency in analysis.

11.7 Linking

11.7.1 Concept of Linked Visualizations

Linked visualizations involve connecting multiple visual representations of the same dataset, allowing interactions in one view to influence and update corresponding elements in other linked views. This interconnectedness creates a cohesive analytical environment where users can explore data from various perspectives simultaneously. For instance, selecting a subset of data points in a scatter plot can automatically highlight related entries in a bar chart or a geographical map. This synchronization facilitates a more comprehensive understanding of the data by revealing relationships and patterns that might remain hidden when visualizations are viewed in isolation. Linked visualizations empower users to perform multivariate analyses seamlessly, enhancing their ability to draw meaningful insights from complex datasets.

11.7.2 Synchronization Techniques

Synchronizing linked visualizations requires effective coordination mechanisms to ensure that interactions in one view are accurately reflected across all connected views. One common technique is the use of shared data sources, where multiple visualizations draw from the same underlying dataset. When a user interacts with one visualization, such as brushing a selection of data points, the shared data source updates accordingly, triggering changes in all linked visualizations. Another technique involves event listeners and callbacks, where specific user actions in one visualization emit events that other visualizations listen for and respond to by updating their displays. Using state management frameworks can help maintain consistency across multiple views by centralizing the state information and ensuring that all visualizations are in sync. Implementing these synchronization techniques ensures that linked visualizations operate harmoniously, providing a unified and interactive data exploration experience.

11.7.3 Benefits for Multivariate Data Exploration

Linking multiple visualizations offers significant advantages for multivariate data exploration by enabling users to analyze multiple dimensions of data concurrently. This approach allows for the simultaneous examination of different variables, facilitating the identification of correlations, trends, and outliers that span across various aspects of the dataset. For example, in a financial dashboard, a line chart depicting stock prices can be linked with a bar chart showing trading volumes and a pie chart representing market sectors. By interacting with one chart, such as selecting a specific time range in the line chart, the other charts automatically update to reflect the corresponding data, providing a holistic view of the financial landscape. This integrated analysis capability enhances the depth and breadth of insights, making it easier for users to uncover complex relationships and make informed decisions based on a comprehensive understanding of the data.

> **⚠ Warning**
>
> Displaying too many small multiples on a single screen can lead to visual clutter and cognitive overload. Limit the number of panels, use pagination, or allow selective display to preserve clarity.

11.7.4   Tools and Frameworks

Several tools and frameworks facilitate the implementation of linked visualizations, each offering unique features and capabilities to support interactive data exploration. `D3.js` is a versatile JavaScript library that provides extensive support for creating dynamic and interactive visualizations, including linking functionalities through its robust event handling and data binding mechanisms. `Plotly` offers high-level interfaces for creating linked plots with minimal code, supporting a wide range of chart types and interactive features. For Python developers, `Bokeh` stands out as a powerful library that enables the creation of linked visualizations with its comprehensive widget and callback system, allowing for seamless interactions between different plots. `Tableau` and `Power BI` are popular business intelligence tools that offer built-in capabilities for linking multiple visualizations, making it easy to create interactive dashboards without extensive coding. These tools and frameworks empower developers and data scientists to build sophisticated linked visualizations that enhance data analysis and insight generation.

11.7.5   Best Practices for Implementing Linking

Linking is a powerful technique in interactive data visualization that synchronizes interactions across multiple visual components. It allows users to select, filter, or highlight data in one view and see corresponding effects in all connected views, thereby enabling multivariate and multidimensional analysis. To be effective, linking must be implemented with clarity, responsiveness, and usability in mind.

Figure 11.10 presents a colorful technical infographic that encapsulates the core design principles for implementing linking in visualizations. The use of minimal text and rich visual metaphors reinforces intuitive understanding of these best practices.

A fundamental design principle is to ensure**synchronized highlighting** across all views. When a user brushes, clicks, or selects data in one chart, related data points in all linked charts should update immediately and visibly, using consistent colors or shapes to indicate correspondence.

Unified color encoding is essential for establishing visual continuity across views. Colors representing categories or clusters must be consistent, enabling users to recognize patterns and relationships at a glance. Inconsistencies can create confusion and misinterpretation.

Another important aspect is the**use of shared data sources or synchronized state management**. Behind the scenes, all views should access the same underlying dataset or a synchronized state model to guarantee coherence across interactions.

Tooltips, animations, or hover effects can further reinforce linking by providing contextual information as the user explores each dimension.**Interactive legends** or controls that highlight corresponding views enhance user agency and understanding.

To prevent overload, the visualization should also provide**clear visual separation of views** and**toggle options** to enable or disable specific linking behaviors. This allows users to isolate or combine views as needed without losing clarity.

Finally, performance is a critical factor. Linking should scale well with data size and number of views, ensuring that interactions remain fluid and immediate across all components.

As visualized in Figure 11.10, these design considerations collectively enable seamless, responsive, and insightful linked interactions that empower users to draw deeper conclusions from complex datasets.

11.7.6 Practical Applications

Linked visualizations are applied across various domains to enhance data analysis and decision-making processes. In **business intelligence**, linked dashboards allow stakeholders to interact with multiple charts and tables simultaneously, enabling comprehensive performance monitoring and strategic planning. For example, a sales dashboard might link a geographic map with sales figures and customer demographics, allowing managers to identify regional trends and target specific market segments effectively.

In **healthcare**, linked visualizations facilitate the analysis of patient data by connecting clinical indicators with treatment outcomes and demographic information, aiding in the identification of effective therapies and population health trends.

In **environmental science**, researchers use linked maps and time-series charts to study climate patterns, enabling the correlation of geographical data with temporal changes.

In **academic research**, linked visualizations support the exploration of complex datasets by connecting various statistical plots and experimental results, fostering a deeper understanding of research findings.

These practical applications demonstrate the versatility and effectiveness of linked visualizations in transforming data into actionable insights across diverse fields.

```python
#Implementing Linked Visualizations with Bokeh
from bokeh.layouts import row
```

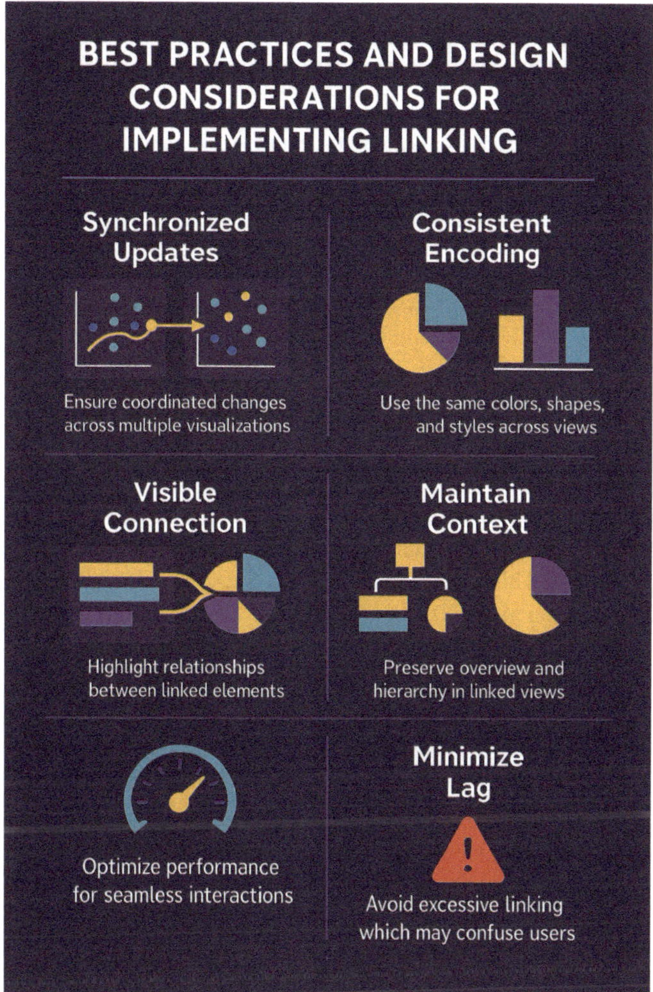

Fig. 11.10: Best Practices and Design Considerations for Implementing Linking. The infographic highlights core strategies including synchronized highlights, unified color encoding, shared data states, contextual tooltips, and toggleable view linking to ensure a cohesive and intuitive multi-view visualization experience.

```python
from bokeh.models import ColumnDataSource, HoverTool
from bokeh.plotting import figure, show

# Sample data
data = {
    'Category': ['A', 'B', 'C', 'D'],
    'Values': [10, 20, 30, 40],
    'Details': ['Detail A', 'Detail B', 'Detail C', 'Detail D']
```

```
11  }
12  source = ColumnDataSource(data=data)
13  selected_source = ColumnDataSource(data={key: [] for key in data})
14
15  # Bar plot
16  p1 = figure(x_range=data['Category'], title="Category Values",
    ↪  toolbar_location="above", tools="hover,tap", width=400, height=400)
17  p1.vbar(x='Category', top='Values', width=0.9, source=source,
    ↪  legend_field="Category",
18        line_color='white', fill_color='navy')
19  hover = p1.select_one(HoverTool)
20  hover.tooltips = [("Category", "@Category"), ("Value", "@Values"),
    ↪  ("Details", "@Details")]
21
22  # Add layout to Bokeh server document
23  layout = row(p1)
24  show(layout)
```

In Figure 11.11, an interactive dashboard showcases linked visualizations where selecting a category in the bar chart dynamically updates the scatter plot to highlight relevant data points. This linkage enables users to explore the relationships between different data dimensions effortlessly, enhancing the overall data analysis experience.

11.8 Annotation

11.8.1 Importance of Annotations in Data Visualization

Annotations play a pivotal role in data visualization by providing contextual information that enhances the interpretability and comprehensibility of visual representations. They serve as guides that help users navigate complex data, highlight significant trends, and draw attention to key insights. By embedding explanatory text, labels, arrows, or markers directly within the visualization, annotations bridge the gap between raw data and meaningful interpretation. This contextualization is essential for ensuring that viewers can quickly grasp the underlying messages and takeaways without extensive analysis. In scientific research, where data can be intricate and multifaceted, annotations facilitate a deeper understanding by clarifying relationships, indicating anomalies, and emphasizing critical findings. Ultimately, annotations transform static visuals into interactive narratives, making complex data more accessible and engaging for a diverse audience.

11.8.2 Types of Annotations

There are various types of annotations that can be employed to enrich data visualizations, each serving a distinct purpose in conveying information effectively.

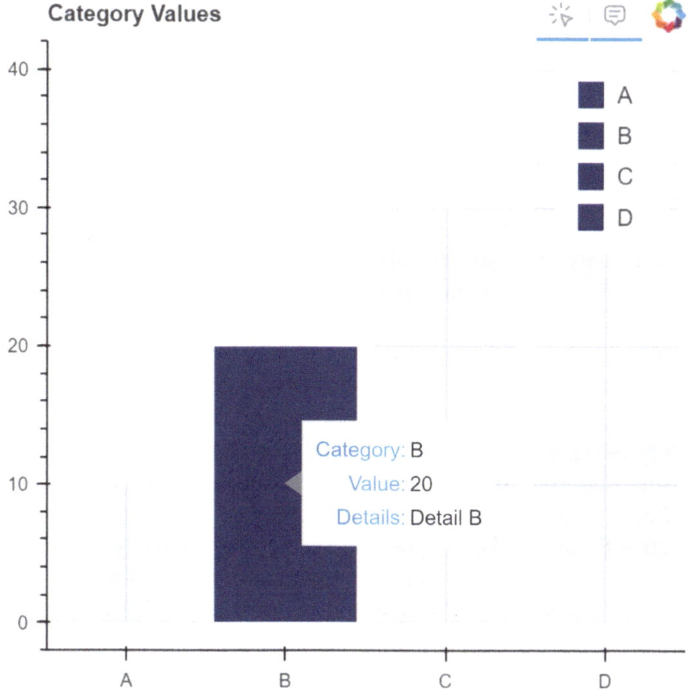

Fig. 11.11: Linked Visualizations in an Interactive Dashboard

Text Annotations involve adding descriptive text to highlight specific data points or regions within a chart. These annotations can provide explanations, definitions, or insights that are not immediately apparent from the data alone.

Label Annotations attach labels to data points, bars, or sections of a visualization, ensuring that important elements are easily identifiable. Labels can include numerical values, categories, or other relevant identifiers.

Arrow Annotations use arrows to draw attention to particular areas of interest, indicating trends, correlations, or deviations within the data. Arrows guide the viewer's focus to specific insights that warrant further exploration.

Shape Annotations incorporate shapes such as circles, rectangles, or highlights around data points to emphasize their significance. These visual markers can differentiate key elements from the rest of the dataset.

Callout Annotations combine text and shapes to create a focused explanation or commentary on a specific aspect of the visualization. Callouts are effective for providing detailed insights without cluttering the overall view.

Each type of annotation contributes uniquely to the storytelling aspect of data visualization, enabling a more nuanced and informative presentation of data.

11.8.3   Best Practices for Designing Effective Annotations

Creating meaningful annotations requires adherence to several best practices to ensure they enhance rather than detract from the visualization.

Clarity and Conciseness: Annotations should be clear and succinct, avoiding unnecessary complexity. The accompanying text should communicate the intended message without overwhelming the viewer.

Relevance: Only annotate elements that add significant value to the interpretation of the data. Irrelevant annotations can create confusion and distract from the primary insights.

Consistency: Maintain a consistent style for annotations throughout the visualization. This includes uniform font sizes, colors, and positioning to create a cohesive look and feel.

Non-Intrusiveness: Annotations should complement the visualization without obscuring important data. Position them strategically to ensure they enhance understanding without blocking critical information.

Interactive Elements: Where possible, incorporate interactivity into annotations, allowing users to engage with them for additional information. This can include tooltips, clickable labels, or expandable sections that provide deeper insights on demand.

Visual Hierarchy: Establish a clear visual hierarchy by differentiating annotations based on their importance. Use varying font weights, colors, or sizes to distinguish primary annotations from secondary ones.

As shown in Figure 11.12, well-designed annotations should balance visibility with subtlety, offering guidance without distracting from the data itself.

By following these best practices, annotations can effectively augment data visualizations, making them more informative and user-friendly.

11.8.4   Tools for Adding Annotations

Several tools and libraries facilitate the addition of annotations to data visualizations, each offering unique features to support interactive and visually appealing annotations.

`Matplotlib` is a widely-used Python library that provides comprehensive annotation capabilities through functions like `annotate`, `text`, and `arrow`. These functions allow for precise placement of text, arrows, and shapes to highlight specific data points or trends.

`Plotly` offers interactive annotation features that enable users to add dynamic text, shapes, and arrows to plots. Its intuitive interface supports the creation of annotations that respond to user interactions, enhancing the interactivity of the visualization.

`Bokeh` excels in creating web-based interactive visualizations with robust annotation support. It allows for the addition of various annotation types, including

Fig. 11.12: Best Practices for Designing Effective Annotations. This infographic visually communicates key principles such as clarity, contextual relevance, consistency, visual hierarchy, unobtrusiveness, and interactivity for annotation design in data visualizations.

labels, arrows, and custom HTML elements, providing a high degree of flexibility and customization.

Seaborn, built on top of Matplotlib, simplifies the creation of statistical visualizations with built-in support for annotations. It integrates seamlessly with Matplotlib's annotation functions, enabling users to enhance their plots with minimal code.

`D3.js` is a powerful JavaScript library that offers extensive annotation capabilities for web-based visualizations. Its flexibility allows for the creation of highly customized and interactive annotations that can be tailored to specific data storytelling needs.

These tools empower data scientists and visualization experts to incorporate annotations that enhance the clarity, engagement, and informativeness of their visualizations.

11.8.5   Examples Highlighting Key Insights

Annotations are instrumental in emphasizing important data points and conveying critical insights within visualizations.

In a **time-series line chart** tracking website traffic over a year. Annotations can mark significant events, such as marketing campaigns or system outages, explaining sudden spikes or drops in traffic. This contextual information helps viewers understand the underlying factors influencing the data trends.

```python
#Adding Annotations to a Time-Series Plot with Plotly
import plotly.graph_objects as go
from datetime import datetime

# Sample time-series data
dates = ['2023-01-01', '2023-02-01', '2023-03-01', '2023-04-01']
traffic = [100, 150, 130, 200]

fig = go.Figure()

# Time-series line plot with custom styling
fig.add_trace(go.Scatter(
    x=dates,
    y=traffic,
    mode='lines+markers',
    name='Website Traffic',
    line=dict(color='royalblue', width=3),
    marker=dict(size=10, symbol='circle', color='lightblue',
    ↪    line=dict(width=2, color='royalblue')),

    ↪    hovertemplate='<b>Date:</b> %{x}<br><b>Visitors:</b> %{y}<extra></extra>'
))

# Highlight event marker \index{Marker} with shaded callout
fig.add_annotation(
    x='2023-02-01',
    y=150,
    xref='x',
    yref='y',
    text='Marketing Campaign Launch',
    showarrow=True,
    arrowhead=2,
    arrowsize=1,
    arrowcolor='crimson',
    ax=0,
    ay=-80,
```

```python
35        bordercolor='crimson',
36        borderwidth=2,
37        borderpad=5,
38        bgcolor='mistyrose',
39        opacity=0.9,
40        font=dict(size=13, color='darkred')
41    )
42
43    # Customize layout for clarity and modern UX
44    fig.update_layout(
45        title=dict(
46            text='Website Traffic Over Time',
47            x=0.5,
48            xanchor='center',
49            font=dict(size=22, family='Segoe UI', color='darkblue')
50        ),
51        xaxis=dict(
52            title='Date',
53            tickangle=-45,
54            tickfont=dict(size=16),
55            showgrid=True,
56            gridwidth=1,
57            gridcolor='lightgray'
58        ),
59        yaxis=dict(
60            title='Number of Visitors',
61            tickfont=dict(size=16),
62            showgrid=True,
63            gridwidth=1,
64            gridcolor='lightgray'
65        ),
66        plot_bgcolor='white',
67        hovermode='x unified',
68        margin=dict(l=60, r=30, t=80, b=80)
69    )
70    fig.show()
```

In this example, the annotation identifies the launch of a marketing campaign, correlating it with a noticeable increase in website traffic. By directly linking events to data trends, annotations provide a clear and immediate understanding of the factors driving changes in the dataset.

> **Design Tip: Annotate Key Events in Animated Visualizations**
>
> When animating time-series data, add annotations or markers to indicate key events or transitions. This guides users' attention and enhances the narrative power of the visualization.

In Figure 11.13, an annotation is added to a specific data point to highlight its significance, while the accompanying text provides context, explaining why this point is noteworthy. This combination of visual and textual cues enhances the overall narrative of the data visualization, making key insights more accessible and memorable.

This example illustrates how annotations can effectively highlight key insights, provide necessary context, and enhance the overall interpretability of data visualizations, making complex information more accessible and actionable.

Fig. 11.13: Annotated Time-Series Plot Highlighting Significant Events

11.9 Animation

11.9.1 Role of Animation in Data Visualization

Animation plays a crucial role in data visualization by bringing static data to life, enabling users to observe changes and transitions over time. By incorporating motion into visual representations, animations help convey complex temporal dynamics and evolving patterns that would be difficult to discern in static charts. This dynamic element enhances data understanding by providing a more engaging and intuitive way to track trends, monitor progress, and identify anomalies as they develop. For example, animating the progression of stock prices over a fiscal year allows viewers to visually follow market fluctuations, making it easier to grasp the volatility and momentum within the data. Animation can guide the viewer's attention to significant events or changes, facilitating a clearer narrative and more impactful storytelling within the visualization. Overall, the integration of animation transforms data visualizations into interactive experiences that promote deeper insights and a more comprehensive comprehension of temporal data.

11.9.2 Techniques for Creating Meaningful Animations

Creating meaningful animations in data visualization involves several key techniques that ensure the motion enhances rather than distracts from the data interpretation. One effective method is the use of smooth transitions, which help maintain continuity and prevent disorientation as the visualization changes states. Smooth interpolations between data points or states allow users to follow the evolution of the data seamlessly. Another technique is the strategic pacing of animations; controlling the speed at which changes occur ensures that viewers have adequate time to absorb and understand each transition without feeling rushed or overwhelmed. Emphasizing critical data points through animated highlights or focal shifts can draw attention to important trends or outliers , reinforcing the narrative of the visualization. Incorporating interactive controls, such as play, pause, and scrubber bars, empowers users to manage the animation flow according to their own pace and interests. Lastly, using consistent visual styles and color schemes throughout the animation helps maintain aesthetic coherence and supports the overall clarity of the data presentation. By applying these techniques, animations can effectively communicate complex information in a clear, engaging, and insightful manner.

> **⚠ Warning**
>
> Animations that are too fast, too slow, or lack control mechanisms (e.g., pause, rewind) may hinder user understanding. Always match animation pacing to the complexity of the data being presented.

11.9.3 Temporal Data Representation

Animation is particularly effective for representing temporal data, as it allows users to visualize changes and developments over time in a dynamic and interactive way. Temporal data encompasses any information that varies with time, such as weather patterns, economic indicators, population growth, or scientific measurements. By animating temporal data, visualizations can illustrate trends, cycles, and sudden shifts that static representations might obscure. For instance, animating the spread of a disease outbreak on a map over several weeks provides a clear depiction of its progression and the effectiveness of containment measures. Similarly, animating economic indicators like GDP growth or unemployment rates over decades can highlight periods of prosperity and recession, offering valuable context for analysis. Time-based animations also facilitate the comparison of different time periods, enabling users to observe how specific variables interact and influence each other over time. This dynamic approach not only enhances the interpretability of temporal data but also makes the exploration of long-term trends more intuitive and engaging for the user.

11.9.4   Interactive vs. Predefined Animations

When incorporating animation into data visualizations, it is essential to distinguish between interactive and predefined animation approaches, as each serves different purposes and user experiences.

Interactive Animations allow users to control the animation flow, offering the ability to play, pause, rewind, or adjust the speed of the animation. This interactivity empowers users to explore the data at their own pace, focusing on specific segments or skipping over less relevant parts. Interactive animations are particularly useful in exploratory data analysis, where users may need to investigate various aspects of the data dynamically. For example, a user might pause an animation to examine a specific data point in detail or rewind to review earlier stages of data progression.

Predefined Animations, on the other hand, follow a set sequence of changes without user intervention. These animations are designed to guide the viewer through a curated narrative, highlighting key insights and transitions in a controlled manner. Predefined animations are ideal for presentations, storytelling, or educational purposes, where the goal is to convey a specific message or illustrate a particular concept clearly and effectively. By limiting user control, predefined animations ensure that the narrative remains focused and that essential information is communicated without distraction.

> **Design Tip: Use Smooth Transitions for Zooming and Panning**
>
> Incorporate animated transitions when zooming or panning to help users maintain context. Abrupt jumps can disorient users and hinder their understanding of spatial relationships in the data.

Both interactive and predefined animations offer unique benefits and can be employed strategically depending on the intended use case and audience. Combining both approaches within a single visualization can also provide a balanced experience, allowing users to engage with the data actively while still benefiting from guided insights.

11.9.5   Best Practices and Design Considerations for Implementing Animations

Animations are a powerful tool in data visualization, enabling users to perceive temporal changes, transitions between data states, and unfolding narratives in an intuitive manner. However, their effectiveness depends heavily on thoughtful design, timing, clarity, and relevance to the data being represented.

Figure 11.14 presents a professional infographic that distills the essential best practices for designing effective animations in scientific and analytical visualizations. This figure uses a flat visual style with minimal text and icon-based guidance, drawing attention to motion-based storytelling strategies that enhance both understanding and user engagement.

A foundational principle is the use of **smooth and coherent transitions**. Abrupt or erratic changes can disorient the viewer, whereas fluid animations guide the eye naturally from one state to the next.**Pacing** is equally critical–animations should progress at a speed that allows for comprehension but avoids sluggishness.

Highlighting change is another best practice; animation should direct attention to evolving values, structures, or trends. Using motion selectively to emphasize key elements prevents visual overload. Designers are encouraged to use **easing functions** to simulate natural movement and adjust motion dynamics (e.g., accelerate-decelerate) to align with cognitive flow.

Another important aspect is **user control**. Providing options to pause, replay, or scrub through animations enhances accessibility and supports self-paced exploration, especially in analytical dashboards or educational tools.

Minimalism in movement is also advised. Animations should serve a clear purpose, not distract from the data. Avoid excessive visual effects unless they carry semantic meaning.

From an accessibility perspective, ensure that animations are**non-disruptive** and offer alternatives for users sensitive to motion. Including **tooltips, legends, or annotations** during animation can further support interpretation.

As visualized in Figure 11.14, a well-designed animation strategy enhances user experience, encourages data exploration, and brings dynamic phenomena into focus without compromising clarity or usability.

11.9.6 Example Demonstrating Temporal Trends

Animation has been effectively utilized in various case studies to highlight temporal trends and enhance data understanding across different domains.

In **climate science**, animations have been used to illustrate changes in global temperature, ice cover, and sea levels over decades. By animating satellite imagery and climate data, scientists have been able to demonstrate the long-term effects of global warming, making the data more accessible and compelling to both the scientific community and the general public. These visualizations have played a crucial role in raising awareness and driving action towards mitigating climate change.

These case studies demonstrate the versatility and effectiveness of animation in highlighting temporal trends, providing deeper insights, and enhancing the overall comprehension of time-based data across diverse fields.

```python
#Creating an Animated Time-Series Plot with Plotly
import plotly.graph_objects as go
import pandas as pd
import numpy as np

# Generate sample time-series data
np.random.seed(0)
dates = pd.date_range(start='2023-01-01', periods=100, freq='D')
```

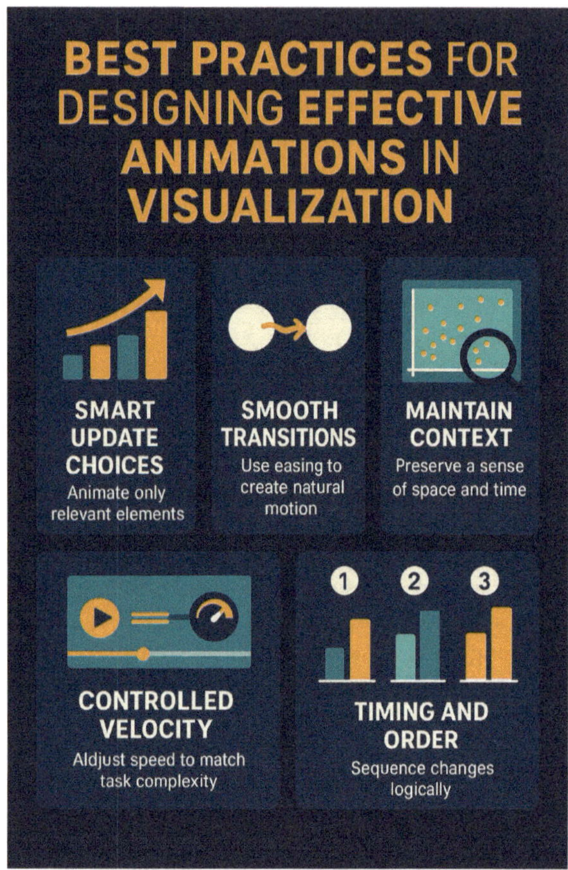

Fig. 11.14: Best Practices for Designing Effective Animations in Visualization. The infographic highlights motion clarity, pacing, focus guidance, user control, minimalism, and accessibility as core principles for integrating animations into data visualizations.

```python
 9  values = np.linspace(100, 300, 100) + np.random.normal(0, 10,
    ↪    100).cumsum()

10
11  df = pd.DataFrame({
12      'Date': dates,
13      'Value': values,
14      'Day': [f'Day {i+1}' for i in range(len(dates))]
15  })

16
17  # Create initial empty figure and define layout
18  fig = go.Figure(
19      layout=go.Layout(
20          title=dict(
```

```python
            text='<b> Progressive Trend in Daily Metrics (2023)</b>',
            x=0.5,
            font=dict(size=22, family='Arial Bold')
        ),
        xaxis=dict(title='<b>Date</b>', tickangle=-45,
        ↪   tickfont=dict(family='Arial Bold'),
                    showgrid=True, gridcolor='lightgray'),
        yaxis=dict(title='<b>Value</b>',
        ↪   tickfont=dict(family='Arial Bold'),
                    showgrid=True, gridcolor='lightgray'),
        plot_bgcolor='white',
        hovermode='x unified',
        updatemenus=[{
            "buttons": [
                {"label": "➡ Play", "method": "animate", "args": [None, {
                    "frame": {"duration": 100, "redraw": True},
                    ↪   "fromcurrent": True}]},
                {"label": "[pause] Pause", "method": "animate", "args":
                ↪   [[None], {
                    "frame": {"duration": 0, "redraw": False}, "mode":
                    ↪   "immediate"}]}
            ],
            "type": "buttons",
            "x": 0.1, "y": 0, "direction": "left", "pad": {"r": 10, "t":
            ↪   87}
        }]
    )
)

# Add initial trace (empty)
fig.add_trace(go.Scatter(
    x=[], y=[], mode='lines+markers+text',
    line=dict(color='royalblue', width=3),
    marker=dict(size=8, color='lightblue', line=dict(width=2,
    ↪   color='royalblue')),
    text=[], textposition='top center',

    ↪   hovertemplate='<b>Date:</b> %{x}<br><b>Value:</b> %{y:.2f}<extra></extra>'
))

# Create frames for each day
frames = []
for i in range(1, len(df) + 1):
    partial_df = df.iloc[:i]
    frame = go.Frame(
        data=[
            go.Scatter(
                x=partial_df['Date'],
                y=partial_df['Value'],
                mode='lines+markers+text',
                line=dict(color='royalblue', width=3),
                marker=dict(size=8, color='lightblue', line=dict(width=2,
                ↪   color='royalblue')),
                text=[f'{v:.1f}' for v in partial_df['Value']],
```

```
66              textposition='top center'
67          )
68      ],
69      name=f'Day {i}'
70  )
71  frames.append(frame)
72
73 fig.frames = frames
74
75 fig.update_layout(
76     xaxis_range=[df['Date'].min(), df['Date'].max()],
77     yaxis_range=[df['Value'].min() - 20, df['Value'].max() + 20]
78 )
79
80 fig.show()
```

In Figure 11.15, an animated line plot illustrates the progression of climate data over time. The animation allows viewers to observe gradual changes and significant shifts in temperature patterns, providing a clear and engaging representation of temporal trends. The accompanying Python code using Plotly demonstrates how to create such an animated visualization, enabling users to track and analyze data dynamics effectively.

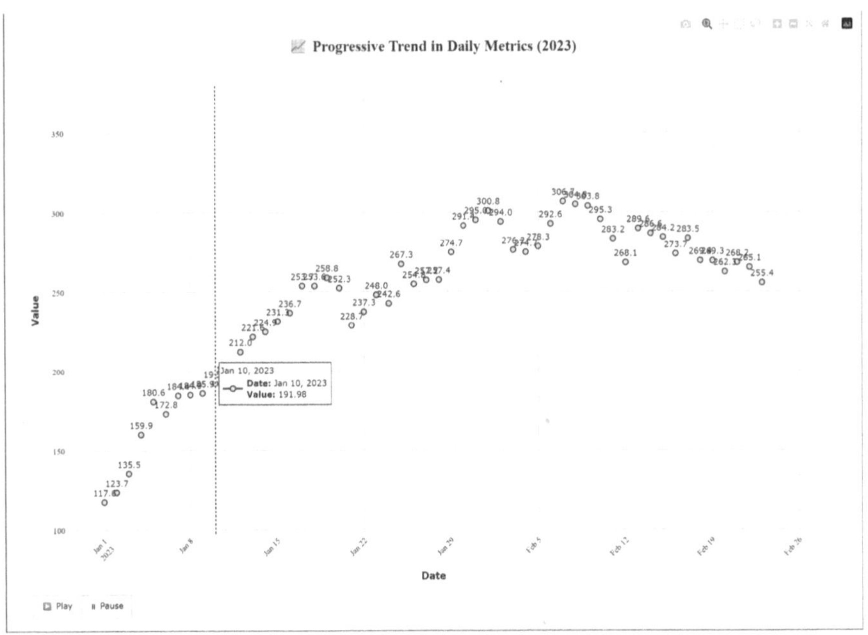

Fig. 11.15: Animated Visualization Highlighting Temporal Trends

11.10 Small Multiples

11.10.1 Concept of Small Multiples

Small Multiples is a powerful visualization technique that involves displaying multiple similar charts or graphs side by side, each representing different subsets of the data. This approach allows for easy comparison across various categories, groups, or time periods without the need for switching between different views. By maintaining consistent scales, axes, and visual encodings across all small multiples, viewers can quickly identify patterns, trends, and discrepancies across the different subsets. This technique enhances the user's ability to perform comparative analysis, making it easier to draw meaningful conclusions from complex and multifaceted datasets. Small Multiples effectively reduce cognitive load by presenting information in a structured and uniform manner, facilitating a clearer and more comprehensive understanding of the data.

11.10.2 Design Principles for Small Multiples

Designing effective Small Multiples requires adherence to several key principles to ensure clarity, consistency, and comparability across all visualizations.

Consistency: Maintain uniform scales, axes, colors, and symbols across all small multiples to ensure that comparisons are accurate and meaningful. Inconsistent design elements can lead to misinterpretation and confusion.

Alignment: Arrange the multiple charts in a grid or aligned layout to facilitate easy scanning and comparison. Proper alignment helps users quickly locate corresponding elements across different charts.

Simplicity: Keep each individual chart simple and uncluttered, focusing on the key variables and relationships. Overcomplicating individual charts can obscure important insights and hinder comparative analysis.

Contextual Information: Provide sufficient context, such as titles, labels, and legends, to ensure that each small multiple is easily interpretable on its own while still contributing to the overall comparative analysis.

Highlighting Differences: Use visual cues like color variations, shading, or annotations to emphasize significant differences or patterns across the small multiples. This helps draw attention to key insights and facilitates quicker identification of trends.

> **Design Tip: Maintain Consistent Scales in Small Multiples**
>
> Ensure all small multiples use identical axes and scales when comparing data subsets. Inconsistent scaling may visually distort differences and lead to incorrect conclusions.

Figure 11.16 illustrates the essential visual design strategies for constructing effective and visually coherent small multiples in scientific visualizations.

By following these design principles, Small Multiples can effectively convey complex information in a clear and comparative manner, enhancing the overall effectiveness of data visualizations.

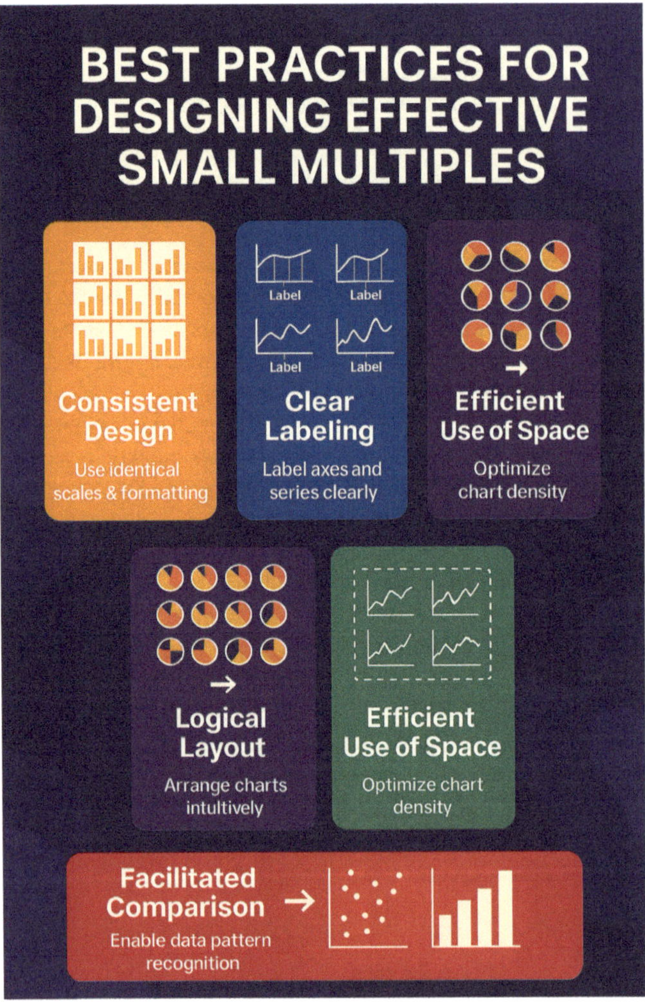

Fig. 11.16: Best Practices for Designing Effective Small Multiples such as consistent scales, aligned axes, minimal chart clutter, grid-based layout, unified color encoding, and contextual labeling to enhance comparative analysis across multiple views.

11.10.3 Facilitating Comparative Analysis

Small Multiples significantly enhance comparative analysis by allowing users to view multiple aspects of the data simultaneously. This side-by-side comparison makes it easier to identify similarities and differences across various subsets, categories, or time periods. For example, displaying monthly sales figures for different regions in separate bar charts arranged in a grid enables sales managers to quickly compare performance across regions and identify areas that require attention. Similarly, in scientific research, Small Multiples can be used to compare experimental results under different conditions, facilitating the identification of patterns and anomalies that inform hypothesis testing and theory development. By presenting multiple perspectives in a cohesive and organized layout, Small Multiples empower users to conduct comprehensive and efficient comparative analyses, leading to more informed and data-driven decisions.

11.10.4 Implementation Techniques

Implementing Small Multiples effectively involves several strategies to ensure that the multiple charts are cohesive and facilitate easy comparison.

Consistent Layout: Use a uniform layout for all small multiples, ensuring that each chart has the same dimensions, scales, and axis orientations. This consistency allows users to effortlessly compare the data across different charts.

Faceted Grids: Arrange the small multiples in a grid or matrix format, where each row or column represents a specific category or group. Faceted grids help in organizing the charts systematically and enhance the visual appeal.

Responsive Design: Ensure that the small multiples are responsive to different screen sizes and devices, maintaining readability and usability across various platforms. This is particularly important for web-based visualizations where users may access the data from different devices.

Interactive Elements: Incorporate interactive features such as tooltips, zooming, and filtering to allow users to engage with each small multiple individually. Interactivity enhances the user experience by providing additional layers of information without overwhelming the primary comparative analysis.

Color Coding: Utilize consistent color schemes across all small multiples to represent the same variables or categories. This uniformity in color coding aids in quickly associating related data points across different charts.

Below is a Python code snippet using Seaborn and Matplotlib to create Small Multiples in the form of facet grids.

```python
#Creating Small Multiples \index{Small Multiples} with Seaborn
import seaborn as sns
import matplotlib.pyplot as plt

# Load example dataset
iris = sns.load_dataset('iris')
```

```
 7
 8   # Create a FacetGrid of scatter plots by species
 9   g = sns.FacetGrid(iris, col="species", height=4, sharex=True, sharey=True)
10   g.map(sns.scatterplot, "sepal_length", "sepal_width", color='teal')
11   g.add_legend()
12
13   plt.subplots_adjust(top=0.8)
14   g.fig.suptitle('Sepal Length vs. Sepal Width by Species')
15   plt.show()
```

In Figure 11.17, a FacetGrid is used to display scatter plots of sepal length versus sepal width for each species of iris. This arrangement allows for easy comparison of the relationships between these two variables across different species, highlighting the distinct patterns and distributions inherent to each group.

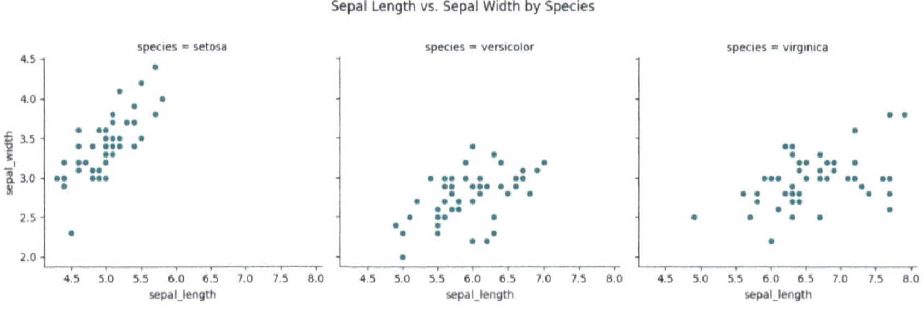

Fig. 11.17: Small Multiples: Facet Grid of Scatter Plots by Category

11.10.5   Examples in Scientific Visualization

Small Multiples are extensively used in scientific visualization to facilitate the comparison of complex datasets across different dimensions or conditions.

In **climate science**, researchers use Small Multiples to compare temperature variations across different regions and time periods. By displaying multiple line charts side by side, scientists can easily identify regional climate trends and anomalies, enabling more accurate climate modeling and forecasting.

In **genomics**, Small Multiples are employed to visualize gene expression levels across various conditions or treatments. By arranging heatmaps or bar charts for each gene or condition in a grid, researchers can quickly compare expression patterns and identify genes that are significantly upregulated or downregulated in response to specific stimuli.

In **astronomy**, Small Multiples are used to display images of celestial objects captured at different wavelengths or times. This allows astronomers to compare the structural and compositional differences of stars, galaxies, and other astronomical entities, aiding in the study of their formation and evolution.

In **ecology**, Small Multiples help visualize species distribution across different habitats or geographic locations. By presenting multiple maps or scatter plots, ecologists can assess biodiversity patterns, habitat preferences, and the impact of environmental changes on various species.

An illustrative example is the use of Small Multiples in `Plotly` to compare multiple statistical plots within a single visualization framework. Below is a Python code snippet demonstrating how to create Small Multiples using Plotly's subplot feature.

```python
#Creating Small Multiples \index{Small Multiples} with Plotly
import plotly.express as px
from plotly.subplots import make_subplots
import plotly.graph_objects as go

# Sample data
tips = px.data.tips()

# Create subplots
fig = make_subplots(rows=1, cols=3, subplot_titles=['Total Bill', 'Tip',
    'Size'])

# Add histograms for each variable
fig.add_trace(go.Histogram(x=tips['total_bill'], name='Total Bill'), row=1,
    col=1)
fig.add_trace(go.Histogram(x=tips['tip'], name='Tip'), row=1, col=2)
fig.add_trace(go.Histogram(x=tips['size'], name='Size'), row=1, col=3)

# Update layout
fig.update_layout(title_text='Small Multiples \index{Small Multiples}
    of Tip Dataset Variables', showlegend=False)
fig.show()
```

In this example, three histograms representing different variables from the tips dataset are displayed side by side using Plotly's subplots. This arrangement allows for straightforward comparison of the distribution of total bills, tips, and party sizes, facilitating a comprehensive analysis of the dataset's characteristics.

These examples underscore the versatility and effectiveness of Small Multiples in scientific visualization, enabling researchers to conduct detailed comparative analyses and derive meaningful insights from multifaceted data.

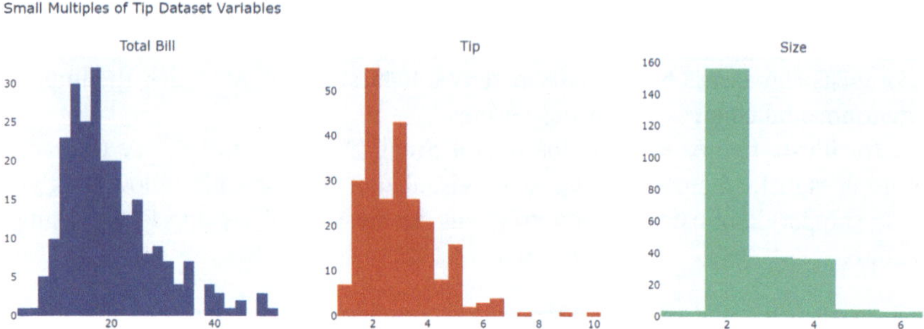

Fig. 11.18: Small Multiples: Histograms of Different Variables in the Tips Dataset

11.11   Focus + Context (Fisheye Views)

11.11.1   Understanding Focus + Context

Focus + Context is a visualization paradigm that aims to provide users with both detailed and overview information within a single display. By employing techniques such as fisheye views, this approach allows a specific area of interest (the focus) to be magnified while simultaneously displaying the surrounding data in a less detailed but still visible manner (the context). This balance enables users to examine fine-grained details without losing sight of the overall data structure, facilitating a more comprehensive and intuitive data exploration experience. The focus + context technique is particularly useful in scenarios where datasets are large or multidimensional, as it helps prevent information overload by guiding the user's attention to relevant sections while maintaining awareness of the broader dataset.

11.11.2   Fisheye View Techniques

Fisheye view techniques distort the visualization in a way that accentuates the focused area while compressing the surrounding regions. This distortion mimics the effect of a fisheye lens in photography, where the center of the image is magnified, and the periphery is reduced. Implementing fisheye views involves applying non-linear transformations to the data coordinates, ensuring that the area of interest remains prominent and easily accessible. Common techniques include curvature-based distortion, where the distance from the focus point increases logarithmically, and tapering, where the edges of the visualization gradually compress. These methods maintain the relative positions of data points, allowing users to navigate through the dataset seamlessly. Below is a Python code snippet using Matplotlib to create a basic fisheye effect by magnifying a specific region of a scatter plot.

11.11.3 Balancing Detail and Overview

Maintaining an effective balance between detail and overview is essential in focus + context visualizations to ensure that users can explore specific data points without losing sight of the larger data landscape. Achieving this balance involves carefully adjusting the degree of magnification and the extent of context preservation. Over-magnifying the focus area can lead to distortion and misrepresentation of data, while insufficient magnification may fail to provide the necessary detail for thorough analysis. The context should be rendered in a way that remains legible and informative, allowing users to correlate the focused data with the broader dataset seamlessly. Techniques such as adaptive scaling, where the magnification level adjusts based on user interaction, and dynamic context resizing, where the context area changes proportionally with the focus area, help maintain this balance. Effective balancing ensures that the visualization remains both detailed and comprehensible, enhancing the user's ability to derive meaningful insights.

> **Design Tip: Avoid Overuse of Fisheye Distortion**
>
> Apply fisheye effects sparingly and clearly indicate areas of distortion. Excessive warping can obscure relationships and reduce interpretability unless well contextualized.

11.11.4 Implementation Strategies

Implementing focus + context visualizations requires a combination of data transformation techniques and thoughtful UI design to create an intuitive and responsive user experience. One common strategy is to use layering, where the focused area is rendered on top of the context layer with higher detail and distinct visual properties. Another approach involves real-time data manipulation, where the dataset is dynamically altered based on the user's focus, ensuring that the magnified region receives the necessary attention. Interactive controls, such as sliders or draggable focus points, allow users to adjust the focus area and explore different sections of the data effortlessly. Incorporating smooth transition animations between different focus states can enhance the visual appeal and usability of the visualization.

As illustrated in Figure 11.19, effective fisheye views depend on a delicate balance between focus magnification and contextual retention, supported by responsive interaction design.

Below is a Python code snippet using Plotly to create an interactive focus + context scatter plot with adjustable focus regions.

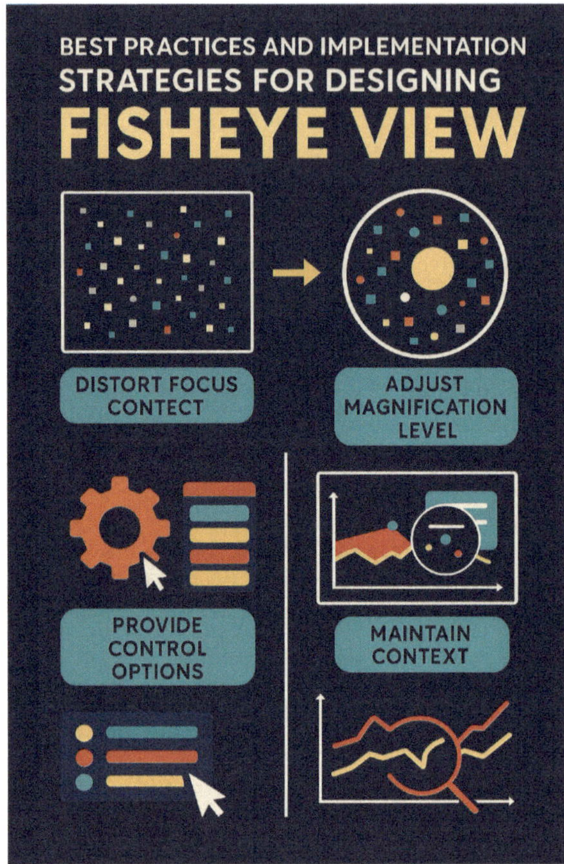

Fig. 11.19: Best Practices and Implementation Strategies for Designing Fisheye View. This infographic uses minimal text and strong visual cues to emphasize principles such as dynamic focus scaling, distortion awareness, context preservation, interaction feedback, and user control for implementing effective fisheye views in data visualizations.

```python
import plotly.graph_objects as go
import numpy as np

# Generate sample data
np.random.seed(0)
x = np.random.normal(50, 15, 1000)
y = np.random.normal(50, 15, 1000)

# Define focus parameters
focus_x, focus_y = 50, 50
focus_radius = 10
```

```python
12   intensity = 1.8  # distortion intensity
13
14   # Calculate distances from focus center
15   dx, dy = x - focus_x, y - focus_y
16   distances = np.sqrt(dx**2 + dy**2)
17
18   # Apply fisheye transformation (Gaussian-like scaling)
19   scale = np.where(
20       distances < focus_radius,
21       1 + intensity * (1 - (distances / focus_radius) ** 2),
22       1
23   )
24   x_fisheye = focus_x + dx * scale
25   y_fisheye = focus_y + dy * scale
26
27   # Create Plotly \index{Plotly} figure
28   fig = go.Figure()
29
30   # Layer 1: Context (original distribution)
31   fig.add_trace(go.Scatter(
32       x=x, y=y,
33       mode='markers',
34       marker=dict(color='lightgray', size=4, opacity=0.2),
35       name='Context',
36       hoverinfo='skip'
37   ))
38
39   # Layer 2: Fisheye magnified points
40   fig.add_trace(go.Scatter(
41       x=x_fisheye, y=y_fisheye,
42       mode='markers',
43       marker=dict(color='orangered', size=6, line=dict(color='black',
          ↪  width=0.5), opacity=0.75),
44       name='Fisheye Focus',
45       hovertemplate='Magnified X: %{x:.2f}<br>Magnified Y: %{y:.2f}'
46   ))
47
48   # Add a dashed circle to indicate focus area
49   theta = np.linspace(0, 2 * np.pi, 100)
50   circle_x = focus_x + focus_radius * np.cos(theta)
51   circle_y = focus_y + focus_radius * np.sin(theta)
52   fig.add_trace(go.Scatter(
53       x=circle_x, y=circle_y,
54       mode='lines',
55       line=dict(color='royalblue', dash='dash'),
56       name='Focus Area',
57       hoverinfo='skip'
58   ))
59
60   # Add annotation
61   fig.add_annotation(
62       x=focus_x, y=focus_y,
63       text='Fisheye Focus Center',
64       showarrow=True,
```

```
65      arrowhead=2,
66      ax=30, ay=-40,
67      font=dict(size=12, color='black'),
68      bgcolor='lightyellow',
69      bordercolor='gray'
70  )
71
72  # Layout formatting
73  fig.update_layout(
74      title=dict(text='Focus + Context Scatter Plot with Fisheye Distortion',
          ↪   font=dict(size=18, family='Arial', color='black')),
75      xaxis=dict(title='X Axis', showgrid=True, zeroline=False),
76      yaxis=dict(title='Y Axis', showgrid=True, zeroline=False),
77      legend=dict(x=0.8, y=1, bgcolor='rgba(255,255,255,0.7)',
          ↪   bordercolor='lightgray'),
78      plot_bgcolor='white',
79      margin=dict(l=40, r=40, t=60, b=40)
80  )
81
82  fig.show()
```

In Figure 11.20, the interactive scatter plot allows users to visualize the focus + context paradigm by distinguishing between the focused data points (in red) and the contextual overview (in blue). Users can modify the focus parameters to explore different regions of the dataset, facilitating a deeper and more flexible data analysis experience.

11.12   Summary and Concluding Thoughts

Advanced visualization techniques play a pivotal role in modern scientific data analysis by transforming static representations into dynamic, interactive, and insightful visual narratives. This chapter has systematically explored a broad range of such techniques–including Zoom-Out, Panning, Brushing, Drill-Down, Filtering, Linking, Annotation, Animation, Small Multiples, Focus + Context (Fisheye Views), and Parallel Coordinates–each offering unique affordances for enhancing data interpretability, user engagement, and analytical depth. These techniques empower researchers to navigate complex and high-dimensional datasets, uncover hidden relationships, detect outliers , and communicate findings more effectively. From providing macro-level context through zooming and panning to facilitating multivariate comparisons via brushing and linking, and enabling temporal or hierarchical exploration through animation and drill-down, each method contributes to a richer, more nuanced understanding of scientific data.

Equally important are the implementation strategies and tool choices, as presented in Tables 11.1, 11.3, and 11.2, which offer practical guidance for choosing appropriate techniques and libraries depending on the dataset type, interaction requirements, and user goals. The use of illustrative Python code throughout the chapter reinforces these strategies by providing actionable templates for integrating advanced visualization into

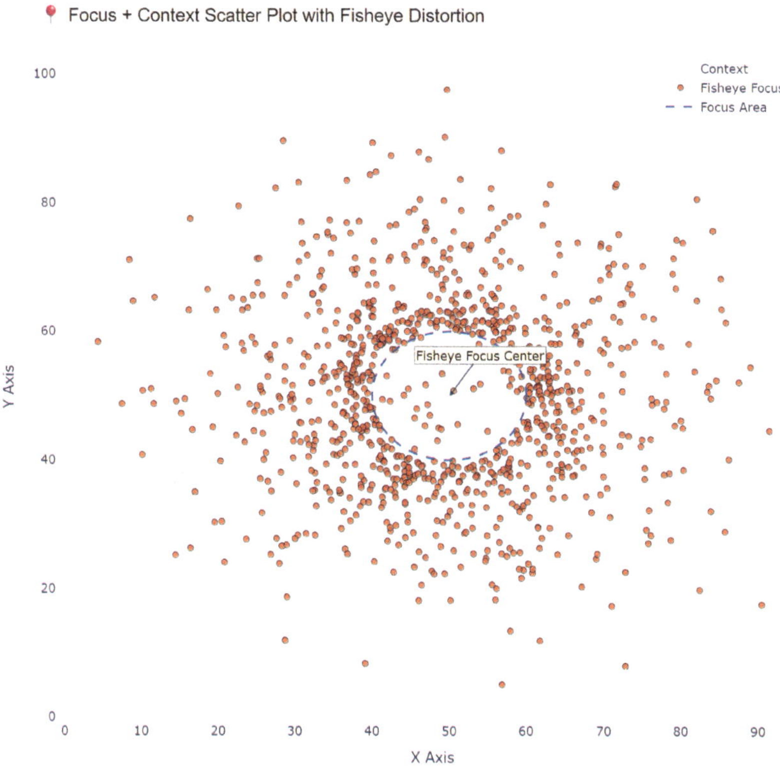

Fig. 11.20: Interactive Focus + Context Scatter Plot with Plotly

scientific workflows. Best practices, such as maintaining visual consistency, ensuring interactivity and accessibility, and optimizing performance, were emphasized as key to designing effective visualizations.

As scientific data continues to grow in volume and complexity, the demand for visualization methods that are both powerful and user-friendly will only increase. By mastering the advanced techniques discussed in this chapter, researchers can not only improve the clarity and impact of their own analyses but also contribute to more transparent, collaborative, and data-driven scientific discourse. Looking forward, the integration of these visualization strategies with emerging technologies–such as explainable AI, real-time dashboards, and immersive environments–offers exciting possibilities for the future of scientific exploration. Ultimately, thoughtful and well-designed visualizations are not just tools for presenting data–they are indispensable instruments for discovery.

Suggested Readings

1. Borhani, Z., Sharma, P., Ortega, F.: Survey of annotations in extended reality systems. IEEE Transactions on Visualization and Computer Graphics **30**(8), 5074–5096 (2024). DOI 10.1109/TVCG.2023.3288869
2. Doerr, N., Lee, B., Baricova, K., Schmalstieg, D., Sedlmair, M.: Visual highlighting for situated brushing and linking. Computer Graphics Forum **43**(3) (2024). DOI 10.1111/cgf.15105
3. Hosseinpour, H., Matzen, L., Divis, K., Castro, S., Padilla, L.: Examining limits of small multiples: Frame quantity impacts judgments with line graphs. IEEE Transactions on Visualization and Computer Graphics **31**(3), 1875–1887 (2025). DOI 10.1109/TVCG.2024.3372620
4. Lan, X., Shi, Y., Wu, Y., Jiao, X., Cao, N.: Kineticharts: Augmenting affective expressiveness of charts in data stories with animation design. IEEE Transactions on Visualization and Computer Graphics **28**(1), 933–943 (2022). DOI 10.1109/TVCG.2021.3114775
5. Lekschas, F., Zhou, X., Chen, W., Gehlenborg, N., Bach, B., Pfister, H.: A generic framework and library for exploration of small multiples through interactive piling. IEEE Transactions on Visualization and Computer Graphics **27**(2), 358–368 (2021). DOI 10.1109/TVCG.2020.3028948
6. Lu, M., Fish, N., Wang, S., Lanir, J., Cohen-Or, D., Huang, H.: Enhancing static charts with data-driven animations. IEEE Transactions on Visualization and Computer Graphics **28**(7), 2628–2640 (2022). DOI 10.1109/TVCG.2020.3037300
7. Shen, Y., Zhao, Y., Wang, Y., Ge, T., Shi, H., Lee, B.: Authoring data-driven chart animations through direct manipulation. IEEE Transactions on Visualization and Computer Graphics **31**(2), 1613–1630 (2025). DOI 10.1109/TVCG.2024.3491504
8. Stokes, C., Bearfield, C., Hearst, M.: The role of text in visualizations: How annotations shape perceptions of bias and influence predictions. IEEE Transactions on Visualization and Computer Graphics **30**(10), 6787–6800 (2024). DOI 10.1109/TVCG.2023.3338451

Appendix A
Visualization Tools and Libraries

In this Appendix, we provide an overview of the most relevant tools and libraries for implementing the visualization techniques discussed throughout this book. While the examples in the main text use Python predominantly, we also highlight popular business-intelligence platforms that facilitate rapid dashboard development. Each library is presented with its core strengths, typical use cases, and pointers to further documentation.

A.1 Python Libraries

A.1.1 Matplotlib

Matplotlib is the foundational plotting library for Python, offering a comprehensive API for creating static, publication-quality figures. It supports a wide range of chart types—from simple line plots and scatter plots to more complex visualizations such as histograms, heatmaps, and 3D surfaces. Its highly customizable nature makes it ideal for fine-tuning every aspect of a figure, and it serves as the basis for higher-level libraries like Seaborn.

A.1.2 Seaborn

Built on top of Matplotlib, Seaborn simplifies the creation of statistically oriented visualizations by providing sensible defaults and high-level functions. It excels at plots such as box plots, violin plots, pair plots, and facet grids, enabling rapid exploratory data analysis with minimal code. Seaborn integrates seamlessly with Pandas DataFrames and is well suited for small multiples and multivariate exploratory workflows.

© The Editor(s) (if applicable) and The Author(s), under exclusive license
to Springer Nature Switzerland AG 2026
R. Damaševičius, *Human-Centred Scientific Data Visualisation*,
Undergraduate Topics in Computer Science,
https://doi.org/10.1007/978-3-032-01606-5

A.1.3 Bokeh

Bokeh is a powerful library for creating interactive, web-ready plots and dashboards directly from Python. It provides widgets (sliders, dropdowns, selectors) and JavaScript callbacks for building linked visualizations, brushing, and real-time updates. Bokeh's server architecture allows Python callbacks to drive UI interactions, making it a strong choice for deploying interactive scientific dashboards.

A.1.4 Plotly

Plotly Express and Plotly Graph Objects offer declarative and object-oriented interfaces, respectively, for building both static and interactive visualizations. Plotly supports a vast array of chart types—including parallel coordinates, treemaps, chord diagrams, and 3D plots—with built-in animation and export to HTML. Its interactivity (hover, zoom, pan) works out of the box, facilitating rapid prototyping and sharing via notebook or web embedding.

A.1.5 NetworkX

NetworkX specializes in the creation, manipulation, and analysis of complex networks. While its native drawing capabilities are basic, integration with Matplotlib and Plotly enables the rendering of network graphs with customizable layouts (force-directed, circular, hierarchical) and interactive features. NetworkX's rich graph algorithms make it indispensable for social network analysis and biological pathway visualization.

A.1.6 Squarify

Squarify implements the squarified treemap algorithm in Python, allowing users to generate area-proportional treemaps with minimal dependencies. When combined with Matplotlib for rendering, Squarify provides a lightweight solution for hierarchical data visualization where space efficiency is paramount.

A.2 JavaScript Frameworks

A.2.1 D3.js

D3.js (Data-Driven Documents) is the de facto standard for web-based, highly customized visualizations. It binds arbitrary data to DOM elements and applies data-driven transformations, enabling developers to implement virtually any visualization technique—from basic charts to fisheye distortion and custom glyphs. Although it has a steep learning curve, D3's flexibility and performance make it ideal for bespoke interactive visualizations.

A.2.2 Vega and Vega-Lite

Vega and Vega-Lite provide a higher-level, JSON-based grammar of graphics for declarative visualization. Vega-Lite simplifies chart creation with concise specifications, while Vega offers full flexibility for custom interactions and layouts. Both compile to D3 under the hood and support advanced features like faceting, interactive selection, and streaming data.

A.2.3 Chart.js and ECharts

Chart.js is a simple yet extensible library for common chart types, optimized for small- to medium-scale visualizations with built-in animations. Apache ECharts offers a rich set of interactive and animated charts—such as streamgraphs, heatmaps, and radial bar charts—with responsive design and performance optimizations for large datasets.

A.3 Business Intelligence Platforms

A.3.1 Tableau

Tableau is a leading commercial platform for drag-and-drop dashboard creation and exploratory analytics. It supports a broad range of data sources, interactive filtering, parameter controls, and storytelling features. Tableau's extension API allows embedding custom D3 or JavaScript visualizations within dashboards, bridging low-code and custom-code workflows.

A.3.2 Power BI

Microsoft Power BI offers integrated data preparation, modeling, and visualization capabilities. With its custom visual SDK, developers can build bespoke visuals using TypeScript and D3, which can then be shared across an organization. Power BI's tight integration with Azure and Excel makes it a popular choice for enterprise reporting.

A.4 Emerging and Specialized Tools

A.4.1 Holoviz (Holoviews, Datashader)

Holoviz provides high-level APIs for building complex visualizations. HoloViews automates the creation of interactive plots, while Datashader renders billions of points by rasterizing data on the server side. Together, they enable visualization of massive datasets with smooth interactivity.

A.4.2 Three.js and WebGL

For true 3D and immersive experiences, Three.js provides a comprehensive API for WebGL rendering. It is used to build 3D scatter plots, volumetric renderings, and virtual reality scenes. Combined with frameworks like A-Frame or React 360, Three.js enables exploration of data in VR and AR contexts.

Further Reading

Readers are encouraged to consult the following resources for in-depth guidance and tutorials on these tools:

- Hunter, J. D. (2007). *Matplotlib: A 2D graphics environment.* Computing in Science & Engineering.
- Waskom, M. (2021). *Seaborn: Statistical data visualization.* Journal of Open Source Software.
- Bostock, M., Ogievetsky, V., & Heer, J. (2011). *D^3 Data-Driven Documents.* IEEE Transactions on Visualization and Computer Graphics.
- Plotly Technologies Inc. (2015). *Introduction to Plotly.js.*

Appendix B
Accessibility and Inclusive Design

Ensuring that scientific visualizations are accessible and inclusive is essential for reaching the broadest possible audience. This appendix outlines key principles, guidelines, and best practices for designing visualizations that accommodate users with diverse abilities, including those with visual, motor, and cognitive impairments.

B.1 Principles of Inclusive Visualization

- **Perceptual Equity**: Design charts so that information is conveyed through multiple channels (color, shape, position, texture) rather than color alone.
- **Operable Interfaces**: Ensure all interactive features (zoom, pan, filtering) are reachable via keyboard and assistive technologies.
- **Understandable Content**: Use clear labels, concise text, and consistent layouts to reduce cognitive load.
- **Robust Compatibility**: Build visualizations that work across browsers, devices, and screen readers, adhering to standards like WCAG 2.1.

B.2 Color and Contrast

B.2.1 High Contrast Ratios

Use a contrast ratio of at least 4.5:1 for text and graphical elements against their background. Tools like the WebAIM Contrast Checker can verify compliance.

© The Editor(s) (if applicable) and The Author(s), under exclusive license
to Springer Nature Switzerland AG 2026
R. Damaševičius, *Human-Centred Scientific Data Visualisation*,
Undergraduate Topics in Computer Science,
https://doi.org/10.1007/978-3-032-01606-5

B.2.2 Color-Blind–Friendly Palettes

- Avoid red–green pairings; opt for palettes such as #0072B2, #D55E00, #F0E442, #CC79A7.
- Use tools like ColorBrewer's ColorBlindSafe palettes for qualitative, sequential, and diverging data.

B.2.3 Redundant Encoding

Pair color with shape, line style, or texture to encode categories. For example:

- Lines: solid vs. dashed
- Markers: circles vs. squares
- Fills: crosshatch vs. dotted

B.3 Keyboard Navigation and Focus Management

B.3.1 Focusable Elements

Ensure all controls (buttons, sliders, dropdowns) have a `tabindex="0"` and visible focus indicators.

B.3.2 Logical Tab Order

Arrange interactive elements in DOM order that matches visual layout. Use landmark roles (e.g., `role="region"`) to group related controls.

B.3.3 ARIA Attributes

- `aria-label`: Provide descriptive labels for controls without visible text.
- `aria-live="polite"`: Announce dynamic updates (e.g., "Filter applied, 32 items displayed").
- `aria-describedby`: Link charts to explanatory text or data summaries.

B.4 Screen Reader Support

B.4.1 Text Alternatives

- Provide `<title>` and `<desc>` elements within SVG for chart titles and descriptions.
- Include data summaries in HTML before or after the visualization (e.g., "Line chart: monthly sales increased from $10k to $25k between January and June").

B.4.2 Data Tables

Offer a hidden or collapsible table representation of chart data to allow screen reader users to navigate individual values.

B.4.3 Annotation Accessibility

Ensure annotations (callouts, tooltips) are reachable via keyboard and read by screen readers using `role="tooltip"` and appropriate ARIA references.

B.5 Accessible Charts and Graphs

B.5.1 Simplified Visuals

- Minimize non-essential decoration (gridlines, background gradients) that may clutter the interface.
- Use clear axis labels and avoid overly small font sizes (minimum 12 px).

B.5.2 Interactive Alternatives

Provide HTML-based controls (e.g., form inputs) as progressive enhancements for users unable to interact with the graphical canvas.

B.5.3 Responsive Design

Ensure charts render legibly on small screens. Use touch-friendly control sizes (minimum 44 × 44 px) and avoid relying solely on hover interactions.

B.6 Testing and Validation

B.6.1 Automated Tools

- **axe-core**, **Pa11y**: Integrate accessibility checks into your build pipeline.
- **Lighthouse**: Use Chrome's auditing tool to catch common accessibility issues.

B.6.2 Manual Testing

- Use screen readers (NVDA, VoiceOver) to verify text alternatives and dynamic announcements.
- Conduct color-blindness simulations (e.g., Coblis) to confirm palette choices.

By applying these guidelines and techniques, you can create scientific visualizations that are not only informative and engaging but also accessible to the widest audience, ensuring that insights are truly available to everyone.

Glossary

2D Density Plot A graphical representation of data concentration across two dimensions. Instead of plotting individual points, it visualizes regions with high or low density, helping to uncover patterns, clusters, and correlations in complex datasets.

Accessibility The practice of designing visualizations that are usable and interpretable by individuals with visual impairments or color vision deficiencies.

Alpha Blending A graphical technique that combines foreground and background colors based on transparency levels. In data visualization, alpha blending helps visualize overlapping data more clearly.

Animation The use of motion in data visualizations to represent changes over time or transitions between different data states, helping users understand dynamic trends and temporal patterns.

Annotation A visualization technique that adds explanatory text, labels, arrows, or shapes directly onto a chart to highlight key data points, clarify trends, and provide contextual information that assists in interpreting the visualization without referring to external notes.

Axes Reference lines on a chart or graph, typically horizontal (x-axis) and vertical (y-axis), used to define the coordinate system and scale for plotting data points. They help users interpret the magnitude, trends, and relationships of the data being visualized.

Axis Label The descriptive text placed along the x-axis or y-axis to indicate what each axis represents. Axis labels should include units where applicable to ensure clarity.

Axis Scaling The process of determining the range and intervals on a plot's axes. Proper axis scaling is critical to ensure accurate and non-misleading representations of data.

Bandwidth A smoothing parameter in kernel density estimation (KDE) that determines the width of the kernel function. A small bandwidth captures fine details (but may introduce noise), while a large bandwidth smooths the data more aggressively.

Bar Chart A chart type that uses rectangular bars to represent and compare categorical data. The height (or length) of each bar corresponds to the value it represents.

Baseline The starting point of an axis scale, often zero. Manipulating the baseline can affect visual interpretation of differences or trends in the data.

Bin Size The resolution of grid cells in a heatmap or histogram-based density plot. Smaller bin sizes show finer detail but may introduce noise, while larger bin sizes offer general trends but reduce granularity.

Bin A range of values used to group continuous or discrete data in a histogram. Each bin collects the frequency of data points that fall within its interval.

Box Plot Also known as a box-and-whisker plot, this chart summarizes the distribution of a dataset using five-number summary statistics: minimum, first quartile (Q1), median, third quartile (Q3), and maximum.

Brushing An interactive method for selecting and highlighting subsets of data points within a visualization—often linked to other views—enabling focused analysis of multivariate relationships and patterns.

Center Point The origin of the spider plot from which all axes radiate. It typically represents the zero or baseline value.

Chord Diagram A circular visualization that shows relationships between entities as arcs connecting outer segments, often used to represent trade flows or network interactions.

Color Contrast The visual difference between two or more colors. High contrast improves segment distinction and readability in charts, especially for users with visual impairments.

Color Mapping The use of color to represent values in a dataset, either as discrete categories or continuous gradients.

Color Palette A predefined or custom set of colors selected for use in a visualization or design. A color palette ensures visual consistency, enhances aesthetics, and can improve data comprehension and accessibility. It may include sequential, diverging, or categorical color schemes depending on the type of data being represented.

Color Transparency A visual technique where overlapping lines or shaded areas are made semi-transparent to reveal underlying data and prevent visual occlusion.

Color-Blind Friendly Palette A set of colors chosen to ensure readability for viewers with color vision deficiencies.

Confidence Interval A shaded area or range around a line that represents the uncertainty or variability in the data, typically shown as a 95% interval in statistical plots.

Contour Plot A 2D plot that uses contour lines to connect areas of equal density or value, resembling topographic maps. Effective for visualizing gradual changes in data, especially for continuous variables.

Data Drilling An interactive feature in charts that allows users to click on a data segment and reveal more detailed, hierarchical information underneath.

Data Gap An interval in the dataset where values are missing or unavailable. Proper handling of data gaps is critical to prevent misleading interpretations.

Data Storytelling The practice of using narrative techniques in combination with data visualizations to communicate insights, highlight trends, and engage audiences more effectively.

Density Histogram A variant of the histogram where the y-axis represents probability density rather than frequency, and the area under the histogram sums to one.

Direct Labeling The practice of placing labels adjacent to the plotted lines rather than in a separate legend. This improves readability, especially in static or print media.

Donut Chart A variation of the pie chart with a central hole. The donut chart offers additional design flexibility, such as placing summary metrics or labels in the center while still conveying part-to-whole relationships.

Drill-Down An exploratory technique that allows users to move from a high-level overview into increasingly detailed layers of data, supporting hierarchical and focused data analysis.

Exploded Pie Chart A pie chart where one or more slices are separated from the center to emphasize specific values or categories. Commonly used to highlight notable data segments.

Filtering The process of including or excluding data based on specified criteria—such as ranges, categories, or boolean conditions—to reduce clutter and focus on relevant subsets of the dataset.

Fisheye View A paradigm that magnifies an area of interest (focus) within a visualization while displaying the surrounding data (context) at reduced scale, maintaining overall awareness alongside detailed inspection.

ggplot2 A powerful data visualization package for R based on the Grammar of Graphics. It is used for creating aesthetically pleasing and highly customizable plots.

ggridges An R package for creating ridgeline plots using ggplot2. It is used to visualize multiple density distributions in a compact form.

Gridlines Horizontal and vertical lines across a plot that help guide the eye to values on the axes. Gridlines should be subtle in appearance to avoid distracting from the data.

Heatmap A two-dimensional graphical representation of data where values are encoded by colors in a grid, useful for visualizing density, correlation matrices, or intensity across two variables.

Hexbin Plot An alternative to scatter plots for large datasets, where data points are aggregated into hexagonal bins and the count or density within each bin is encoded by color, reducing overplotting.

Histogram A graphical representation of the distribution of numerical data where data is grouped into contiguous, non-overlapping intervals (bins), and the height of each bar corresponds to the frequency or density of observations within that interval.

Human-Centered Design An approach to visualization that focuses on the usability, accessibility, and needs of the end user, ensuring the design communicates effectively across diverse audiences.

Interactive Dashboard A digital interface that integrates multiple visualizations and interactive controls, allowing users to explore, filter, and analyze data in real-time.

Interpolation A technique used to estimate missing values by constructing new data points within the range of known data. Common forms include linear and spline interpolation.

Jittering A method of adding random noise to the position of data points in a scatterplot to prevent overlap, particularly for discrete or categorical values.

Kernel Density Estimation (KDE) A non-parametric method to estimate the probability density function of a continuous variable. KDE is commonly used in ridgeline plots to create smooth curves representing data distributions.

Layering The technique of overlaying multiple data polygons in a single spider plot to compare datasets across the same set of categories.

Legend A chart component that explains the meaning of visual encodings such as colors or patterns.

Line Plot A chart type that displays data as a series of points connected by straight lines. It is ideal for showing trends over time or across continuous variables.

Linking A technique used to synchronize multiple visualizations such that interactions in one view (e.g., selection or brushing) are reflected across other related views, enhancing multivariate data exploration.

Logarithmic Scale A type of scale where each unit represents a constant multiplicative factor. Useful for displaying exponential growth or data spanning several orders of magnitude.

Marker The symbol used to represent a single data point in a scatterplot. Common marker shapes include circles, squares, triangles, and crosses, often used in combination with color or size encodings.

MATLAB A numerical computing environment and programming language developed by MathWorks. It includes built-in functions for plotting and visualizing data across engineering and scientific domains.

Matplotlib A widely-used Python library for creating static, animated, and interactive visualizations. It provides fine-grained control over plots and is commonly used for line plots, bar charts, histograms, and more.

Multivariate Scatterplot An extension of a basic scatterplot that uses additional visual encodings such as color, size, or shape to represent more than two variables.

Network Graph A diagram that depicts relationships between entities (nodes) connected by links (edges), often used to visualize social networks, biological systems, or communication structures.

Normalization The process of rescaling different variables to a common range (e.g., 0–1 or 0–100) to enable accurate comparison on a spider plot.

Outliers Data points in a box plot that fall outside the whiskers, typically beyond 1.5 times the IQR from Q1 or Q3. They are plotted individually and may indicate variability, data entry issues, or unique cases.

Overplotting A visual clutter caused by displaying too many lines or data series in a single plot, making it hard to distinguish between them. It can be mitigated with color transparency or small multiples.

Panning An interactive navigation technique in which users move the view horizontally or vertically across a dataset without changing the zoom level, allowing continuous exploration of large visualizations.

Parallel Coordinates A high-dimensional data visualization technique in which each variable is represented by a parallel axis and data points are displayed as lines intersecting the axes, allowing users to detect patterns, correlations, and anomalies across multiple dimensions.

Perceptually Uniform Color Map A color map where equal steps in data are perceived as equal steps in color. Examples include `viridis`, `plasma`, and `cividis`, which improve interpretability and accessibility of visual data representations.

Pie Chart A circular statistical graphic divided into slices to illustrate numerical proportions. Each slice's angle and area represent a part of a whole, typically used for visualizing small sets of categorical data.

Plotly An interactive graphing library for Python (and other languages) that supports zooming, panning, and hover effects. It is used for creating dashboards and web-based visualizations.

Radar Chart A synonymous term for spider plot, emphasizing the radial symmetry and spoke-like structure of the chart design.

Radial Bar Chart A circular version of a bar chart where bars extend outward from the center, often used for cyclical data such as months or seasons.

Ridgeline Plot A data visualization technique used to display the distribution of a continuous variable across multiple categories by layering density plots vertically. Also known as a joy plot.

Sankey Diagram A flow diagram where the width of the arrows is proportional to the flow quantity, used to represent energy, material, or cost transfers between processes or entities.

Scatterplot A graphical representation of data using Cartesian coordinates, where each point represents the value of two continuous variables.

Seaborn A Python visualization library built on top of Matplotlib that provides a high-level interface for drawing attractive and informative statistical graphics. It simplifies complex plots like violin plots, box plots, and heatmaps.

Small Multiples A layout technique where multiple similar charts are displayed side by side, each depicting a different subset of data, enabling straightforward comparison across categories, groups, or time periods.

Spider Plot Also known as a radar chart, a spider plot is a circular graph used to display multivariate data by plotting values along multiple axes radiating from a central point.

Stacked Bar Chart A bar chart where subcategories are stacked within a single bar for each category, allowing comparison of total values and proportions simultaneously.

Stacked Histogram A type of histogram that visualizes multiple datasets in a cumulative form, with each dataset's contribution stacked atop the previous within the same bins.

Storytelling The fusion of narrative elements with data visualization techniques to guide viewers through a structured story, emphasizing insights and supporting retention by weaving context, emotion, and data-driven evidence.

Streamgraph A variation of a stacked area chart where layers flow around a central axis, used to visualize time-series data across multiple categories in a smooth, organic format.

Tooltip A small interactive box that appears when a user hovers over or clicks a data point in an interactive scatterplot, providing additional information without cluttering the view.

Transparency A visual property that controls the opacity of plot elements. Used in density plots to reduce overplotting and improve the readability of overlapping data.

Treemap A space-filling visualization technique used to represent hierarchical data as a set of nested rectangles, where the size and color of each rectangle reflect quantitative attributes.

Trend Line A visual representation of the general direction or pattern in a dataset, often added using linear regression or smoothing techniques to highlight relationships between variables.

Truncated Axis A visual design choice where the axis does not start at zero, often leading to exaggerated differences between bars and misleading interpretations.

Under-clustering A histogram design issue resulting from too few bins, which can over-simplify the data and mask significant details or features.

User Experience (UX) The overall experience and satisfaction of a person using a histogram, influenced by clarity, design, accessibility, and interactivity.

Violin Plot A chart that combines a box plot with a kernel density estimate to visualize the distribution, spread, and modality of the data more richly than a box plot alone.

Visual Hierarchy A principle in design that structures visual elements in a way that guides the viewer's attention, typically by varying size, contrast, or layout to highlight important data first.

Zoom-Out A visualization technique that enables users to view a broader portion or the entirety of a dataset by reducing the level of detail, thereby gaining an overall perspective and identifying macro-level patterns and trends.

Index

2D Density Plot, 99, 117, 265–276, 278, 282, 283, 286, 288, 290–294

Accessibility, 1, 6–8, 10, 15–19, 32, 34, 35, 39, 87, 90, 93–95, 97, 110, 111, 116–121, 123, 124, 132, 137, 146, 147, 149–151, 155, 163, 164, 167, 175, 187, 188, 190, 191, 193, 215–218, 223–225, 227, 228, 235, 241, 242, 246, 247, 250–253, 261, 263, 265, 269, 272, 274, 286, 287, 290–294, 296, 300, 305, 311, 316, 321, 327, 332, 336, 341, 354, 358, 365, 381, 382, 395
Alpha Blending, 28, 39, 45, 91, 98
Animation, 54, 103, 105, 167, 169, 183–185, 192, 378–381, 384, 395
Annotation, 27, 107, 109, 122, 178, 213, 248, 251, 332, 345, 373–375, 377, 385
Axes, 24, 25, 27, 28, 36, 37, 39, 43, 44, 53, 54, 56, 63, 66, 67, 79, 81, 89, 92, 96, 99, 110, 111, 117, 118, 120, 124, 141, 145, 164, 195, 197–200, 211, 217, 218, 220–222, 224, 284, 329, 385
Axis Label, 36, 58, 64, 72, 73, 79, 84, 89, 90, 94, 109–112, 116–118, 139, 149, 155, 206, 210, 211, 245, 248, 267–269, 273–275, 281, 287, 293, 330
Axis Scaling, 24, 44, 58, 84, 145, 150, 164, 202, 289, 293

Bandwidth, 75, 238, 259, 262, 265, 268, 283, 285, 286
Bar Chart, 50, 61, 64–66, 72, 73, 78–83, 85, 127, 336, 361, 368, 372
Baseline, 47, 48, 54, 197, 229, 268
Bin, 99, 125, 128, 134, 141, 143, 271, 332
Bin Size, 128, 129, 144, 145, 159, 164, 271

Box Plot, 68–76, 78, 80–83, 85, 127, 239
Brushing, 351–355, 368, 395

Color Contrast, 10, 175, 264, 305
Color Mapping, 96, 298, 332
Color Palette, 70, 82, 110, 139, 173, 174, 186, 201, 235, 238, 243, 246, 247
Color Transparency, 45
Color-Blind Friendly Palette, 27, 28, 32, 39, 215, 216, 218, 223–228, 235, 245, 247, 251–254, 262–265, 272, 274, 282, 286–288, 290–294
Contour Plot, 278

Data Drilling, 184, 345, 358
Data Storytelling, 13, 24, 27, 61, 72, 84, 105, 107, 114, 115, 122, 167, 180–183, 187, 191–193, 338, 373, 374, 378, 380
Density Histogram, 137
Direct Labeling, 40, 178
Donut Chart, 169, 171, 173, 178, 180–184, 187
Drill-Down, 186, 187, 190, 192, 305, 338, 340, 342, 343, 353, 357–361

Exploded Pie Chart, 176

Filtering, 66, 79, 80, 102, 288, 315, 318, 321, 323, 326, 334, 335, 362–365, 367, 387
Fisheye View, 390

ggplot2, 241, 244–246, 248, 250
ggridges, 241, 242, 244, 246–249
Grid Lines, 25, 34, 60, 80, 94, 96, 99, 100, 112, 139, 140, 163, 201, 222, 275

Heatmap, 156, 258, 267, 281
Hexbin Plot, 91, 99, 117, 337, 338

© The Editor(s) (if applicable) and The Author(s), under exclusive license
to Springer Nature Switzerland AG 2026
R. Damaševičius, *Human-Centred Scientific Data Visualisation*,
Undergraduate Topics in Computer Science,
https://doi.org/10.1007/978-3-032-01606-5

Histogram, 124–130, 132, 134, 136, 137, 139, 142–150, 152–155, 157, 160–163, 165, 166, 257, 277
Human-Centered Design, 19, 60, 89, 121–123

Interactive Dashboard, 41, 55, 72, 79, 166, 180, 370, 372
Interpolation, 26, 30, 31, 46, 47

Jittering, 71, 92, 99

Kernel Density Estimation, 74, 75, 265, 268–271, 273, 283–286, 292, 293

Layering (spider plot), 107, 181, 194, 212, 213, 219, 223, 225, 228, 234, 257, 265, 282, 283, 286
Legend, 26, 28, 40, 82, 94, 110, 118, 135, 178, 186, 209, 217, 218, 220, 223, 226, 248, 251, 261, 268, 291–294, 298, 300, 305, 311, 327
Line Plot, 22–26, 29, 31, 33–36, 40, 41, 43–45, 47–49, 51, 53, 58–60
Linking, 27, 341, 352, 356, 369–371, 377
Logarithmic Scale, 29, 44

Marker, 70, 87, 89, 91, 93, 96–98, 101, 107, 110, 113, 115, 121, 206, 223, 226, 227
MATLAB, 163
Matplotlib, 53–57, 78, 79, 87, 96, 103, 111–114, 118, 132, 137, 146, 147, 154, 157–159, 161, 188, 345, 353, 387, 390
Multivariate Scatterplot, 94, 100, 101, 121

Network Graph, 297, 306, 307, 310, 337, 338
Normalization, 92, 126, 199, 220, 222, 243, 258–260, 262, 329, 331

Outliers, 25–27, 30, 62, 63, 65, 68–76, 79–81, 83, 84, 88, 91–94, 96, 102, 103, 105–107, 116, 118, 120, 126–129, 143, 153, 154, 164, 206, 215, 229, 231, 232, 258, 266, 282, 299, 318, 334, 340, 351, 357, 362, 368, 379, 394
Overplotting, 91, 92, 98–100, 117, 120, 272, 288, 290, 293, 294, 297, 332, 338

Panning, 43, 102, 112, 160, 309, 329, 330, 347–350, 380, 394
Parallel Coordinates, 101, 198, 352
Perceptually Uniform Color Map, 56, 132, 290, 292
Pie Chart, 168–173
Plotly, 54–57, 78, 79, 85, 111, 112, 114, 152, 154, 161, 165, 166, 188, 202, 203, 207, 208, 265, 353, 384, 391

Radar Chart, 101, 194–196, 206, 209
Radial Bar Chart, 331, 332, 334, 335
Ridgeline Plot, 228–234, 236, 238–242, 244, 246–248, 251, 253–264

Sankey Diagram, 297, 311, 313, 315
Scatterplot, 88–93, 95, 98, 100–105, 107–115, 117–123
Seaborn, 78, 79, 159–161, 353, 387
Small Multiples, 28, 341–343, 353, 385–389
Spider Plot, 194–204, 206, 207, 209–212, 214, 215, 217–219, 222, 223, 225–227
Stacked Bar Chart, 63–66
Stacked Histogram, 135, 136
Streamgraph, 298, 323, 325–327

Tooltip, 36, 43, 102, 277, 311
Transparency, 27, 28, 31, 39, 45, 58, 70, 106, 107, 111, 112, 117, 120, 121, 132, 134, 135, 139–142, 152
Treemap, 298–300, 302–305
Trend Line, 94
Truncated Axis, 44, 81, 84

Under-clustering, 124, 143–145, 164
User Experience (UX), 14, 150, 360

Violin Plot, 65, 74–76, 78, 85
Visual Hierarchy, 12, 15, 34, 87, 90, 93, 106, 107, 119, 120, 164, 300, 338, 359, 374, 375

Zoom-Out, 340–345, 353, 394